# SONAR SYSTEMS

Edited by **Nikolai Kolev**

**Sonar Systems**
Edited by Nikolai Kolev

**Published by InTech**
Janeza Trdine 9, 51000 Rijeka, Croatia



**Publishing Process Manager** Davor Vidic

**Technical Editor** Teodora Smiljanic

**Cover Designer** InTech Design Team

Additional hard copies can be obtained from orders@intechopen.com

Sonar Systems, Edited by Nikolai Kolev
p. cm.
ISBN 978-953-307-345-3

For EU product safety concerns:
IN TECH d.o.o., Prolaz Marije Krucifikse Kozulić 3, 51000 Rijeka, Croatia,
info@intechopen.com or visit our website at intechopen.com.

# Meet the editor

Dr. Nikolai Kolev graduated from Bulgarian Naval Academy in 1985 and served as an Electronics Officer in the Navy. Since 1989 he was consecutively PhD student, assistant, associate professor and Head of the Department of Electronics between 2003 - 2007. Currently, he is an associate professor at the same department. He has been lecturing courses in radio waves, antennas and microwave devices, radar and sonar. His scientific interests are in the area of radar and sonar signal processing and lately in sensor networks. Dr. Kolev has more than 30 scientific publications and is an author and coauthor of three text books and 3 patents in Bulgarian. He is also active in analog and microcontroller systems embedded design and programming in C, LabView and Matlab. Dr. Kolev has been a member of the program committee of EUSAR conference since 2002 and a reviewer of IET journals. Since 2007 he has been respectively a project coordinator and leader of two NSF projects about the development of a sonar research laboratory and sensor network for underwater surveillance.

# Contents

# Preface

Most of Earth's surface is covered with water. These water areas have been used for fishing and transportation by mankind since ancient times but only with the invention of sonar systems an opportunity was created for in depth exploration and use of the ocean because the mechanical acoustic waves are the only physical phenomena applicable for underwater sensing. This book is an edited collection of research articles covering the current state of sonar systems, the signal processing methods and their applications prepared by experts in the field.

The book is grouped into three major sections: Innovative Sonar Systems, Modeling and Simulation; Sonar Signal Processing; Biology and Marine Science Sonar Applications. However, there are interrelations between sections and chapters because it is impossible to strictly divide the sonar system engineering level from the used signal pre- and post-processing methods and the applications.

Even though the synthetic aperture has been a known method for increasing the angular resolution in sonar and radar sensors for more than half a century, operational sonar systems have been developed only in the last years. The chapter of R. Hansen is an excellent introduction in synthetic aperture sonar and shapes the future use of interferometric principles for 3D reconstruction of underwater ocean scenes. The current state of the sensor network technology fosters the application of the longstanding bistatic and multistatic sonar concepts. The chapter of D. Orlando and F. Ehlers presents the advances in multistatic signal processing for tracking of low target strength targets in noisy environment. The sonar systems simulation methodology is covered in the chapter of D. Gueriot and C. Sintes. Modeling and simulation are an important area of investigation leading to enhancing the design of current sonar systems, algorithm performances estimation and new system models development.

Several authors in their chapters present important signal pre- and post-processing principles and applications. The chapter of J. Piper gives the fundamentals of sonar array beamforming and direction of arrival estimation. The quality and capability of passive sonar array systems are maybe of greatest concern for naval submarines sonar designers in order to make them undetectable for the enemy sensor systems and give an operational advantage. Next follows the contribution of Nunes de Moura et al. on passive sonar signal detection and classification based on independent component analysis covering DEMON and LOFAR signal processing principles with application in the experimental results presented on submarine sonar system operator displays. Modern sonar passive array systems are using adaptive matched field signal processing techniques. The submission of N. Kolev and G. Georgiev presents the basics of vertical sonar array matched field signal processing for underwater source localization in depth and range. Sonar video signal post processing is important to improve the image quality, data interpretation and applies image recognition techniques. The chapter of E. Dura presents image processing techniques for detection and classification of artificial mine like objects on side scan sonar images. Current state of sonar images denoising techniques is presented in the chapter by Isar et al. The chapter of Peyvandi et al. bridges the gap between classical passive sonar signal processing methods and some newly introduced probabilistic concepts of signal processing and classification as the Hidden Markov Model and Artificial Neural Networks. An application in underwater sensor network acoustic communication is outlined.

Nature has created air passive hearing mechanism for humans but also a sophisticated biosonar for some earth and water species which combines active and passive acoustic principles, subject to a scientific investigation and a challenge for technical implementation. The chapters of Gudra et al. and Mora et al. give an interesting insight for readers on these topics. H. Manik has developed bottom surface and fish target strength models and a multi frequency acoustics methodology to characterize and quantify the fish and sea bottom scattering by an echo sounder. Experimental results are presented for the area of Jawa Island. Georgakarakos et al. have made a review of acoustic biomass estimation methods. Experimental results for a fish-school species identification using a neural network with multibeam sonar data are presented. The side scan sonar is the most affordable and used system for constructing images of the sea bottom with applications in the marine geology. A. Savini has analyzed the influence of side scan sonar parameters on bottom sonar images from geomorphological point of view. The chapter represents a valuable source of seafloor images and reference documentation about the main features and sediment characteristics of the Mediterranean continental shelf.

On one hand the book will be interesting for scientists, engineers and students specializing in the area of sonar systems, signal processing and marine science because they will find in depth investigations and practical solutions of a broad range of problems presented by the authors. On the other hand, it will be also useful for readers with general knowledge as a guide and an introduction in sonar systems principles of operation, signal processing methods and many interesting applications nowadays.

I am grateful to contributing authors, InTech team and especially to Mr. Vidic for the continuous help and support during the editorial process.

**Assoc. Prof., Ph.D, Nikolai Kolev,**
Department of Electronics,
NV Naval Academy, Bulgaria

# Part 1

# Innovative Sonar Systems, Modeling and Simulation

# 1

# Introduction to Synthetic Aperture Sonar

Roy Edgar Hansen
*Norwegian Defence Research Establishment*
*Norway*

## 1. Introduction

SONAR is an acronym for **SO**und **N**avigation **A**nd **R**anging. The basic principle of sonar is to use sound to detect or locate objects, typically in the ocean. Sonar technology is similar to other technologies such as: RADAR = **RA**dio **D**etection **A**nd **R**anging; ultrasound, which typically is used with higher frequencies in medical applications; and seismic processing, which typically uses lower frequencies in the sediments. There are many good books that cover the topic of sonar (Burdic, 1984; Lurton, 2010; Urick, 1983). There are also a large number of books that cover the theory of underwater acoustics more thoroughly (Brekhovskikh & Lysanov, 1982; Medwin & Clay, 1998).

The principle of Synthetic aperture sonar (SAS) is to combine successive pings coherently along a known track in order to increase the azimuth (along-track) resolution. A typical data collection geometry is illustrated in Fig. 1. SAS has the potential to produce high resolution images down to centimeter resolution up to hundreds of meters range. This makes SAS a suitable technique for imaging of the seafloor for applications such as search for small objects, imaging of wrecks, underwater archaeology and pipeline inspection.

SAS has a very close resemblance with synthetic aperture radar (SAR). While SAS technology is maturing fast, it is still relatively new compared to SAR. There is a large amount of SAR literature (Carrara et al., 1995; Cumming & Wong, 2005; Curlander & McDonough, 1991; Franceschetti & Lanari, 1999; Jakowatz et al., 1996; Massonnet & Souyris, 2008).

This chapter gives an updated introduction to SAS. The intended reader is familiar with sonar but not SAS. The only difference between traditional sonar and synthetic aperture is

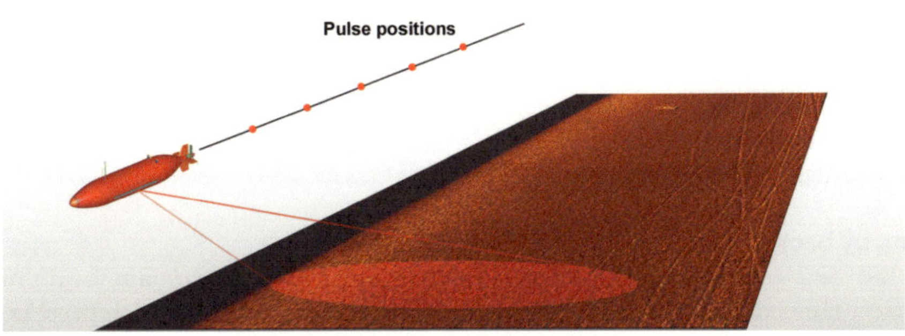


Fig. 1. Data acquisition geometry for synthetic aperture sonar.

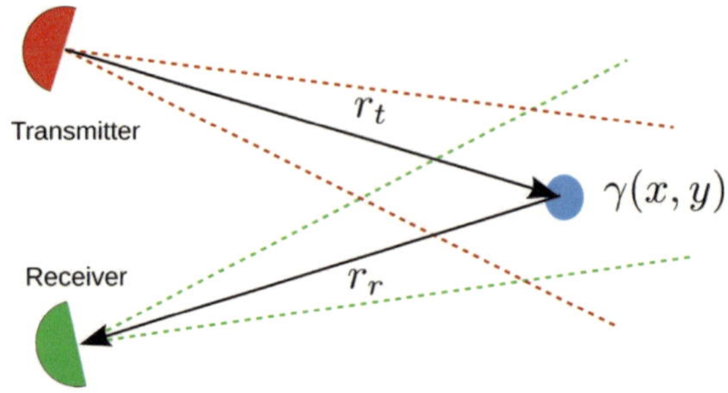


Fig. 2. Basic imaging geometry.

the construction of the aperture (or array). We start by giving a review of sonar imaging and describe the backprojection method. Then, we derive the angular resolution. We describe the multi-element receiver concept and calculate the area coverage rate for SAS. We explain the frequency dependence in SAS, and discuss some of the choices and trade-offs in SAS design. We list some of the specific challenges in SAS, and suggest possible solutions. We give an overview of the signal processing involved in a SAS processor, and we discuss properties of SAS images. Finally, we show numerous examples of SAS images from a particular system, the HISAS 1030 interferometric SAS.

## 2. Sonar Imaging

Assume a transmitter that insonifies a scene with acoustic reflecting material represented by a reflectivity function $\gamma(x, y)$. A part of the scattered acoustic field is recorded by one or more receiver hydrophones. This is illustrated in Fig. 2. Sonar imaging is the *inverse problem*, namely to estimate the reflectivity function from the received data from one or more transmitted pulses.

Sonar imaging can be separated into range processing of the data and angular processing (beamforming) of the data. *Beamforming* is defined as the method or processing algorithm that focuses the signal from several receivers in a particular direction. Beamforming can be applied to all types of multi-receiver sonars: active, passive, towed array, bistatic, multistatic, and synthetic aperture. Sonar beamforming is well covered in (Burdic, 1984; Johnson & Dudgeon, 1993; Nielsen, 1991).

Range processing is signal processing applied to each time series individually. The processing is a function of transmit waveform. There are several types of signals that are used as transmit waveforms. Classical sonar often uses gated continuous-wave (CW) pulses, sometimes referred to as pings. Modern sonar, and virtually all SAS systems, use phase coded transmit signals where the signal bandwidth is determined by the phase coding (or frequency spread of the signal modulation). The reason to use phase coded waveforms is to increase the transmit signal energy while maintaining large signal bandwidth. The range processing done on phase coded signals are referred to as *pulse compression* or *matched filtering*. This is covered in a number of excellent books such as (Franceschetti & Lanari, 1999; Levanon, 1988).

### 2.1 Imaging by backprojection

The simplest and most intuitive type of beamforming is time-domain beamforming by *backprojection*. This is done by back propagating the received signal via each pixel in the scene to be imaged and into the transmitter. This is illustrated in Fig. 3. Backprojection is also known as Delay-And-Sum (DAS) (Johnson & Dudgeon, 1993, pages 117-119). Functionally,

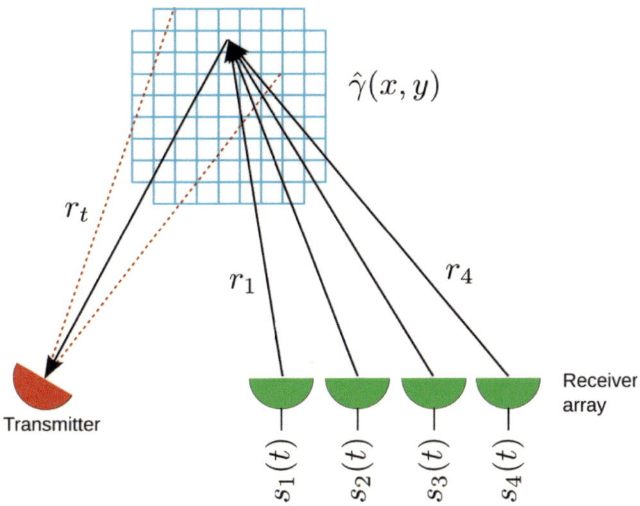


Fig. 3. Backprojection geometry.

backprojection is also closely related to Kirchhoff migration in seismic processing (Claerbout, 1995). Formally, backprojection is straightforward and can be done as follows. First the signals in each receiver are filtered and pulse compressed. Then the signal from each receiver is delayed to the pixel of interest

$$t_n(x,y) = (r_t + r_n)/c. \tag{1}$$

where $r_t$ is the distance from the transmitter to the pixel, $r_n$ is the distance from the pixel to receiver $n$, and $c$ is the sound velocity. Then the recorded, pulse compressed signals $s_n$ are summed coherently over all receivers to produce an estimate of the reflectivity function

$$\hat{\gamma}(x,y) = \frac{1}{N}\sum_{n=1}^{N} A_n s_n(t_n(x,y)). \tag{2}$$

The amplitude $A_n$ is generally a function of geometry and frequency $A_n = A_n(r_t, r_r, \omega)$. In addition, the direction dependent sensitivity of the receivers can be taken into account. This amplitude factor can also be used to control the sidelobe suppression vs the angular resolution (Van Trees, 2002). For sampled time series $s_n(t_n(x,y))$ may have to be interpolated to obtain accurate values in the image. A simple algorithmic description of the backprojection algorithm is shown in Listing 1. We immediately realise why this method is called *Delay-And-Sum*. The signals are delayed to the correct pixel, and then summed.

### 2.2 Angular resolution

The *angular resolution* can be defined as the minimum angle for which two reflectors can be separated in the sonar image. A very simple and intuitive way to derive this is as follows. Assume a phased array receiver of length $L$ consisting of a number of elements as illustrated in Fig. 4. The angular resolution for this receiver is the angle difference for which the echo from two reflectors gives destructive interference in the receivers. Consider a reflector at broadside and a reflector at an angle $\beta/2$. As the maximum range difference will always be on the ends of the array, we only consider the center element and the end element displaced $L/2$ apart. The distance difference between these two reflectors is

$$\delta R = R_0 - R_1. \tag{3}$$

```
for all directions
   for all ranges
      for all receivers
         Calculate the time delay
         Interpolate the received time series
         Apply appropriate amplitude factor
      end
      sum over receivers and store in gamma(x,y)
   end
end
```

Listing 1. Backprojection Pseudo-code

For range differences larger than $\delta R = \lambda/4$ destructive interference will start to occur. The difference in range is

$$\delta R = R_0 - R_1 = L/2 \sin(\beta/2). \tag{4}$$

We define $\beta$ as the *angular resolution* and solve for when destructive interference will start

$$\delta R = L/2 \sin(\beta/2) = \lambda/4. \tag{5}$$

For small angles, we can approximate the sin-function

$$L/2\beta/2 = \lambda/4 \Rightarrow \boxed{\beta = \frac{\lambda}{L}}. \tag{6}$$

Hence, the angular resolution (in radians) equals the inverse length of the array measured in wavelengths. A longer array or a higher frequency (shorter wavelength) gives better angular resolution. More details and a more accurate derivation of angular resolution can be found in (Van Trees, 2002).

### 2.3 Imaging techniques for phased arrays

Fig. 5 shows the basic imaging geometry for an array of receivers (a phased array). The field of view is defined by the beamwidth of each of the receivers and their look direction. The

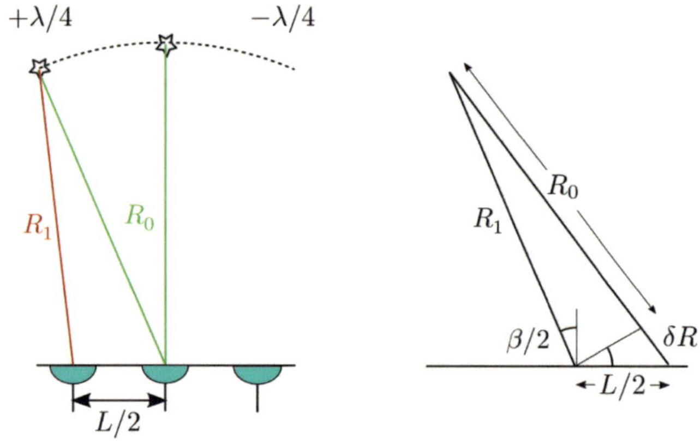


Fig. 4. Geometrical derivation of angular resolution.

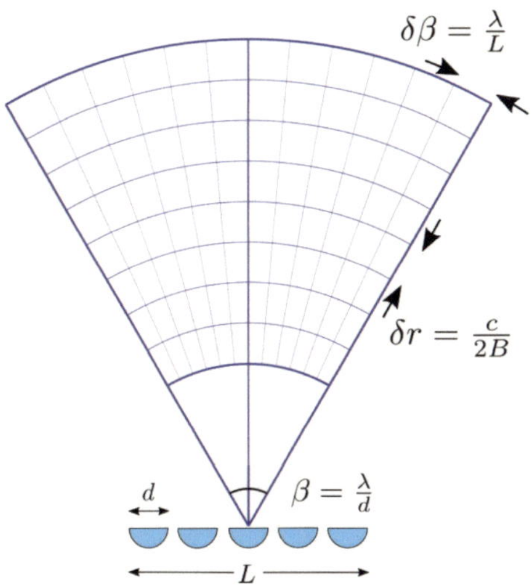


Fig. 5. Field of view and resolution in imaging sonar.

angular resolution is given by the array length and the wavelength. If the individual elements are more than $\lambda/2$ apart, the directivity of each element must be small enough to mitigate alias lobes (see section 3.1). For an active system, the range resolution is defined by the bandwidth of the system. The maximum range is determined by the pulse repetition interval and/or the sensitivity of the receiver elements. For a linear array it can be fruitful and efficient for the imaging to define a polar coordinate system fixed to the sonar.

In underwater applications, beamforming can be used to produce different types of products. For high frequency sonar imaging, this can be categorized into three different types, as illustrated in Fig. 6:

**Sectorscan sonar** produces a two-dimensional image for each pulse. These images are usually shown on a display pulse by pulse. Sectorscanning sonar is often used as hull mounted sonars for forward looking imaging or wide swath imaging. In fisheries acoustics, some cylindrical arrays actually produce full 360 degrees view.

**Sidescan sonar** is a particular type of sonar that uses the platform motion to cover different parts of the seafloor. A sidescan sonar produces one or a few beams around broadside, and an image of the seafloor is produced by moving the sonar and using repeated pulses. This is a very popular technology, it has fairly low hardware complexity and can therefore be more affordable. There are several books covering sidescan sonar very thoroughly (Blondel, 2009; Fish & Carr, 2001).

**Synthetic aperture sonar** uses multiple pulses to create a large synthetic array (or aperture). From this, an image of the seafloor is produced such that the information from multiple pulses goes into each pixel on the seafloor. It is, from a certain point of view, the combination of sidescan sonar and sectorscan sonar.

## 3. SAS sampling and coverage rate

In all types of array signal processing, the sampling of the array, or the spacing of the elements and their directivity, is critical. If the array is too sparsely sampled, grating or alias lobes will

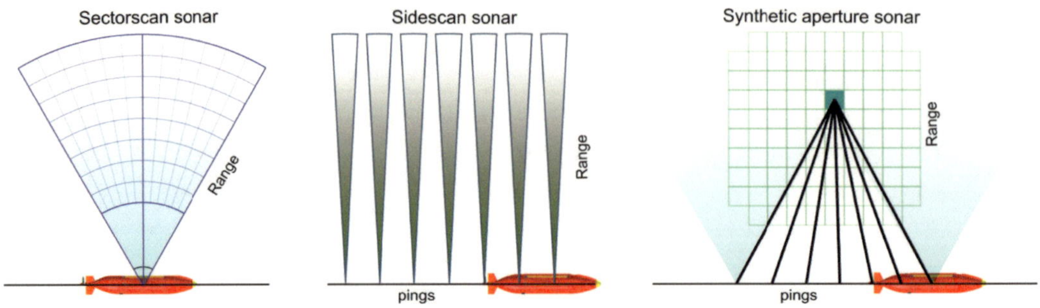


Fig. 6. Phased array imaging concepts for sonar.

occur, and the image quality will be reduced. In this section we describe the sampling criterion and show how this affects SAS design.

### 3.1 Undersampling of arrays

Consider a simple array with two elements displaced $D$ apart, as illustrated in Fig. 7. Assume an inbound plane wave at broadside $\theta = 0$ with wavelength $\lambda$ from a reflector in the far field. Any plane wave coming in the direction

$$\sin\theta_n = \pm n\frac{\lambda}{2D} \tag{7}$$

where $n$ is an integer, will be perfectly in phase with the plane wave at broadside. Note that the factor two, i.e. $\lambda/2$, comes from assuming two-way propagation (transmission and reception). This ambiguity will cause grating lobes or alias lobes in the beampattern (Manolakis et al., 2000; Van Trees, 2002). To avoid grating lobes, the beamwidth of each individual element must be smaller than the distance to the first grating lobe. Assuming that each element is of size $d$ with a beamwidth of $\beta \approx \lambda/d$, we get the sampling criterion

$$D \leq d/2. \tag{8}$$

Hence, for a well sampled array, the effective distance between each element must be half the size of each element. This is fulfilled for a densely populated array without gaps between elements (see below).

Grating lobes are such that the problem they cause cannot be fixed in signal processing afterwards. For linear uniform arrays (ULAs), there is a deterministic and simple one-to-one

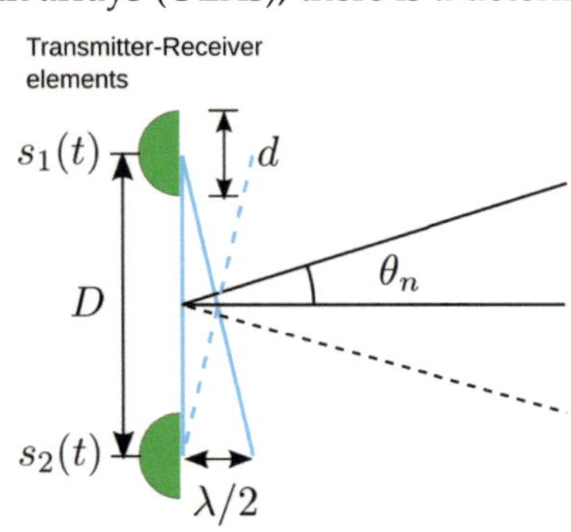


Fig. 7. Along track sampling and grating (alias) lobes.

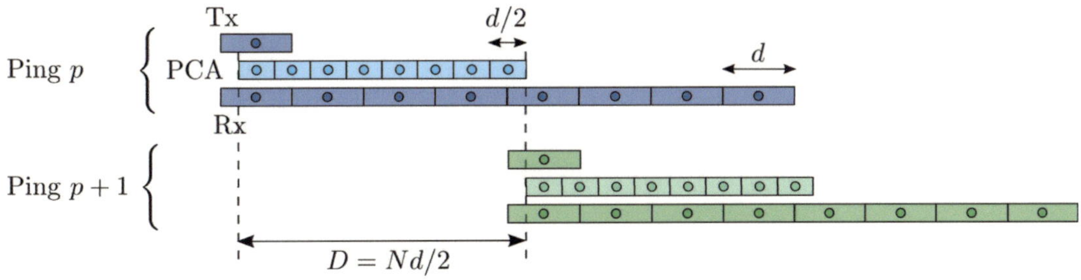


Fig. 8. Multi-element receiver array and the Phase Center Approximation.

relation between the sampling of the array and the angular position of the grating lobes. Grating lobes are frequency dependent. For large bandwidth systems, the grating lobes will be less defined due to the fact that their positions change with frequency.

### 3.2 Multi-element receiver arrays in SAS

In SAS, the transmitter-receiver pair is moved along-track and pulsed repeatedly to form a long synthetic antenna. The sampling criterion (8) imposes a maximum allowed displacement between pings, which again, transforms into a maximum range for a given speed. The maximum range $R_{max}$ is given by

$$2R_{max}/c = T_{pri} = D/v \quad \Rightarrow \quad R_{max} = \frac{cD}{2v} \tag{9}$$

where $v$ is the platform speed and $T_{pri}$ is the pulse repetition interval. This relation imposes a very strong limitation to SAS. Consider a platform speed of $v = 2\,\text{m/s}$. For a displacement of $D = 5\,\text{cm}$, the maximum range then becomes $R_{max} = 18.75\,\text{m}$ for a sound velocity of $c = 1500\,\text{m/s}$. This is an unacceptable small range for most applications. To overcome this, a receiver array of multiple elements can be used (Bruce, 1992; Cutrona, 1975). The multi-element receiver array is used in almost all existing SAS systems today.

A simple description of how a single transmitter multiple receiver system can be used to form a synthetic aperture is as follows. Consider a single transmitter $Tx$ and $N$ receivers of size $d$ in a linear array $Rx$, as illustration shown in Fig. 8. In stead of treating this as a bistatic system with one transmitter and many receivers, we assume a virtual array where each element is placed at the middle position between each transmitter-receiver pair (indicated in light green). In this array, each element is a transmitter-receiver (monostatic). This is the Phase Center Approximation (PCA) (Bellettini & Pinto, 2002). The virtual array (or PCA array) is half the length of the receiver array, with element size $d/2$, and each of the elements in the virtual array can be placed directly in a synthetic array (similar to SAR). The maximum displacement between two pulses becomes the length of the PCA array $D = Nd/2 = L/2$, and the maximum range becomes

$$R_{max} = \frac{cL}{4v} \tag{10}$$

where $L$ is the length of the receiver array. The maximum range can then be increased by increasing the length of the receiver array. Fig. 9 shows the maximum range as function of vehicle speed for three different receiver array lengths. Note that the area coverage rate (speed times range) is a constant proportional to the array length.

In airborne SAR, aperture undersampling is not a problem, since the phase velocity for electromagnetic waves in air is relatively high compared to the platform speed. In large

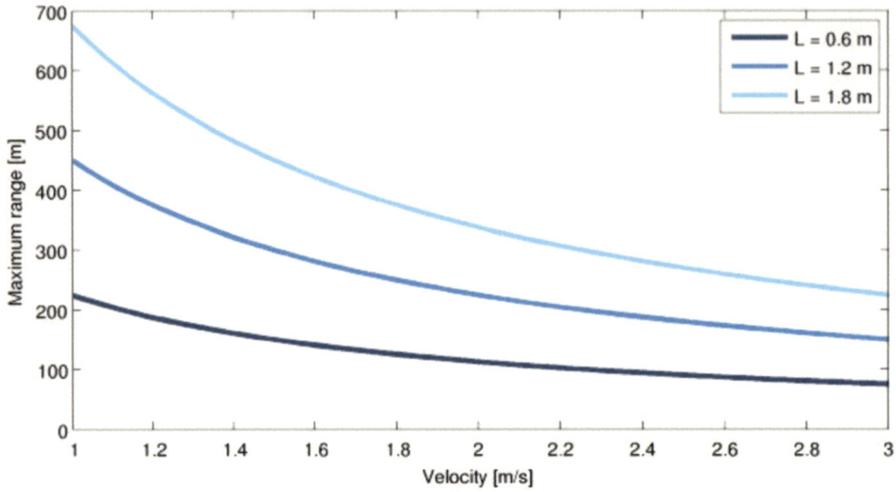


Fig. 9. Maximum range for given platform speed for three different receiver array lengths.

swath high resolution space-borne SAR, there is a limitation related to the sampling criterion. Advanced SAR systems have 2D phased arrays and use electronic steering of the transmitter and receiver arrays for different SAR modes. ScanSAR (Franceschetti & Lanari, 1999) and *Terrain Observation by Progressive Scans* (TOPS) (Gebert et al., 2010) are used to increase the area coverage rate while lowering the resolution. Spotlight mode (Jakowatz et al., 1996) is used to increase the resolution while lowering the area coverage rate.

## 4. Design of Synthetic Aperture Sonar

SAS is different from traditional sonar in several ways. In this section we describe some of the significant differences that affect the performance and application areas of SAS.

### 4.1 Resolution

While traditional imaging sonar has constant angular resolution, and thereby range dependent along-track resolution, SAS produces range independent along-track resolution. This is done by increasing the length of the synthetic array as function of range. Consider the left illustration of Fig. 10. The maximum length of the synthetic aperture is given by the field of view of each transmitter/receiver element. At range $R_1$, the length of the synthetic aperture becomes

$$L_1 \approx \beta R_1 \tag{11}$$

where $\beta = \lambda/d$ is the field of view. The along-track resolution is given by the synthetic antenna

$$\delta x \approx R_1 \frac{\lambda}{2L_1} \tag{12}$$

where the factor 2 comes from the fact that both the transmitter and receiver is moved along the synthetic aperture (creating a focused receiver and transmitter). Inserting for $L_1$ and $\beta$ we get a resolution of

$$\delta x \approx R_1 \frac{\lambda}{2L_1} = R_1 \frac{\lambda}{2\beta R_1} = R_1 \frac{\lambda}{2\lambda/dR_1} = \frac{d}{2}. \tag{13}$$

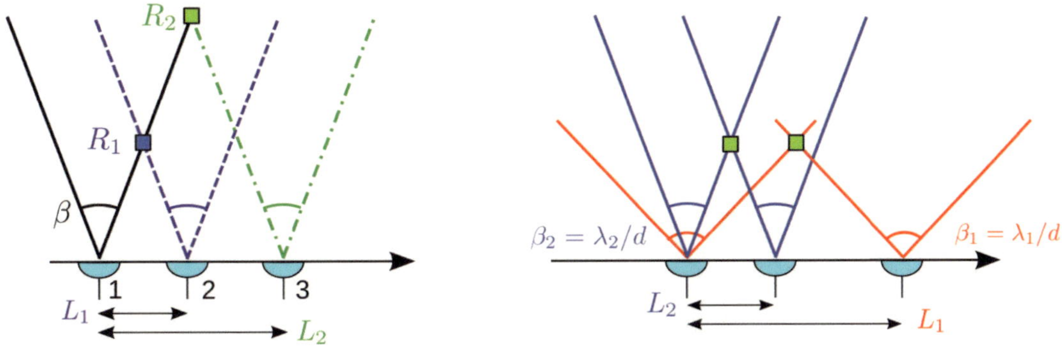


Fig. 10. Left: Range independent along-track resolution in SAS. Right: Frequency independent along-track resolution in SAS.

Hence, the along-track resolution for full length synthetic apertures is given by the element size alone, independent of range. This is a well known result, and there are numerous books covering this (Curlander & McDonough, 1991; Franceschetti & Lanari, 1999). This gives a degree of freedom not available in traditional sonar, where the cross-range resolution inevitably decreases with increasing range.

The length of the synthetic aperture is range dependent. Consider again the left drawing in Fig. 10. At range $R_2$, the length of the synthetic aperture becomes $L_2$ to obtain the same along-track resolution as for range $R_1$. The image quality becomes range dependent, since the transmission loss (Lurton, 2010) and thereby the Signal to Noise Ratio (SNR) is range dependent. Since the synthetic aperture is longer at longer range, the required accuracy in navigation, sound velocity and topography becomes thougher (see section 5). Hence, SAS at long range is inherently more difficult than SAS at short range.

### 4.2 Frequency dependence

Another intriguing fact is that SAS provides frequency independent along-track resolution. This is achieved by increasing the length of the synthetic aperture for decreasing frequency. The right drawing in Fig. 10 illustrates this. The lower frequency (red curve) requires a longer synthetic aperture than the higher frequency (blue curve) for a fixed resolution. The angular spread, however, becomes frequency dependent. For lower frequency SAS, larger angular spread is needed for a given along-track resolution. There is a lower limit to the resolution, where (13) does not hold. The best possible along-track resolution that can be achieved is $\delta x = \lambda/4$, for 180° field of view.

The frequency independence gives a degree of freedom compared to traditional sonar. Seawater is a dissipative medium for acoustic waves through viscosity and chemical processes (Brekhovskikh & Lysanov, 1982, pages 9-11), (Lurton, 2010, pages 23-27), (Medwin & Clay, 1998, pages 104-110). Acoustic absorption in seawater is frequency dependent, such that lower frequencies reach longer than higher frequencies. Fig. 11 shows the two-way transmission loss for some typical frequencies used in high frequency sonar imaging. For a given reception threshold (sensitivity), the maximum achievable range changes substantially. In traditional sonar, a typical design criterion is to choose the highest possible frequency for the desired range. Then design the transmitter and receiver arrays to obtain best possible azimuth resolution. In SAS, the along-track resolution is independent of frequency. This allows the center frequency to be chosen for other reasons than resolution.

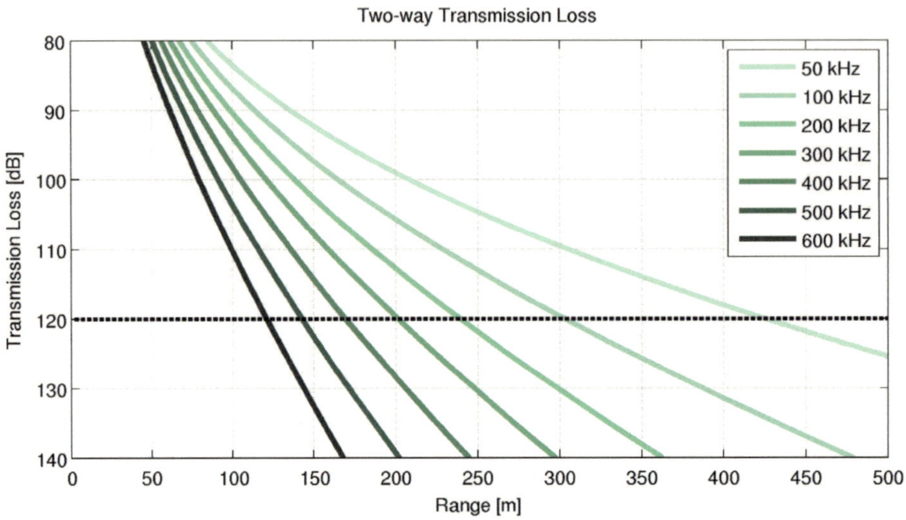


Fig. 11. Two-way transmission loss for typical frequencies in SAS.

### 4.3 Design parameters

There are many different SAS designs based on different philosophies. Here, we list some systems to illustrate the variety in existing designs. In (Bellettini & Pinto, 2009), the sonar has a center frequency of 300 kHz, and a receiver array length of 1.2 m with 36 elements. This sonar was designed to have a maximum range of 200 m at 2 m/s, good shallow water performance and as little angular spread as possible on the targets. In (Larsen et al., 2010), the center frequency is 60 kHz, and the receiver array is 4.3 m with 32 elements and a maximum range is 1500 m at 1 m/s speed. This system was specifically designed for deep tow with very large swath. The system described in (Jean, 2008) has a center frequency of 100 kHz, a receiver array of length 2 m with 24 elements, giving a maximum range of 300 m at 2.5 m/s speed. The system described in (Glover & Campell, 2010) is a dual frequency SAS at 17.5 kHz and 150 kHz, with a receiver array of 0.6 m length and 24 receiver elements. This gives a maximum range of 100 m at 2 m/s speed. The system described in section 8.1 has a center frequency of 100 kHz and a receiver length of 1.2 m with 32 elements, giving 200 m range at 2 m/s speed.
In Table 1 we list some design parameters (left column) and what controls these (right column). The general rule of thumb is that higher performance in resolution and area coverage rate requires more hardware, more processing capability and thereby also more space and power consumption (as expected). A slightly less obvious conclusion is that a high resolution system inherently is more robust. The future will surely bring new designs as the frequency agility of transducers improves, and the ability for large number of channels increases.

## 5. Challenges in SAS

The principle of SAS has been known in more than 40 years (Gough & Hawkins, 1998; Hayes & Gough, 2009), but it is only the last few years SAS has become commercially available. The main reason for this is the specific challenges that has to be solved for successful SAS imagery. In this section, we list some fundamental challenges and indicate how these can be solved. More details can be found in (Hagen & Hansen, 2008; 2009; Hansen, 2010; Hansen, Callow, Sæbø & Synnes, 2010).

| Design Parameter | Relation |
|---|---|
| Range resolution | System Bandwidth |
| Along-track resolution | Element size |
| Area coverage rate | Receiver array length |
| Maximum range | Frequency, receiver array length and speed |
| Signal to Noise Ratio | Frequency, geometry and maximum range |
| Robustness | Relative bandwidth and redundancy |
| Complexity | Beamwidth, number of elements and relative bandwidth |
| Throughput | Bandwidth times number of receiver elements |

Table 1. Design parameters and their relation in SAS.

### 5.1 Navigation

The sonar has to be positioned with accuracy better than a fraction of a wavelength along the synthetic aperture. This is the same requirement as for any other array sensor, but inherently more difficult to obtain since a synthetic antenna is formed by a moving platform. A sharp image depends on accurate positioning of the elements in the array. How much error is tolerated is dependent on the scale of the error and the required image quality (Carrara et al., 1995). Navigation of underwater platforms is more difficult than navigation of airborne and terrestrial platforms because GPS is not available. This challenge is the most important, and the first that has to be overcome for successful SAS imagery.

One solution is to use the sonar data itself for navigation (referred to as *micronavigation*). Compare the illustration in Fig. 12 with Fig. 8. By moving the SAS system a distance $D = Md/2$ such that $M < N$, there will be redundancy (or overlap) in the synthetic aperture. The overlap in the phase center antenna (PCA) is indicated in orange. Estimating which channels overlap can be used to estimate the displacement along-track and cross-track. This can be done by cross correlating the data recorded by elements from ping $p$ with data recorded by other elements from ping $p + 1$. The method is named displaced phase center antenna (DPCA) or redundant phase center (RPC). DPCA was first used in Moving Target Indicator (MTI) radars in the 1950s (Dickey Jr et al., 1991). In SAS, early descriptions of DPCA can be found in (Pinto et al., 1997; Sheriff, 1992). A very good overview of the method is given in (Pinto, 2002), and a detailed study of the performance is given in (Bellettini & Pinto, 2002). In (Hansen et al., 2003) different strategies to combine DPCA with traditional inertial navigation is described.

Micronavigation using DPCA causes a trade-off between navigation performance and efficiency (area coverage rate). This is due to the fact that the displacement between pings has to be less than the maximum displacement $L/2$. More overlap gives better navigation but lower area coverage rate. An exception to this is if several transmitters, specific bandwidth, and a specific pulse regime is used (Billon & Fohanno, 2002).

### 5.2 Topographic errors

When running a vehicle on a non-straight track, a non-straight synthetic aperture is formed and the imaging geometry becomes dependent on the full three-dimensional geometry. This means that the position of the sonar has to be known and the topography (or bathymetry) of the scene to be imaged has to be known (Jakowatz et al., 1996, pages 187-197). Only then, successful SAS processing can be done. This is critical for robust SAS, and a significant problem since the topographic changes in rough terrain may impose severe non-linear tracks (Hansen et al., 2009). There are two solutions to this challenge: either run on a straight line or

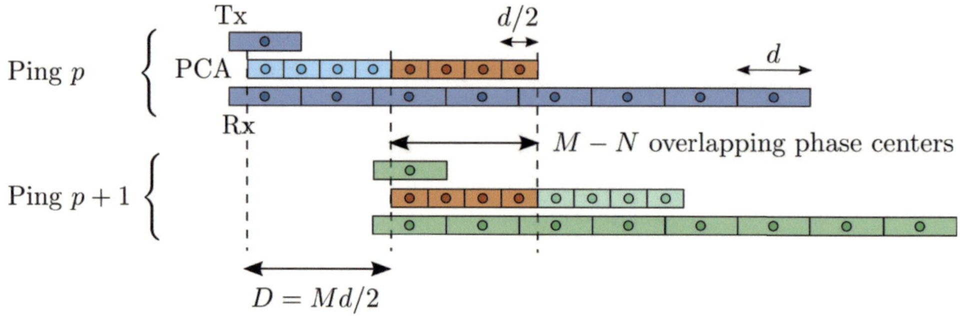


Fig. 12. Multi-element receiver array and the principle for Displaced Phase Center Antenna.

obtain a map of the area before synthetic aperture processing. The former is impractical (or impossible) on small platforms such as autonomous underwater vehicles (AUVs) or towfish systems. The latter can be obtained by using an interferometric sonar to map the scene prior to SAS processing.

### 5.3 Ocean environment

SAS is near-field imaging. This implies that the sound velocity has to be accurately estimated for well focused imaging. The sound velocity in the ocean varies with depth (Brekhovskikh & Lysanov, 1982, pages 1-9), (Lurton, 2010, chapter 2.6). There might also be local horizontal and temporal variations. This can cause variation in the sound velocity of up to 2% along the acoustic path. The effect of incorrect sound velocity on SAS is described in (Hansen et al., 2007; Hansen, Callow, Sæbø & Synnes, 2010). To overcome this, the sound velocity can be measured directly or calculated using a Conductivity, Temperature, Depth (CTD) sensor. Another possibility is to auto correct for incorrect sound velocity by using an autofocus technique (Hansen et al., 2007; Jakowatz et al., 1996).

### 5.4 Vehicle stability

SAS systems are generally multi-element receiver systems (see section 3.1). This affects synthetic aperture processing in several ways. For vehicles operated in a crabbing environment (where the heading is not aligned with the track), a baseline occurs between overlapping elements. Even if the track is perfectly linear, the synthetic aperture becomes non-linear and the image quality in SAS processing becomes dependent on the topography accuracy (see section 5.2). Large crab-induced baselines is challenging for imaging and also micronavigation (Callow, 2010).

### 5.5 Multipath environment

When operating in shallow waters, multiple reflections (or multipath) via the sea surface might affect the performance of sonar (Lurton, 2010, chapter 2.4), (Bellettini et al., 2003). This can affect the SAS data threefold: Multipath can potentially lower the temporal coherence between pings in the micronavigation; multipath can reduce the spatial coherence in interferometry; and multipath can add unwanted signal to the SAS images, causing loss of shadow contrast and fidelity. The latter two problems applies to all sonars in shallow waters, not only SAS. How much the shallow water environment affects SAS is actually dependent of the seafloor conditions, the sound velocity profile and the sea surface roughness (Synnes et al., 2009).

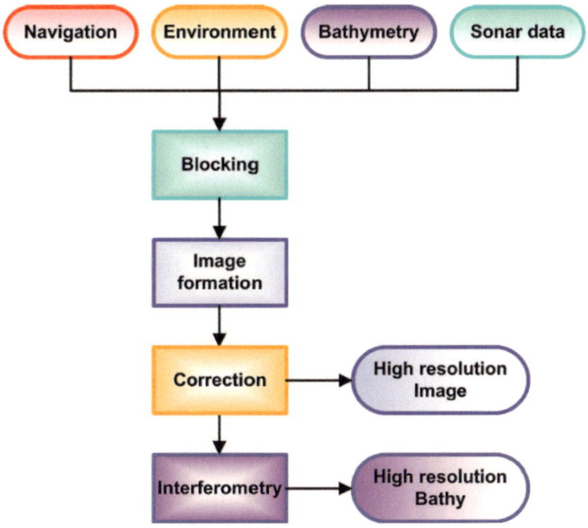


Fig. 13. Overview of SAS signal processing flow.

## 6. Signal processing of SAS data

Signal processing of SAS data can be done in many different ways. Fig. 13 shows the shcematic overview of a processing flow. Signal processing of SAS data is similar to SAR processing, with a few exceptions. Accurate knowledge of the vehicle navigation (track position and sensor orientation), the ocean environment and in particular the sound velocity, and the seafloor topography must be obtained for successful SAS imagery (see section 5).

The first step (blocking) in the processing is to divide the data into portions suitable for synthetic aperture processing. The blocking selects which pings to put into one synthetic aperture, and decides what to do with the data. If micronavigation is required, this must take place before imaging. There are two classes of image formation algorithms: time domain imaging and frequency domain imaging. In general, the frequency domain algorithms are more efficient but require more controlled vehicle behavior (close to straight line tracks). Candidates are the wavenumber algorithm (Soumekh, 1994),(Cumming & Wong, 2005, chapter 8), (Carrara et al., 1995, chapter 10) (also referred to as range migration algorithm or Omega-K algorithm), and the chirp scaling algorithm (Cumming & Wong, 2005, chapter 7), (Franceschetti & Lanari, 1999). Alternatively, the image formation can be done in time domain (see section 2.1). An overview of different algorithms used in SAS imaging can be found in (Gough & Hawkins, 1997).The benefit of time domain beamforming is that the imaging grid can be arbitrary and the data acquisition (the synthetic aperture) can be severely distorted. This gives an added flexibility not available in frequency domain imaging (Massonnet & Souyris, 2008, chapter 2.4). The disadvantage of time domain imaging is the computational load. After image formation, blind correction for residual errors (known as autofocus) in the image can be performed (Jakowatz et al., 1996, chapter 4), (Callow, 2003). It should be noted that most autofocusing techniques are local methods - they have a certain success in correcting local errors, but do not perform well on a large image with different sources for errors in different locations in the image. For interferometric systems, the final stage in the SAS processing flow is bathymetry estimation using interferometry (Hanssen, 2001; Sæbø, 2010).

## 7. Properties of SAS images

The goal of sonar imaging is to estimate the acoustic reflectivity in the best possible manner, given the sensor and geometry (see section 2). (Oliver & Quegan, 1998) gives an excellent overview of the properties of SAR images (highly relevant for SAS), related to the signal processing and the scene content. In the following, we list common measures of system performance, applicable to any imaging system.

**Geometrical resolution** or detail resolution. This is the minimum distance between two reflectors where they can be resolved in the image. The theoretical geometric resolution is given by the bandwidth for the range dimension and the element size for the along-track dimension (see section 4.1). The true geometrical resolution is also dependent on the image quality. Defocus will reduce the geometrical resolution (Oliver & Quegan, 1998, chapter 3.2).

**Radiometric resolution** or contrast / value resolution, echogenicity or target strength accuracy. This is the accuracy of the estimated value in each pixel. All coherent imaging systems suffers from speckle (Goodman, 2007; Oliver & Quegan, 1998). Speckle is random variability caused by constructive and destructive interference between individual scatterers in each geometrical resolution cell. Speckle causes a variance in the pixel value and thereby a reduced radiometric resolution. There are different methods to despeckle images and to estimate the scattering cross section. The traditional approach in SAR has been to apply multilook processing to reduce speckle (Jakowatz et al., 1996, chapter 3.3). There also exist more advanced methods to estimate the scattering cross section (Massonnet & Souyris, 2008, chapter 3.11), (Oliver & Quegan, 1998, chapter 6). The radiometric resolution strictly depends on a fully calibrated system, where the whole system has to be energy preserving. This is non-trivial to obtain, and it implies that all the terms except the Target Strength in the sonar equation has to be accounted for (Ainslie, 2010; Lurton, 2010), (Curlander & McDonough, 1991, chapter 7).

**Dynamic range** or resolvability of small targets in the presence of large targets. This is a function of the sidelobe levels or the shape of the point spread function. There is a trade-off between geometrical resolution and dynamic range. Large dynamic range requires large sidelobe suppression which causes poorer geometrical resolution (Franceschetti & Lanari, 1999, chapter 3.1), (Carrara et al., 1995, chapter 8). In addition to the sidelobes, the alias lobes (or grating lobes) must also be controlled in order to obtain the desired dynamic range. This can only be achieved by oversampling the synthetic aperture (Bellettini & Pinto, 2009; Gough & Hawkins, 1997).

**Sensitivity** or detection ability of low level targets. This is determined by several of the terms in the sonar equation (Lurton, 2010; Urick, 1983), such as system self noise, transmit power, transmission loss (which is a function of acoustic frequency) and processing gain. See (Ainslie, 2010) for a detailed description of the terms in the sonar equation. A system with larger pulse compression gain will have improved sensitivity.

**Temporal resolution** or framerate. This is the number of independent images on the scene per unit time. Since SAS is based on space-time processing (spatial movement uses time to generate aperture), SAS only has one frame on the scene when using full aperture length. Multi-aspect imaging (Hansen et al., 2008) can be applied to produce multiple aspects from different apertures and thereby different time intervals. There is a trade-off between looks and along-track resolution (since the aperture becomes shorter).

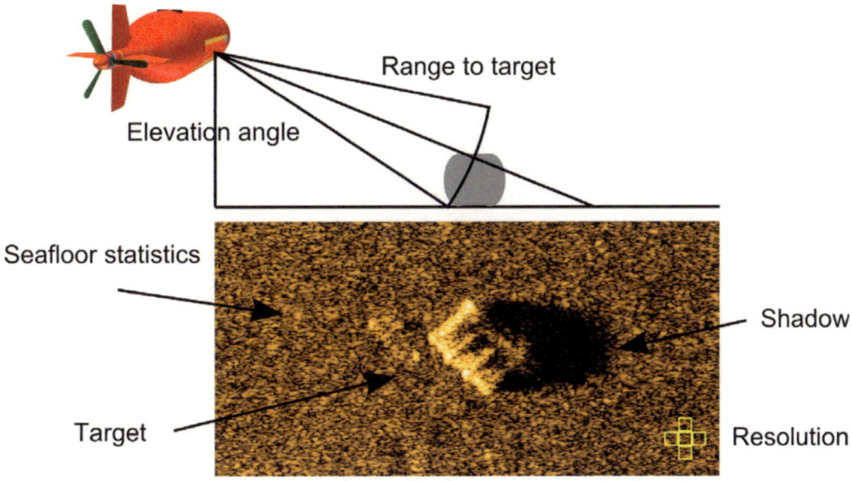


Fig. 14. The ability to retrieve relevant information from a SAS image is dependent on a number of different factors.

The ability to extract the relevant information from a SAS image depends on a number of different factors, as illustrated in Fig. 14. Observation geometry given by range and elevation angle is important for interpretation of the highlight structure and shadow. Backscattered target signals can contain elements of specular reflections, diffuse scattering, transparency, resonant scattering and multiple scattering, all of which complicate the process of retrieving the relevant information. Finally, image resolution is critical in resolving or classifying objects, shadow shape and the surrounding scatterers on the seafloor.

## 8. Applications of SAS

There are many applications where SAS is suitable. Very high resolution acoustic imaging can be achieved on traditional sonar using very high frequencies, and there are sonar systems today using up to 2 MHz frequency. They are, hovewer, very limited in range. When large area coverage *and* very high resolution is needed at the same time, SAS is really the only technology that can provide a solution. In this section, we describe a particular SAS system, the HISAS 1030, and show example images from different applications.

### 8.1 The HISAS 1030 interferometric SAS

HISAS 1030 is a wideband widebeam interferometric SAS developed by Kongsberg Maritime and FFI (Fossum et al., 2008; Hagen et al., 2008). The sonar contains two along-track receiver arrays of 1.2 m length with 32 elements in each array, and a vertical baseline approximately 30 cm which equals 20 wavelengths. The transmitter is a vertical phased array with 16 elements, and the transmit beam can be electronically steered and shaped to obtain the best possible performance in shallow waters (see Section 5.5). The transmitter can also be used as a receiver, giving 16 individual receiver channels along a vertical array. Fig. 15 shows the sonar mounted on a HUGIN 1000-MR AUV. Typical HISAS 1030 specifications are listed in Table 2. The SAS processing is done in a software suite named *FOCUS Toolbox* (Hansen et al., 2005; 2003).

Fig. 16 shows an example SAS image made by the HISAS 1030 on the HUGIN 1000 AUV. The image shows the wreck of the 1500 dwt oil tanker Holmengraa lying on a slanted seabed at

Fig. 15. The HISAS 1030 interferometric SAS on the HUGIN autonomous underwater vehicle. The picture was taken just before launch during a scientific mission on board the research vessel H U Sverdrup II in Norwegian waters in April 2010.

| | |
|---|---|
| Center frequency | 100 kHz |
| Wavelength | 1.5 cm |
| Bandwidth | 30 kHz |
| Total frequency range | 50-120 kHz |
| Along-track resolution | 3 cm |
| Cross-track resolution | 3 cm |
| Maximum range @ 2 m/s | 200 m |
| Area coverage rate | 2 km$^2$/h |

Table 2. Typical system specifications for the HISAS 1030 interferometric SAS.

depth 77 m. The distance to the center of the image is about 95 m. The length of the wreck is about 68 m and width about 9 m.
Fig. 17 captures the essence of SAS: Long range *and* high resolution at the same time. The large (middle) image shows a SAS image where the range is 25 m (left) to 325 m (right), and the dynamic range is 32 dB. The data was collected with HISAS 1030 on HUGIN AUV running at 2.3 knots at 40 m altitude, outside Horten, Norway in approximately 200 m water depth. The upper image shows a section from 200 m range to 250 m range with the wreck of the German WWII submarine U735. The lower image shows a section from 260 m range to 290 m range with a 1 m$^3$ cube. The length of the synthetic aperture is (see section 4.1)

$$L_{\text{sa}} \approx R\eta \frac{\lambda}{d} \tag{14}$$

Fig. 16. SAS image of the wreck of the Norwegian tanker Holmengraa that was sunk during WWII in 1944. Courtesy of Kongsberg Maritime.

where $R$ is the range, $\lambda$ is the wavelength at center frequency, $d$ is the along-track element size in the array. $\eta$ is a programmable parameter controlling the *process beamwidth*, that is, the beamwidth actually processed. In this particular case, $\eta = 2/3$, and the length of the synthetic aperture at maximum range becomes $L_{sa} \approx 90\,\text{m} = 6000\lambda$. The SAS resolution-gain, defined as the ratio between along-track resolution in real aperture $\delta x_{ra}$ and synthetic aperture $\delta x_{sa}$ is

$$Q_{sa} = \frac{\delta x_{ra}}{\delta x_{sa}} = \frac{L_{sa}}{L/2} \approx R\eta \frac{2\lambda}{Ld}, \tag{15}$$

where and $L$ is the array length. In Fig. 17, the SAS resolution-gain is $Q_{sa} \approx 150$ at maximum range. This is a considerable resolution improvement, and the equivalent along-track resolution is very difficult to obtain using real aperture techniques.

### 8.2 Underwater archaeology

SAS is a candidate technology in searching for wrecks and other objects of historic interest. Fig. 18 shows a SAS image collected by the Royal Norwegian Navy in a training mission close to the town of Tromsø in the winter of 2009. The data was collected at 34 m water depth, close to shore. The image shows the wreck of a German WWII Heinkel He 115 seaplane. The original length of the plane is 17 m and the wingspan was 22 m. Note the small part outside the right wing of the seaplane. This is probably one of the floats. The object in the lower left part of the image is probably the tail-section of another plane of the same type.

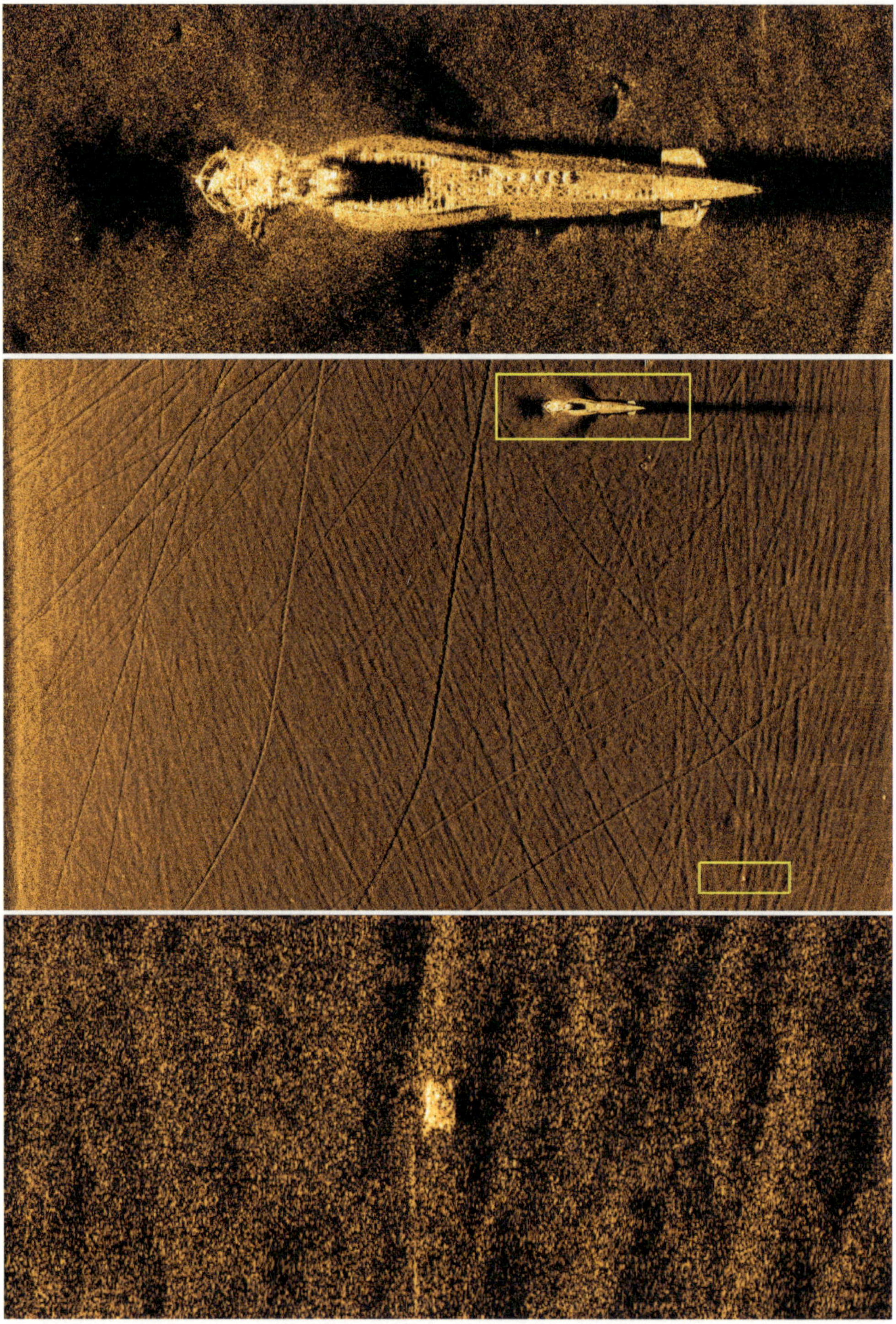

Fig. 17. Middle: SAS image where the range is 25 m (left) to 325 m (right). Courtesy of Kongsberg Maritime.

Fig. 18. SAS image of a German WWII Heinkel He 115 seaplane lying on the seafloor. Courtesy of the Royal Norwegian Navy. Photograph is from wikipedia.

### 8.3 Search for small objects

When searching for small objects over very large areas, SAS is an excellent tool. Fig. 19 shows a SAS image from an area with debris. The sonar has traveled along the vertical direction of the image, and the sonar look direction is towards right. The size of the image is 190 m (along-track) and 30 m to 165 m cross-track, and the water depth is around 70 m. The yellow boxes indicates three objects, and the small images below show zoomed images of the objects. These are two drums of approximate size 0.9 m length and 0.6 m diameter, and a cylinder of approximate length 2.5 m. The lower row of images shows optical images of the same objects. The optical images were collected with another HUGIN AUV. The altitude was 5 m on the data collection of the optical images. We see several interesting features in the objects. The drum at 73 m has clear indications of partial transparency. The back end of the barrel is clearly visible, and there is acoustic pollution in the shadow region. The drum at 112 m has a less defined back end and deeper shadow contrast, indicating that this object is less transparent. In the optical images, indeed, we see that the drum at 73 m has severe damage and holes, while the drum at 112 m looks to be more intact. In the optical image of the cylinder, we see a small cavity in the lower right end. This appears as a highlight in the SAS image.

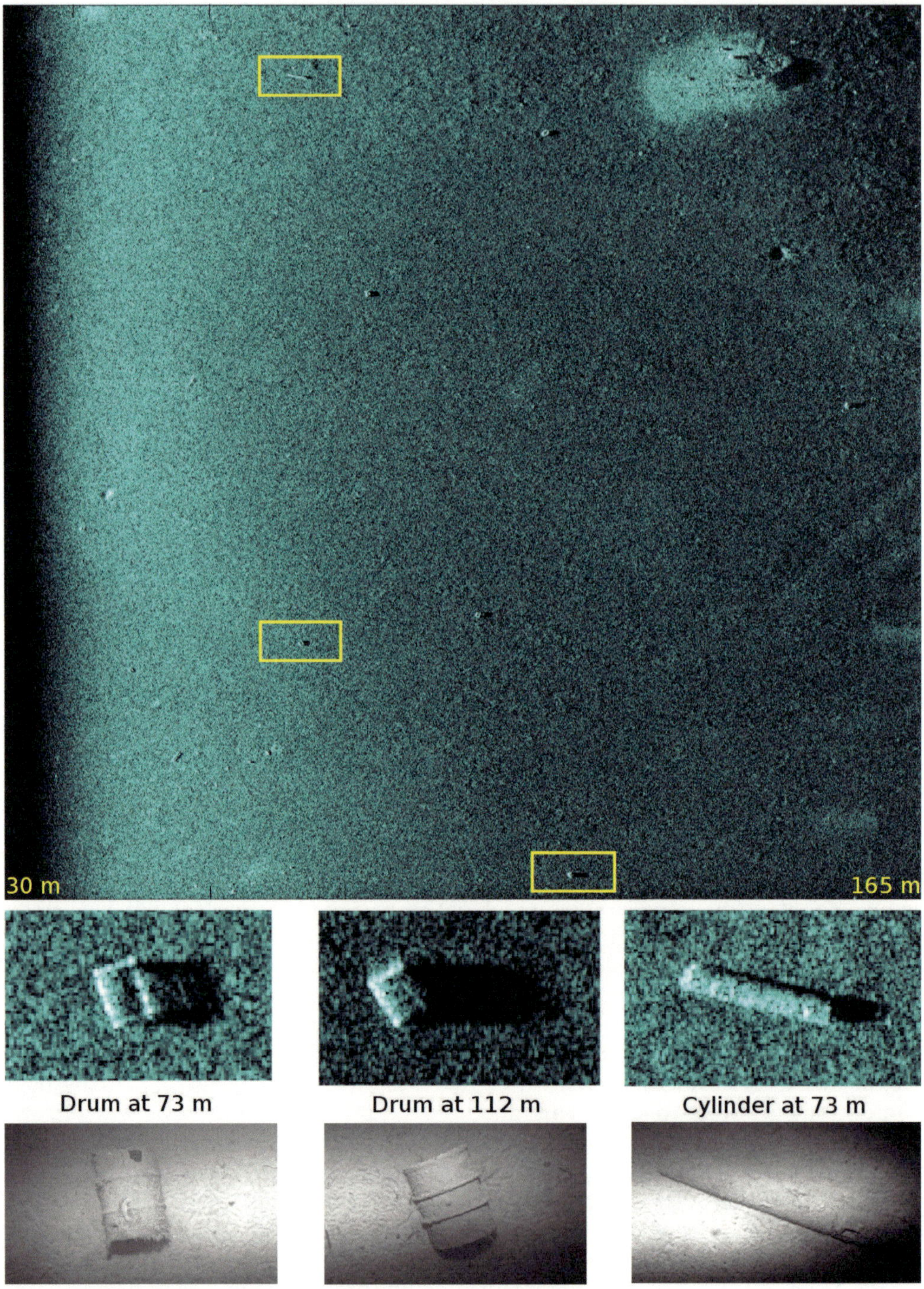


Fig. 19. Upper: SAS image of an area with small objects. The range is 30 m (left) to 165 m (right), and the along-track (vertical) is 190 m. The small cut-outs shows zoomed images of three different objects. The lower row shows optical images of the same targets. Courtesy of Kongsberg Maritime / FFI.

### 8.4 Inspection of man made constructions

External inspection of underwater constructions such as pipelines is an important task. The objective of these inspections is to detect burial, exposure, free spans and buckling of the pipeline, as well as possible damages due to trawling, anchoring and debris near the pipeline. SAS may be well suited technology for some of these tasks (Hagen et al., 2010), (Hansen, Sæbø, Callow & Hagen, 2010). Fig. 20 shows an example SAS image collected by a HUGIN AUV during a demonstration in San Diego, USA, in 2010. The image shows a sewer pipeline, and a rope or wire on the seafloor.

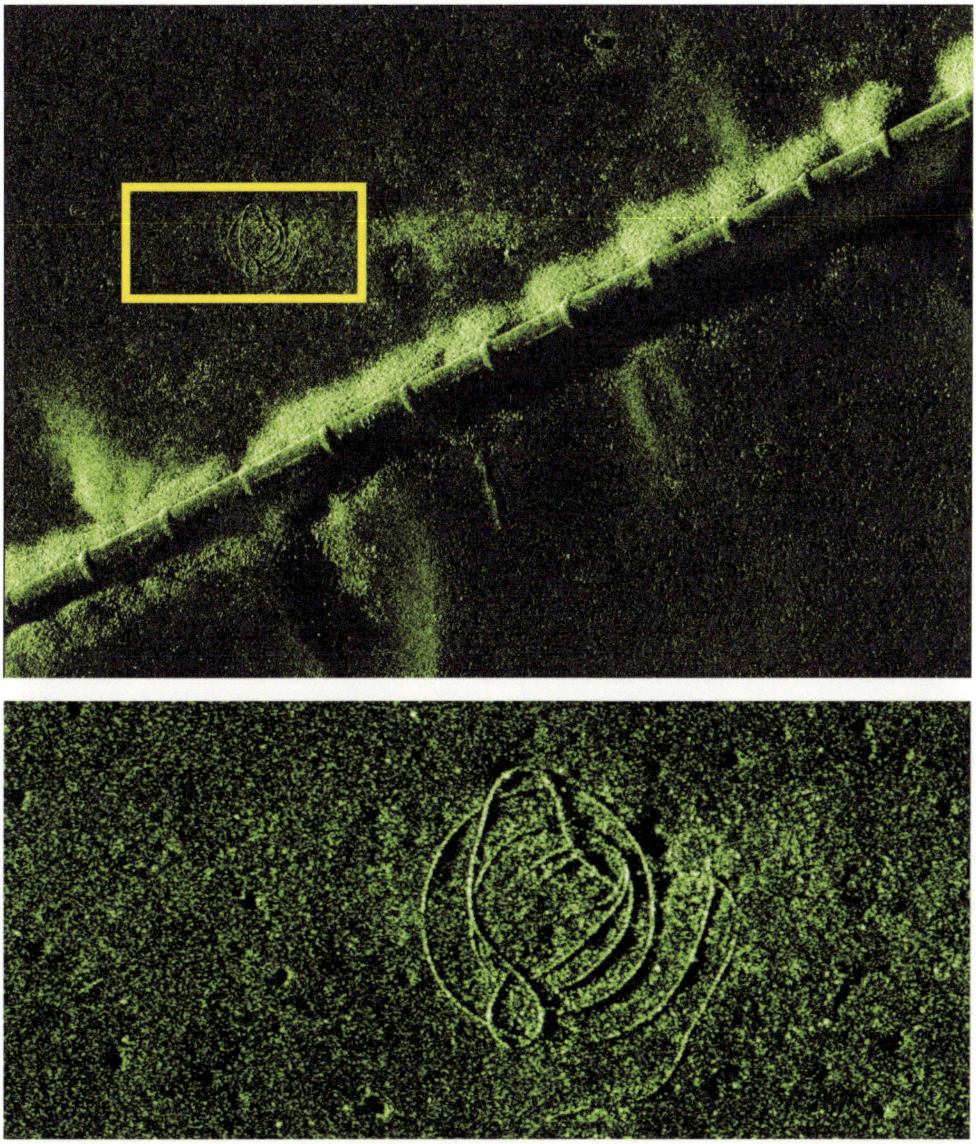

Fig. 20. Upper: SAS image of a pipeline outside San Diego. The range in the upper image is 65 m - 145 m. Lower: 20 m times 10 m zoomed area at range 85 m of a rope or wire on the seafloor. Courtesy of Kongsberg Maritime.

## 9. Conclusion

Synthetic aperture sonar (SAS) is an advanced signal processing technique to improve resolution in sonar imagery. The main application is detailed documentation of the seafloor, in areas such as search for small objects, underwater archaeology, detailed seabed mapping and documentation of underwater installations. Successful SAS is dependent on accurate knowledge about the sonar position, the ocean environment and the seabed topography.
SAS is substantially more mature now than 10 years ago. To illustrate the maturity of commercially available systems, we show a final example of SAS data collected by the HUGIN AUV carrying the HISAS 1030. Fig. 21 shows SAS images and interferometric SAS relative bathymetries of a German WWII Focke Wulf 190 A-3 aeroplane that was found by the Royal Norwegian Navy mine warfare flotilla. The length of the plane is 9 m and the wingspan is 10.5 m. The tail was damaged as we see in the SAS images and the bathymetries. The motor fell off during the recovery. These images was produced at sea by the Royal Norwegian Navy personnel during the search operation. The images was constructed using micronavigation and sidescan bathymetry as a preprocessing step, then backprojection in three dimensions for image formation, and finally bathymetry estimation using a maximum likelihood phase estimator in the interferometric processing.

## 10. Acknowledgments

The author thanks the very good colleagues Hayden J Callow, Torstein O Sæbø, and Stig A V Synnes at the Norwegian Defence Research Establishment. The author also thank Kongsberg Maritime, and in particular Per Espen Hagen and Bjørnar Langli, for a long standing collaboration and providing data for the analysis. Finally the author wish to thank the Royal Norwegian Navy Mine Warfare Service for the long fruitful collaboration and kind permission to use data recorded during Navy operations with their HUGIN 1000-MR AUV.

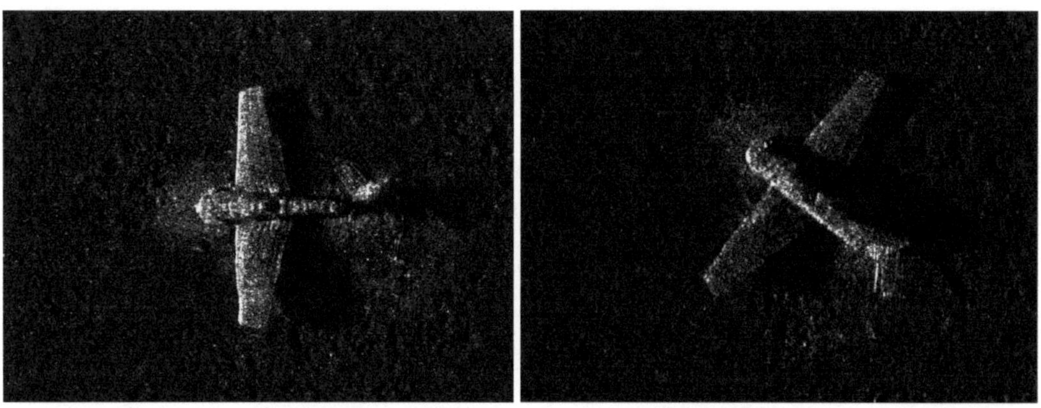

(a) SAS image 1 (b) SAS image 2

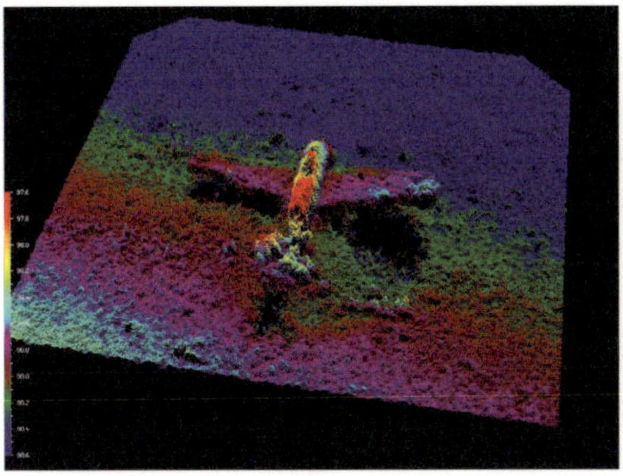

(c) Bathymetry by SAS interferometry 1

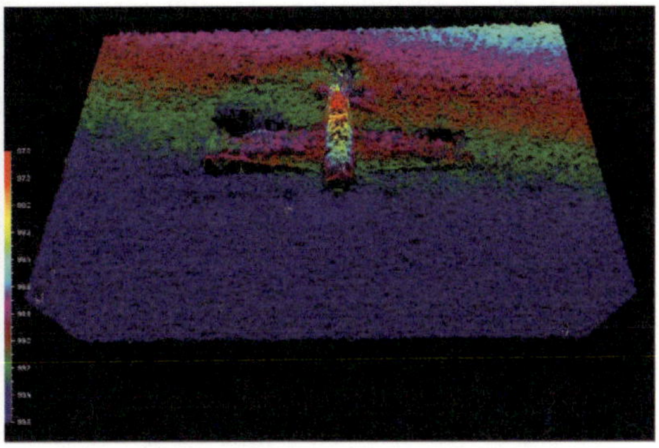

(d) Bathymetry by SAS interferometry 2

(e) Photograph taken during the recovery

(f) Photograph taken during the recovery

Fig. 21. The German WWII Focke Wulf 190 A-3 aircraft. The plane was found by the Royal Norwegian Navy Mine warfare flotilla at 98 m water depth during an underwater archaeology mission. Courtesy of the Royal Norwegian Navy.

## 11. References

Ainslie, M. (2010). *Principles of Sonar Performance Modelling*, Springer Verlag.

Bellettini, A. & Pinto, M. A. (2002). Theoretical accuracy of synthetic aperture sonar micronavigation using a displaced phase-center antenna, *IEEE J. Oceanic Eng.* 27(4): 780–789.

Bellettini, A. & Pinto, M. A. (2009). Design and Experimental Results of a 300-kHz Synthetic Aperture Sonar Optimized for Shallow-Water Operations, *IEEE J. Oceanic Eng.* 34(3): 285–293.

Bellettini, A., Pinto, M. A. & Wang, L. (2003). Effect of multipath on synthetic aperture sonar, *Proc. 5th World Congr. Ultrasonics*, Paris, France, pp. 531–534.

Billon, D. & Fohanno, F. (2002). Two improved ping-to-ping cross-correlation methods for synthetic aperture sonar: theory and sea results, *OCEANS'02 MTS/IEEE*, Vol. 4, IEEE, pp. 2284–2293.

Blondel, P. (2009). *The Handbook of Sidescan Sonar*, Geophysical Sciences, Springer Praxis Books.

Brekhovskikh, L. & Lysanov, Y. (1982). *Fundamentals of Ocean Acoustics*, Vol. 8 of *Springer Series in Electrophysics*, Springer-Verlag, Berlin, Germany.

Bruce, M. P. (1992). A processing requirement and resolution capability comparison of side-scan and synthetic-aperture sonars, *IEEE J. Oceanic Eng.* 17(1): 106–117.

Burdic, W. S. (1984). *Underwater acoustic system analysis*, Prentice Hall.

Callow, H. J. (2003). *Signal Processing for Synthetic Aperture Sonar Image Enhancement*, PhD thesis, University of Canterbury, Christchurch, New Zealand.

Callow, H. J. (2010). Comparison of SAS processing strategies for crabbing collection geometries, *Proceedings of Oceans 2010 MTS/IEEE*, Seattle, US.

Carrara, W. G., Goodman, R. S. & Majewski, R. M. (1995). *Spotlight Synthetic Aperture Radar: Signal Processing Algorithms*, Artech House.

Claerbout, J. (1995). *Basic earth imaging*, Stanford Exploration Project.
URL: *http://sepwww.stanford.edu/*

Cumming, I. G. & Wong, F. H. (2005). *Digital Processing of Synthetic Aperture Radar Data: Alogirthms and Implementation*, Artech House.

Curlander, J. C. & McDonough, R. N. (1991). *Synthetic Aperture Radar: Systems and Signal Processing*, John Wiley & Sons, Inc., 605 Third Avenue, New York, NY.

Cutrona, L. J. (1975). Comparison of sonar system performance achievable using synthetic-aperture techniques with the performance achievable by more conventional means, *J. Acoust. Soc. Am.* 58(2): 336–348.

Dickey Jr, F., Labitt, M. & Staudaher, F. (1991). Development of airborne moving target radar for long range surveillance, *Aerospace and Electronic Systems, IEEE Transactions on* 27(6): 959–972.

Fish, J. P. & Carr, H. A. (2001). *Sound reflections: Advanced Applications of Side Scan Sonar*, LowerCape Publishing.

Fossum, T. G., Hagen, P. E., Langli, B. & Hansen, R. E. (2008). HISAS 1030: High resolution synthetic aperture sonar with bathymetric capabilities, *Shallow survey*, Portsmouth, NH, USA.

Franceschetti, G. & Lanari, R. (1999). *Synthetic Aperture Radar Processing*, CRC Press.

Gebert, N., Krieger, G. & Moreira, A. (2010). Multichannel azimuth processing in ScanSAR and TOPS mode operation, *IEEE Trans. Geosci. Remote Sensing.* 48(7): 2994–3308.

Glover, N. P. & Campell, I. (2010). Simultaneous low and high frequency high resolution SAS and a statistical method of quantifying the resolutions obtained, *Proceedings of Synthetic Aperture Sonar and Radar 2010*, Lerici, Italy.

Goodman, J. W. (2007). *Speckle Phenomena in Optics: Theory and Applications*, Roberts and Company.

Gough, P. T. & Hawkins, D. W. (1997). Imaging algorithms for a strip-map synthetic aperture sonar: Minimizing the effects of aperture errors and aperture undersampling, *IEEE J. Oceanic Eng.* 22 (1): 27–39.

Gough, P. T. & Hawkins, D. W. (1998). A short history of synthetic aperture sonar, *Geoscience and Remote Sensing Symposium Proceedings, 1998. IGARSS'98. 1998 IEEE International*, Vol. 2, IEEE, pp. 618–620.

Hagen, P. E., Børhaug, E. & Midtgaard, Ø. (2010). Pipeline inspection with interferometric SAS, *Sea Technology* pp. 37–40.

Hagen, P. E., Fossum, T. G. & Hansen, R. E. (2008). HISAS 1030: The next generation mine hunting sonar for AUVs, *UDT Pacific 2008 Conference Proceedings*, Sydney, Australia.

Hagen, P. E. & Hansen, R. E. (2008). Synthetic aperture sonar challenges ... and how to meet them, *Hydro International* pp. 26–31.

Hagen, P. E. & Hansen, R. E. (2009). Robust synthetic aperture sonar operation for AUVs, *Proceedings of Oceans '09 MTS/IEEE Biloxi*, Biloxi, MS, USA.

Hansen, R. E. (2010). Robust synthetic aperture sonar for autonomous underwater vehicles, *Proceedings of Synthetic Aperture Sonar and Radar 2010*, Lerici, Italy.

Hansen, R. E., Callow, H. J. & Sæbø, T. O. (2007). The effect of sound velocity variations on synthetic aperture sonar, *Proceedings of Underwater Acoustic Measurements 2007*, Crete, Greece.

Hansen, R. E., Callow, H. J., Sæbø, T. O., Hagen, P. E. & Langli, B. (2008). High fidelity synthetic aperture sonar products for target analysis, *Proceedings of Oceans '08 Quebec*, Quebec, Canada.

Hansen, R. E., Callow, H. J., Sæbø, T. O., Synnes, S. A., Hagen, P. E., Fossum, T. G. & Langli, B. (2009). Synthetic aperture sonar in challenging environments: Results from the HISAS 1030, *Proceedings of Underwater Acoustic Measurements 2009*, Nafplion, Greece.

Hansen, R. E., Callow, H. J., Sæbø, T. O. & Synnes, S. A. V. (2010). Challenges in seafloor imaging and mapping with synthetic aperture sonar, *Proceedings of EUSAR 2010*, Aachen, Germany, pp. 540–543.

Hansen, R. E., Sæbø, T. O., Callow, H. J. & Hagen, P. E. (2010). Interferometric synthetic aperture sonar in pipeline inspection, *Proceedings of Oceans 2010 MTS/IEEE*, Sydney, Australia.

Hansen, R. E., Sæbø, T. O., Callow, H. J., Hagen, P. E. & Hammerstad, E. (2005). Synthetic aperture sonar processing for the HUGIN AUV, *Proceedings of Oceans '05 Europe*, Vol. 2, Brest, France, pp. 1090–1094.

Hansen, R. E., Sæbø, T. O., Gade, K. & Chapman, S. (2003). Signal processing for AUV based interferometric synthetic aperture sonar, *Proceedings of Oceans 2003 MTS/IEEE*, San Diego, CA, USA, pp. 2438–2444.

Hanssen, R. F. (2001). *Radar Interferometry: Data Interpretation and Error Analysis*, Kluwer Academic Publishers.

Hayes, M. P. & Gough, P. T. (2009). Synthetic aperture sonar: A review of current status, *IEEE J. Oceanic Eng.* 34(3): 207–224.

Jakowatz, J. C. V., Wahl, D. E., Eichel, P. H., Ghiglia, D. C. & Thompson, P. A. (1996). *Spotlight-Mode Synthetic Aperture Radar: A Signal Processing Approach*, Kluwer Academic Publishers.

Jean, F. (2008). Shadows, synthetic aperture sonar and forward looking gap-filler: different imaging algorithms, *Proceedings of OCEANS 2008 - MTS/IEEE Kobe Techno-Ocean*, Kobe, Japan.

Johnson, D. H. & Dudgeon, D. E. (1993). *Array signal processing: Concepts and Techniques*, Signal processing series, Prentice Hall, Englewood Cliffs, NJ, USA.

Larsen, L. J., Wilby, A. & Stewart, C. (2010). Deep ocean survey and search using synthetic aperture sonar, *Proceedings of Oceans 2010 MTS/IEEE*, Seattle, US.

Levanon, N. (1988). *Radar Principles*, Wiley Interscience.

Lurton, X. (2010). *An Introduction to Underwater Acoustics: Principles and Applications*, second edn, Springer Praxis Publishing.

Manolakis, D. G., Ingle, V. K. & Kogon, S. M. (2000). *Statistical and Adaptive Signal Processing*, McGraw-Hill.

Massonnet, D. & Souyris, J. (2008). *Imaging with synthetic aperture radar*, EFPL Press.

Medwin, H. & Clay, C. S. (1998). *Fundamentals of Acoustical Oceanography*, Academic Press, San Diego, CA, USA.

Nielsen, R. O. (1991). *Sonar signal processing*, Artech House.

Oliver, C. & Quegan, S. (1998). *Understanding Synthetic Aperture Radar Images*, Artech house, Inc.

Pinto, M. A. (2002). High resolution seafloor imaging with synthetic aperture sonar, *IEEE Oceanic Eng. Newsletter* pp. 15–20.

Pinto, M. A., Fohanno, F., Trémois, O. & Guyonic, S. (1997). Autofocusing a synthetic aperture sonar using the temporal and spatial coherence of seafloor reverberation, *in* O. Bergem & A. P. Lyons (eds), *High Frequency Acoustics in Shallow Water*, SACLANTCEN Conference Proceedings, NATO SACLANT Undersea Research Centre, La Spezia, Italy, pp. 417–424.

Sæbø, T. O. (2010). *Seafloor Depth Estimation by means of Interferometric Synthetic Aperture Sonar*, PhD thesis, University of Tromsø, Norway.

Sheriff, R. (1992). Synthetic aperture beamforming with automatic phase compensation for high frequency sonars, *Autonomous Underwater Vehicle Technology, 1992. AUV'92., Proceedings of the 1992 Symposium on*, IEEE, pp. 236–245.

Soumekh, M. (1994). *Fourier Array Imaging*, Prentice Hall, Englewood Cliffs, NJ, USA.

Synnes, S. A., Hansen, R. E. & Sæbø, T. O. (2009). Assessment of shallow water performance using interferometric sonar coherence, *Proceedings of Underwater Acoustic Measurements 2009*, Nafplion, Greece.

Urick, R. J. (1983). *Principles of Underwater Sound*, Mcgraw-Hill Book Company.

Van Trees, H. L. (2002). *Optimum Array Processing (Detection, Estimation, and Modulation Theory, Part IV)*, Wiley-Interscience.

2

# Advances in Multistatic Sonar

Danilo Orlando[1] and Frank Ehlers[2]
[1]*Dipartimento di Ingegneria dell'Innovazione - Università del Salento*
[2]*NATO Undersea Research Centre*
*Italy*

## 1. Introduction

Although the concept of multistatic active sonar (MAS) has been around for over 50 years, new trends have brought this technology to the forefront of anti-submarine warfare research. These trends include advancements in sonar sensors and signal processing, advancements in submarine stealth, and a desire to track targets in noisy and reverberant environments, such as near-shore or shallow waters. The latest trend is to exploit the really game-changing capabilities of unmanned and autonomously operating underwater vehicles.

The focus of this chapter is on advances in signal processing enabling especially the tracking of low signature targets, namely targets with low signal-to-noise ratio (SNR), in a multisensor environment. In particular, the track-before-detect (TBD) approach and its adaption to pre-selected contact-based tracking are addressed. The TBD approach is designed to track low SNR targets. TBD-based procedures jointly process several consecutive pings and, relying on target kinematics or, simply, exploiting the physically admissible target transitions, declare the presence of a target and, eventually, its track (Orlando et al., IEEE-TSP 2010). A TBD algorithm is typically fed by unthresholded data or thresholded data with significantly lower thresholds than the ones used by conventional trackers. Moreover an important feature of a TBD scheme is the so-called constant false track acceptance rate (CFTAR) property: if a TBD scheme ensures the constant false alarm rate property with respect to the unknown statistics of the disturbance, then it allows controlling the overall false track acceptance rate. The TBD algorithm herein presented considers a bistatic sonar architecture and is capable of handling raw hydrophone data (Orlando et al., CIP 2010). Remarkably, it guarantees the CFTAR property. Performance analysis highlights its potential to implement automatic track continuation and to prepare automatic classification for temporarily weak targets as these tasks are usually the challenges that MAS systems have to overcome.

In the context of multistatic sonar, a batch algorithm is also introduced, that jointly processes measurements provided by multiple sensors over a certain number of consecutive pings (Orlando et al., FUSION 2010). These measurements are time differences of arrival and bearing information of a target maneuvering in the surveillance region. This approach is tested on a benchmark data set provided by METRON in the context of collaborative international multi-laboratory research that is ongoing in the ISIF Multi-Static Tracking Working Group (Orlov, Metron Data set 2009).

The remainder of this chapter is organized as follows: the next section is devoted to the description of bistatic and multistatic sonar systems. Section III focuses on the derivation of the TBD (or TBD-based) processors. In Section IV a node selection strategy for multistatic sonar systems is described on a conceptual level while Section V provides some illustrative examples. Finally, Section VI contains concluding remarks and hints for future work.

## 2. Multistatic sonar

When low frequency active sonar was first used to find and track underwater targets, the required hardware for both the acoustic source and the acoustic receiver was installed onboard a single unit. The deployed part (or wet-end) of the system is quite heavy and large in extend and there has to be a synchronization between the activating source and the receiving hydrophones. Hence, the natural approach is to use such a monostatic system (source and receiver collocated) in order to avoid the necessity and cost for multiple units involved and data synchronization issues. However, looking at the cylindrical structure of the target it turns out that approaching or opening aspects of the target result in very low detection probabilities for those monostatic systems.
An enormous improvement in performance is possible if multiple acoustic sources and receivers are deployed in a spatially distributed manner. Multiple units have to be used and data synchronization has to be solved in order to implement these so called multistatic active sonar systems. A multistatic architecture can provide:

- short latency (due to effective Doppler processing);
- high precision (due to triangulation);
- fewer false alarms;
- anti-stealth.

These benefits can only be exploited by automatic fusion and tracking of the multistatic data. Automatic tracking algorithms use the information from consecutive measurements to discriminate between physically feasible movements and random movements of prospective targets. By this discrimination the overall false alarm rate can be reduced, especially when different bistatic aspects on the target are combined within the tracking algorithm. It is possible to summarize the major issues of MAS as follows:

1. estimation task: non-linear and noisy measurements and variable sound channels;
2. multi-target issue: false alarms, clutter targets, and bottom reflections;
3. non-cooperative evader: stealthy target with evading actions;
4. multi-agent coordination whereby communication between sensors needed, but not always present.

In this chapter, two TBD-based techniques are introduced in order to overcome limitations due to the issues 1 and 2.
Before proceeding further, notice that a multistatic system consists of multiple sources and multiple receivers and, hence, it can be viewed as composed of different bistatic subsystems. For this reason, the bistatic configuration is first introduced in Subsection 2.1, then the description of the multistatic architecture is given in Subsection 2.2.

### 2.1 System description: the bistatic architecture

The bistatic sonar scenario involves a projector and a receiver, placed at a different location and equipped with an array of sensors, a point-like target at a certain "distance" from the array (range), and a signal that travels from the projector to the target and from the target to the receiver. The system under consideration utilizes a planar array (see Figure 1) consisting of

$N_a$ arms; each of them contains $N_s$ identical sensors spaced $d$ meters apart from one another. The total number of sensors is

$$N = N_a(N_s - 1) + 1, \quad (1)$$

where we are accounting for the fact that the arms share the central sensor. Moreover, let $\theta_a = 2\pi/N_a$ be the separation angle between two consecutive arms, $\theta_r$ the azimuthal angle of the impinging target echo, measured clockwise from the reference arm, and

$$\theta_i = \theta_r - (i-1)\theta_a, \quad i = 1, \ldots, N_a, \quad (2)$$

the angle of arrival with respect to the $i$th arm.
Suppose that the projector omnidirectionally transmits the following pulse waveform

$$\Re e \left\{ A e^{j\phi} p(t) e^{j2\pi f_c t} \right\}, \quad t \in [0, T_p[, \quad (3)$$

where

$$p(t) = \frac{1}{\sqrt{T_p}} e^{j\pi k t^2} \quad (4)$$

is a frequency modulated waveform of duration $T_p$, $A > 0$ is an amplitude factor related to the transmitted power, $\phi \in [0, 2\pi[$ is the initial phase of the carrier signal, $f_c$ is the carrier frequency, $\Re e\{z\}$ denotes the real part of the complex number $z$, and $k = B/T_p$, with $B$ the pulse bandwidth. In the next subsection, we derive the discrete-time form for the received signal.

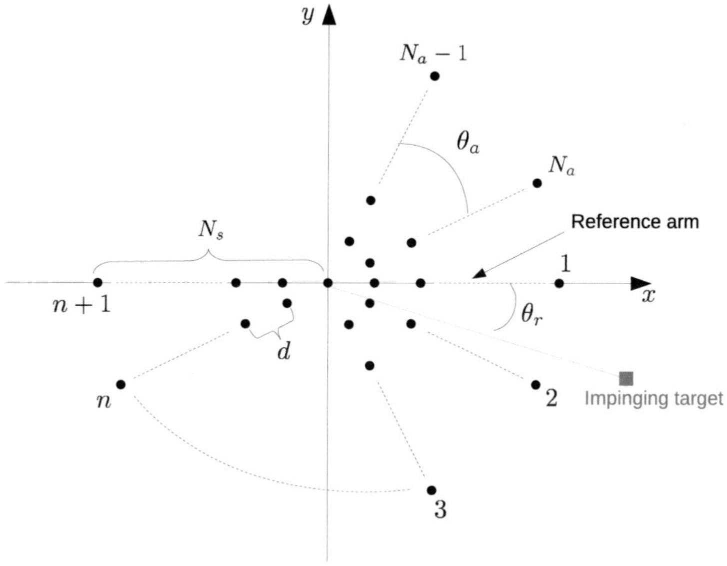


Fig. 1. The planar array and the associated reference system: each sensor is represented by a filled circle.

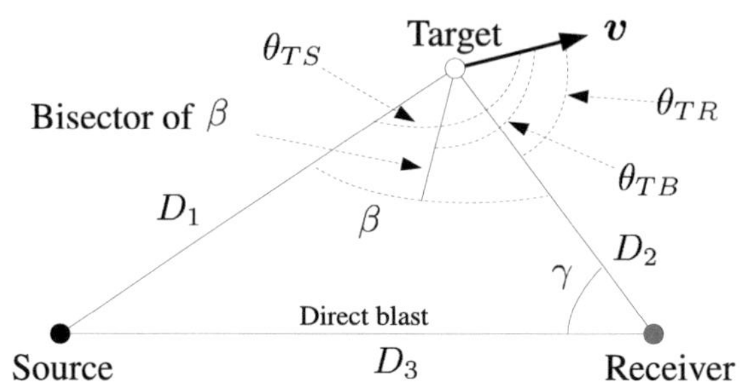


Fig. 2. Bistatic geometry

### 2.1.1 Discrete-time sensor model

Consider the geometry depicted in Figure 2 and suppose that a prospective (point-like) target is moving within the surveillance area with a velocity vector $v \in \mathbb{R}^{2\times 1}$. The received signal at the $n$th sensor of the $i$th arm is given by

$$r_{n,i}(t) = \Re e\left\{\frac{\alpha}{\sqrt{T_p}} e^{j\pi k(t-\tau_{n,i}(t))^2} e^{j2\pi f_c(t-\tau_{n,i}(t))}\right\} + n_{n,i}(t), \tag{5}$$

where

- $\alpha \in \mathbb{C}$ is a complex and unknown (deterministic) factor accounting for the transmitted power, the transmitting and receiving gain, the two-way path loss, and the sonar cross-section of the target; hereafter $\alpha$ denotes any proper modification of the previous constant;
- $\tau_{n,i}(t)$ is the time delay between the transmission by the source and the arrival of the target echo at the $n$th sensor of the $i$th arm;
- $n_{n,i}(t)$ is the overall disturbance component; in the sequel we use the above symbol to denote any modification of the noise components.

Now observe that $\tau_{n,i}(t)$ consists of three components

$$\tau_{n,i}(t) = \tau_1(t) + \tau_2(t) - \frac{d(n-1)}{c}\cos\theta_i, \tag{6}$$

where[1] $c$ ($\gg \|v\|$) is[2] the speed of sound in water, $\tau_1(t)$ is the travel time of the transmitted signal from source to target, and $\tau_2(t)$ is the travel time of the target echo from target to receiver[3].
For the sake of convenience, it is worth defining the following time difference of arrival

$$\tau'(t) = \tau_1(t) + \tau_2(t) - \tau_3, \tag{7}$$

where $\tau_3 = D_3/c$ is the travel time from source to receiver ($D_3$ is the distance between the source and the receiver). It follows that equation (5) can be recast as

$$r_{n,i}(t) \approx \Re e\left\{\alpha e^{j\pi k(t-\tau'(t)-\tau_3)^2} e^{j2\pi f_c t} e^{j2\pi(n-1)\nu_i} e^{-j2\pi f_c\tau_3} e^{-j2\pi f_c\tau'(t)}\right\} + n_{n,i}(t), \tag{8}$$

---

[1] Recall that $d$ is the interelement spacing
[2] $\|\cdot\|$ denotes the Euclidean norm.
[3] To be more precise, $\tau_2(t)$ is the travel time of the target echo from the target to the origin of the reference system which coincides with the center of the receiver (see Figure 1).

where

$$\nu_i = \frac{d}{\lambda}\cos\theta_i \tag{9}$$

is the *spatial frequency* with respect to the $i$th arm. An equivalent monostatic range $D'(t)$, corresponding to $\tau'(t)$, can be defined as follows

$$D'(t) = \frac{c\tau'(t)}{2} = D' - v'\left(t - \frac{\tau'(t)}{2}\right), \tag{10}$$

where $D'$ is the initial equivalent monostatic range and (Willis, Scitech Pub. 2005)

$$v' = \|v\| \cos\theta_{TB} \cos\frac{\beta}{2}(< \|v\| \ll c). \tag{11}$$

Equation (11) can be obtained by writing $D'(t)$ as follows

$$D'(t) = \left[\frac{D_1(t) + D_2(t) - D_3}{2}\right], \tag{12}$$

where $D_1(t)$ is the source-target distance and $D_2(t)$ is the receiver-target distance, and evaluating its first derivative with respect to $t$, namely

$$\frac{d}{dt}D'(t) = \frac{1}{2}\|v\|(\cos\theta_{TR} + \cos\theta_{TS}), \tag{13}$$

where

$$\theta_{TR} = \theta_{TB} - \beta/2, \tag{14}$$

$$\theta_{TS} = \theta_{TB} + \beta/2. \tag{15}$$

It is not difficult to show that knowledge of $D'$ and $\theta_r$ allows to evaluate the initial receiver-target distance, $D_2$ say. Indeed, from Figure 2 the following relation among the initial source-target distance, $D_1$ say, $D_2$, and $D_3$ holds true

$$D_1^2 = D_2^2 + D_3^2 - 2D_2D_3\cos\gamma. \tag{16}$$

Then, solving the next equation

$$D' = \frac{D_1 + D_2 - D_3}{2} \tag{17}$$

with respect to $D_1$ and substituting the solution into (16) yields

$$D_2 = D'\left[\frac{1 + D_3/D'}{1 + (D_3/D')(1 - \cos\gamma)/2}\right]. \tag{18}$$

The above equation allows us to recast the expression of the received signal in terms of the equivalent monostatic quantities $v'$, $D'$, and $\tau'(t)$. More specifically, from equation (10) we obtain

$$\tau'(t) \approx \frac{2D'}{c} - \frac{2v't}{c} = \tau' - \frac{2v't}{c}, \tag{19}$$

where[4]

$$\tau' = \frac{2D'}{c}. \tag{20}$$

[4] Remember that $c \gg v'$.

Then, inserting (19) into (8) yields

$$r_{n,i}(t) \approx \Re e \left\{ \alpha e^{j\pi k(t-\tau'-\tau_3)^2} e^{j2\pi f_c t} e^{j2\pi(n-1)\nu_i} e^{j2\pi f'_D t} \right\} + n_{n,i}(t), \quad (21)$$

where

$$f'_D = 2\frac{f_c v'}{c}. \quad (22)$$

After complex baseband conversion, the output of a filter matched to $p(t)$ is given by (see (Bandiera at al., M&C 2009))

$$y_{n,i}(t) = \alpha \int_{-\infty}^{+\infty} e^{j\pi k(u-\tau'-\tau_3)^2} e^{j2\pi(n-1)\nu_i} e^{j2\pi f'_D u} e^{-j\pi k(t-u)^2} du + n_{n,i}(t). \quad (23)$$

Let $u_1 = u - \tau' - \tau_3$, then

$$\begin{aligned} y_{n,i}(t) &= \alpha e^{j2\pi(n-1)\nu_i} \int_{-\infty}^{+\infty} e^{j2\pi f'_D u_1} e^{j\pi k u_1^2} e^{-j\pi k[u_1-(t-\tau'-\tau_3)]^2} du_1 + n_{n,i}(t) \\ &= \alpha \chi_p(t-\tau'-\tau_3, f'_D) e^{j2\pi(n-1)\nu_i} + n_{n,i}(t), \end{aligned} \quad (24)$$

where again $\alpha$ and $n_{n,i}(t)$ denote a proper modification of the target response and of the noise component, respectively, and $\chi_p(\cdot,\cdot)$ is the (complex) ambiguity function of $p(t)$.
In order to generate the vector of the noisy returns corresponding to the $l$th range gate, $l = 1,\ldots,L$, the output of the matched filter $y_{n,i}(t)$ is sampled at

$$t_l = t_{\min} + (l-1)T_B, \quad (25)$$

where $t_{\min}$ denotes the beginning of the sampling process and $T_B = 1/B$. The time samples are grouped to form an $N$-dimensional vector (recall that $N = N_a(N_s - 1) + 1$), as follows

$$\begin{aligned} z_l &= [y_{1,1}(t_l) \cdots y_{N_s,1}(t_l)\; y_{2,2}(t_l) \cdots y_{N_s,2}(t_l) \cdots y_{2,N_a}(t_l) \cdots y_{N_s,N_a}(t_l)]^T \\ &= \boldsymbol{s}_l + \boldsymbol{n}_l, \end{aligned} \quad (26)$$

where

- $^T$ denotes transpose;
- $\boldsymbol{s}_l \in \mathbb{C}^{N\times 1}$ is the useful signal vector[5];
- $\boldsymbol{n}_l \in \mathbb{C}^{N\times 1}$ is the noise vector.

It is important to stress here that the bistatic range resolution is conventionally taken to be

$$\delta = c\frac{T_B}{2}. \quad (27)$$

However, two targets lying on the isorange contours must be at least physically separated by a distance given by

$$\frac{cT_B}{2\cos(\beta/2)} \quad (28)$$

to generate a separation $\delta$ at the receiver (Willis, Scitech Pub. 2005).

[5] $\boldsymbol{s}_l$ incorporates the complex amplitude $\alpha$

Now, assuming that

$$\tau' + \tau_3 \in [t_{\min}, t_{\min} + (L-1)T_B], \quad (29)$$

it follows that

$$\exists \bar{l} \in \{1, \ldots, L-1\} : \tau' + \tau_3 = t_{\min} + (\bar{l}-1)T_B + \epsilon', \quad (30)$$

$0 \le \epsilon' \le T_B$, and, hence, that

$$\mathbf{s}_l = \begin{cases} \alpha \chi_p(-\epsilon', f'_D)\mathbf{v}(\theta_r), & l = \bar{l}, \\ \alpha \chi_p(T_p - \epsilon', f'_D)\mathbf{v}(\theta_r), & l = \bar{l}+1, \\ \mathbf{0}, & l \neq \bar{l}, \bar{l}+1, \end{cases} \quad (31)$$

where $\mathbf{0}$ is a null vector of proper dimensions and

$$\mathbf{v}(\theta_r) = \left[1 \; e^{j2\pi\nu_1} \; \ldots \; e^{j2\pi(N_s-1)\nu_1} \; \ldots \; e^{j2\pi\nu_{N_a}} \; \ldots \; e^{j2\pi(N_s-1)\nu_{N_a}}\right]^T \quad (32)$$

is the *spatial steering vector*.
In the following we assume that the target is located at the center of the $l$th range gate, namely that there is no spillover of target energy to adjacent matched filter samples. Thus, equation (31) becomes

$$\mathbf{s}_l = \begin{cases} \alpha \mathbf{v}(\theta_r), & l = \bar{l}, \\ \mathbf{0}, & l \neq \bar{l}. \end{cases} \quad (33)$$

A remark is now in order. Observe that the $\mathbf{z}_l$'s (equation (26)) are usually processed to generate the contacts used by conventional trackers. An alternative to the traditional approach consists in feeding the $\mathbf{z}_l$'s to a TBD processor that operates with raw data (see Subsection 3.1).

### 2.2 Contact-based multistatic architecture

This subsection is devoted to the description of a multisensor surveillance sonar system that processes thresholded data to track prospective targets. In particular, we focus our attention on a sensor network which provides a set of time difference of arrival (TDOA) and bearing information on $M$ consecutive pings. Notice that each network node makes hard decisions and transmits these results to the fusion center for track estimate. In other words, we consider a parallel fusion network employing centralized fusion architecture (Varshney, Springer 1997).
The considered system consists of multiple acoustic sound sources and multiple receivers, more precisely

- $N_{sr}$ sources that alternately illuminate the surveillance region with a ping occurring every $T_{pg}$ seconds; the $N_{sr}$-source ping schedule is

$$n_{sr} = [(m-1) \mod N_{sr}] + 1, \quad (34)$$

  where $n_{sr}$ is the index of the transmitting source at the $m$th ping and $[a \mod b]$, $a, b \in \mathbb{N}$, is the remainder of the division $a/b$.

- $N_r$ synchronous omnidirectional receivers.

Source and receiver positions are known. Each receiver provides a set of measurements (or contacts), denoted as follows

$$Z_{i,m} = \{\mathbf{z}_{i,1,m}, \ldots, \mathbf{z}_{i,N_{i,m},m}\}, \quad i = 1, \ldots, N_r, \quad m = 1, \ldots, M, \quad (35)$$

where $M$ is the total number of consecutive illuminations and $N_{i,m}$ is the number of measurements collected by the $i$th receiver at the $m$th ping. As to the $z_{i,j,m}$'s, they are 2-dimensional vectors defined as

$$z_{i,j,m} = \begin{bmatrix} d_{i,j,m} \\ \phi_{i,j,m} \end{bmatrix} \in \mathbb{R}^{2\times 1}, \tag{36}$$

where $\phi_{i,j,m}$ is the bearing measured clockwise from the receiver north and $d_{i,j,m}$ denotes the TDOA between the direct blast and the contact.

### 2.2.1 Sensor model

In the sequel, we assume that when a measurement has originated from the target, the errors for TDOA and bearing are independent zero-mean Gaussian random variables, namely

$$\Delta d_{i,j,m} \sim \mathcal{N}_1(0,\sigma_d^2) \quad \text{and} \quad \Delta\phi_{i,j,m} \sim \mathcal{N}_1(0,\sigma_\phi^2), \quad \forall i,j,m, \tag{37}$$

where $\sigma_d > 0$ and $\sigma_\phi > 0$ are the respective standard deviations and are supposed known. On the other hand, when the contacts are due to the noise, we assume that they are uniformly distributed in the measurement space, i.e.,

$$d_{i,j,m} \sim \mathcal{U}(d_{i,m}^{\min}, d_{i,m}^{\max}) \quad \text{and} \quad \phi_{i,j,m} \sim \mathcal{U}(\phi_{i,m}^{\min}, \phi_{i,m}^{\max}), \tag{38}$$

where

- $d_{i,m}^{\min}$ and $\phi_{i,m}^{\min}$ denote the minimum values for TDOA and bearing, respectively;
- $d_{i,m}^{\max}$ and $\phi_{i,m}^{\max}$ are the maximum values for TDOA and bearing, respectively.

These parameters are tied to the application in question and to the employed technology, more precisely they can be chosen on the basis of: the surveillance area size, the peculiarities of the sought target, the quality of the used sensors, etc.

In Subsection 3.2 a tracking algorithm which takes advantage of the above multistatic system is introduced. In addition, such an algorithm relies on the main idea behind the TBD paradigm, namely, it jointly processes the available measurements over several consecutive pings.

## 3. Target tracking: the track-before-detect approach

Traditional tracking algorithms are designed assuming that the sensor provides a set of discrete measurements at each scan (or ping). In an activated surveillance system such measurements could be obtained by thresholding the output of a matched filter fed by a baseband version of collected data. Then, the tracking algorithm links measurements across time and estimates the parameters of interest. The threshold value must be low enough to guarantee a sufficiently high probability of target detection. However, a low threshold gives rise to a high rate of false alarms. It follows that to avoid false tracks it is necessary to effectively solve the data association problem (Bar-Shalom & Fortmann, Academic Press 1988). A reliable means of validating the track estimate as a target-originated one is also required.

An alternative approach, referred to as track-before-detect (TBD), consists of feeding the processor with unthresholded data or thresholded data with significantly lower thresholds than the ones used by conventional trackers. TBD-based procedures jointly process several consecutive pings (or scans) and, relying on a target kinematics or, simply, exploiting the physically admissible target transitions, jointly declare the presence of a target and, eventually, its track. A TBD algorithm can improve track accuracy and follow low SNR targets at the price of an increase of the computational complexity. Moreover, a TBD scheme ensuring the constant false alarm rate property with respect to the unknown statistics of the disturbance controls the overall false track acceptance probability and, hence, it is capable of guaranteeing the constant false track acceptance rate (CFTAR property).

Most of the TBD algorithms existing in the open literature are conceived to be implemented in optical and radar systems. Their use in connection with sonar systems has received less attention. For a description of existing results see (Orlando et al., IEEE-TSP 2010; Buzzi et al., IEEE-TAES 2005; S. Buzzi et al., IEEE-TAES 2008; Kramer & Reid, Radar Conference 1990; Wallace, Radar Conference 2002). A family of low-complexity power-efficient TBD procedures has been presented in (Buzzi et al., IEEE-TAES 2005). Therein, the continuous-time continuous-amplitude signal collected by a pulse Doppler radar is discretized to reflect the sectorization of the coverage area and the range gating operation, and the generalized likelihood ratio test (GLRT) is solved resorting to a Viterbi-like tracking algorithm. The proposed algorithm has a complexity linear in the number of integrated scans and in the time on target. The emphasis is on detection performance more than tracking: in fact, the GLRT does not rely on the target kinematics; it simply takes into account a maximum target velocity in order to define the admissible target transitions in range and azimuth (the Doppler is dealt with as a nuisance quantity due to the considered system and target parameters). However, a rough estimate of the target parameters is obtained as a by-product of the construction of the target statistic. (Orlando et al., IEEE-TSP 2010) extends the derivation of (Buzzi et al., IEEE-TAES 2005) to the context of space-time adaptive processing.

In the next subsection, we focus on a bistatic architecture and describe a proper modification of the algorithms proposed in (Orlando et al., IEEE-TSP 2010) to handle raw hydrophone data. In Subsection 3.2 we apply the main idea behind the TBD approach to thresholded data provided by a sensor network over several consecutive pings.

### 3.1 Track-before-detect strategies for bistatic sonars

The system considered here has a bistatic configuration and integrates the returns from $M$ consecutive pings before deciding whether or not a target is present in the surveillance area. Relying on the mathematical model for the received signal derived in Subsection 2.1.1, we utilize design criteria based upon the GLRT to derive a class of adaptive detectors which guarantee the CFTAR property under design assumptions with respect to the overall spectral properties of the noise. Moreover, at the design stage, we assume that the unknown clutter covariance matrix can possibly change from ping to ping. The performance assessment is provided in Section 5. It is carried out resorting to real sonar data collected by the deployable underwater surveillance system, called DEMUS, of NATO Undersea Research Centre (NURC).

#### 3.1.1 Detector designs

As stated above, the receiver jointly processes data from $M$ consecutive pings before discriminating between the noise-only hypothesis and the signal-plus-noise hypothesis. Thus, the track-before-detect problem at hand can be formulated in terms of a binary hypothesis testing problem as follows (Orlando et al., IEEE-TSP 2010)

$$\begin{cases} H_0 : \boldsymbol{z}_{l,m} = \boldsymbol{n}_{l,m}, & l = 1,\ldots,L, \quad m = 1,\ldots,M \\ H_1 : \begin{cases} \boldsymbol{z}_{l,m} = \alpha_m \boldsymbol{v}(\theta_{r,m}) + \boldsymbol{n}_{l,m}, & l = l_m, \quad m = 1,\ldots,M, \\ \boldsymbol{z}_{l,m} = \boldsymbol{n}_{l,m}, & l \neq l_m, \quad m = 1,\ldots,M, \end{cases} \end{cases} \tag{39}$$

where

- $l \in \{1,\ldots,L\}$ and $m \in \{1,\ldots,M\}$ are integers indexing the range cells and the pings, respectively;
- $\boldsymbol{z}_{l,m} \in \mathbb{C}^{N\times 1}$, $l = 1,\ldots,L$, $m = 1,\ldots,M$, are the vectors containing the noisy returns;
- $\boldsymbol{n}_{l,m} \in \mathbb{C}^{N\times 1}$, $l = 1,\ldots,L$, $m = 1,\ldots,M$, are independent and identically distributed complex normal random vectors with zero mean and unknown covariance matrices $\boldsymbol{R}_m \in \mathbb{C}^{N\times N}$;
- $\theta_{r,m}$ denotes the azimuthal angle of the impinging target echo at the $m$th ping.

The above assumptions imply that the probability density function (pdf) of the overall data matrix, denoted by

$$\boldsymbol{Z} = [\boldsymbol{z}_{1,1} \ \cdots \ \boldsymbol{z}_{L,M}], \tag{40}$$

can be written as

$$f_0(\boldsymbol{Z};\boldsymbol{R}_1,\ldots,\boldsymbol{R}_M) = \prod_{m=1}^{M} \frac{1}{\pi^{NL}\det(\boldsymbol{R}_m)^L} \exp\left\{-\mathrm{Tr}\left[\boldsymbol{R}_m^{-1}\left(\boldsymbol{z}_{l_m,m}\boldsymbol{z}_{l_m,m}^{\dagger} + \boldsymbol{S}_{l_m,m}\right)\right]\right\} \tag{41}$$

under $H_0$ and

$$f_1(\boldsymbol{Z};\boldsymbol{R}_1,\ldots,\boldsymbol{R}_M,\boldsymbol{\alpha}) = \prod_{m=1}^{M} \frac{1}{\pi^{NL}\det(\boldsymbol{R}_m)^L} \exp\left\{-\mathrm{Tr}\left[\boldsymbol{R}_m^{-1}\left(\boldsymbol{u}_{l_m,m}\boldsymbol{u}_{l_m,m}^{\dagger} + \boldsymbol{S}_{l_m,m}\right)\right]\right\} \tag{42}$$

under $H_1$, where

- $\det(\cdot)$ and $\mathrm{Tr}[\cdot]$ denote the determinant and the trace of a square matrix, respectively;
- $\boldsymbol{S}_{l_m,m} \in \mathbb{C}^{N\times N}$ is $(L-1)$ times the sample covariance matrix of the noise based on the available data over the $m$th ping, but for the $l_m$th range cell, namely

$$\boldsymbol{S}_{l_m,m} = \sum_{l \neq l_m} \boldsymbol{z}_{l,m}\boldsymbol{z}_{l,m}^{\dagger}; \tag{43}$$

- $\boldsymbol{\alpha} = [\alpha_1 \ \cdots \ \alpha_M]^T$ is the vector of target responses and[6]

$$\boldsymbol{u}_{l_m,m} = \boldsymbol{z}_{l_m,m} - \alpha_m \boldsymbol{v}_m(\theta_{r,m}). \tag{44}$$

In the following we apply the so-called two-step GLRT-based design procedure (Kelly & Nitzberg, IEEE-TAES 1992) in order to come up with a class of fully-adaptive detectors. To be more precise, the following rationale is adopted

- first assume that the covariance matrices of the noise are known $\forall m$ and implement the GLRT maximizing the likelihood functions over the unknown parameters;

[6] Recall that $\boldsymbol{v}_m$ is defined by (32)

- then, replace the unknown matrices $\boldsymbol{R}_m$, $m = 1, \ldots, M$, with proper estimates.

Thus, the GLRT under the assumption that the $\boldsymbol{R}_m$'s are known is given by

$$\frac{\max_{\mathcal{D} \in \mathcal{S}} \max_{\boldsymbol{\alpha}} f_1(\boldsymbol{Z}; \mathcal{D}, \boldsymbol{\alpha}, \boldsymbol{R}_1, \ldots, \boldsymbol{R}_M)}{f_0(\boldsymbol{Z}; \boldsymbol{R}_1, \ldots, \boldsymbol{R}_M)} \underset{H_0}{\overset{H_1}{\gtrless}} \eta, \tag{45}$$

where

- $\mathcal{D} = \{(l_1, \theta_{r,1}), \ldots, (l_M, \theta_{r,M})\} \in \mathcal{S}$ is the sequence of points occupied by a prospective target in the Range-Azimuth domain, with $\mathcal{S}$ the set of all physically admissible target trajectories;
- $\eta$ is the threshold value to be set in order to ensure the desired probability of false alarm ($P_{fa}$).

Performing the maximization of (45) with respect to $\boldsymbol{\alpha}$ yields

$$\max_{\mathcal{D} \in \mathcal{S}} \sum_{m=1}^{M} \frac{\left| \boldsymbol{v}(\theta_{r,m})^\dagger \boldsymbol{R}_m^{-1} \boldsymbol{z}_{l_m,m} \right|^2}{\boldsymbol{v}(\theta_{r,m})^\dagger \boldsymbol{R}_m^{-1} \boldsymbol{v}(\theta_{r,m})} \underset{H_0}{\overset{H_1}{\gtrless}} \eta. \tag{46}$$

An adaptive version of decision scheme (46) can be obtained by replacing the $\boldsymbol{R}_m$'s with proper estimates, $\widehat{\boldsymbol{R}}_m$'s say. The most common estimate of $\boldsymbol{R}_m$ is the sample covariance matrix based upon the available data at the $m$th ping, namely

$$\widehat{\boldsymbol{R}}_m = \frac{1}{L} \sum_{l=1}^{L} \boldsymbol{z}_{l,m} \boldsymbol{z}_{l,m}^\dagger. \tag{47}$$

The above estimate can be alternatively modified in order to exclude range cells contaminated by useful echoes as follows

$$\widehat{\boldsymbol{R}}_m = \frac{1}{K} \sum_{l \in \Omega_R \setminus \Omega_T} \boldsymbol{z}_{l,m} \boldsymbol{z}_{l,m}^\dagger, \tag{48}$$

where

- $\Omega_R = \{1, \ldots, L\}$;
- $\Omega_T$ is the set of consecutive integers indexing the range cells contaminated by target returns;
- $K$ is the cardinality of $\Omega_R \setminus \Omega_T$, namely the set containing the elements of $\Omega_R$ that do not belong to $\Omega_T$.

### 3.1.2 Implementation issues

This subsection is aimed at discussing some implementation issues concerning the above TBD algorithm. In fact, an adaptive version of (46) is time consuming since it requires maximization with respect to the sequence $\mathcal{D}$ of pairs $(l_m, \theta_{r,m})$. In order to overcome this limitation, observe that maximization over $\mathcal{D}$ cannot be conducted separately with respect to the $(l_m, \theta_{r,m})$ pairs: indeed, physical constraints on the target trajectory imply that $(l_{m+1}, \theta_{r,m+1})$ depends upon

$$\{(l_1, \theta_{r,1}), \ldots, (l_m, \theta_{r,m})\}, \quad m = 1, \ldots, M-1. \tag{49}$$

Moreover, let us partition the azimuth region as follows

$$\Omega_A = \left\{ \frac{1}{q}2\pi, \frac{2}{q}2\pi, \ldots, 2\pi \right\}, \tag{50}$$

where $q \in \mathbb{N}$ is a parameter which controls the accuracy of the azimuthal angle estimate. With these simplifications in mind, it is possible to recast detector (46) as follows

$$\max_{\substack{\mathcal{D} \in (\Omega_R \times \Omega_A)^M \\ (l_m, \theta_{r,m}) \in \mathcal{P}((l_{m-1}, \theta_{r,m-1}))}} \sum_{m=1}^{M} \frac{\left| \boldsymbol{v}_m(\theta_{r,m})^{\dagger} \boldsymbol{R}_m^{-1} \boldsymbol{z}_{l_m,m} \right|^2}{\boldsymbol{v}_m(\theta_{r,m})^{\dagger} \boldsymbol{R}_m^{-1} \boldsymbol{v}_m(\theta_{r,m})} \underset{H_0}{\overset{H_1}{\gtrless}} \eta, \tag{51}$$

where

- $\times$ denotes the Cartesian product;
- $(\Omega_R \times \Omega_A)^M$ is the $M$th Cartesian power of the set $(\Omega_R \times \Omega_A)$;
- $\mathcal{P}((l_{m-1}, \theta_{r,m-1}))$ denotes the set of elements of $(\Omega_R \times \Omega_A)$ that can be reached from $(l_{m-1}, \theta_{r,m-1})$ under the upper bounds $V_R$ and $V_A$ on the radial velocity and on the tangential velocity of the target, respectively.

The above optimization problem can be solved constructing an expanded trellis diagram, whose states are the elements of $\Omega_R \times \Omega_A$ and hence using a Viterbi-like procedure to find the best path metric in this expanded trellis (Buzzi et al., IEEE-TAES 2005; Forney, IEEE Proc. 1973). The Viterbi algorithm would require determining (at most) $Lq$ paths in the expanded trellis diagram of depth $M$, with a consequent maximum complexity[7] $\mathcal{O}(LqM)$ (linear in the number of ping). Finally, it is easy to prove that (51) coupled with (47) or (48) ensures the CFTAR property with respect to the unknown noise parameters.

### 3.2 A batch tracking algorithm for multistatic sonars

In this subsection, a tracking algorithm for multistatic sonars is derived, that borrows the main idea of the TBD approach by jointly processing thresholded data over several consecutive pings. The considered system is described in Subsection 2.2. In order to estimate the target trajectory, the likelihood function of the measurements (that depends on the target position at each ping) is maximized by resorting to a Viterbi-like procedure (see Subsection 3.1). For the reader's ease, recall that

$$Z_{i,m} = \{ z_{i,1,m}, \ldots, z_{i,N_{i,m},m} \} \tag{52}$$

denotes the set of measurements transmitted to the fusion center by the $i$th receiver at the $m$th ping, where

$$z_{i,j,m} = \begin{bmatrix} d_{i,j,m} \\ \phi_{i,j,m} \end{bmatrix} \tag{53}$$

with $d_{i,j,m}$ the TDOA and $\phi_{i,j,m}$ the bearing. In the following, we assume the presence of one maneuvering target with deterministic motion in the surveillance region.
Let us start denoting by

$$Z = \{ Z_{1,1}, \ldots, Z_{1,M}, \ldots, Z_{N_r,1}, \ldots, Z_{N_r,M} \} \tag{54}$$

[7] We resort to the usual Landau notation.

the overall data set and defining a grid $G$ of $N_x N_y$ points[8], that covers the entire surveillance region. At the $m$th ping the nominal position of the target is given by

$$\boldsymbol{x}_m = \begin{bmatrix} x_m \\ y_m \end{bmatrix}, \quad (x_m, y_m) \in G. \tag{55}$$

Now, assuming that, given the sequence of the nominal target positions over $M$ consecutive pings, denoted by

$$X = \{\boldsymbol{x}_1, \boldsymbol{x}_2, \dots, \boldsymbol{x}_M\}, \tag{56}$$

the measurements are independent of each other, it is not difficult to show that the pdf of $Z$ can be written as follows

$$f_1(Z; X, J) = \prod_{m=1}^{M} \prod_{i=1}^{N_r} \left\{ \frac{(1 - P_{i,m})}{(U_{i,m})^{N_{i,m}}} + P_{i,m} \frac{f_z(\boldsymbol{z}_{i,j_{i,m},m}; \boldsymbol{x}_m)}{(U_{i,m})^{N_{i,m}-1}} \right\}, \tag{57}$$

where $P_{i,m}$ is the probability of detection of the $i$th receiver at the $m$th ping[9], $j_{i,m} \in \mathbb{N}$ is an integer indexing the target originated contact for the $i$th receiver at the $m$th ping,

$$J = \{J_1, \dots, J_M\}, \quad \text{with} \quad J_m = \{j_{1,m}, \dots, j_{N_r,m}\}, \tag{58}$$

$$U_{i,m} = (d_{i,m}^{\max} - d_{i,m}^{\min})(\phi_{i,m}^{\max} - \phi_{i,m}^{\min}). \tag{59}$$

Finally, $f_z(\cdot;\cdot)$ is given by

$$f_z(\boldsymbol{z}_{i,j_{i,m},m}; \boldsymbol{x}_m) = \frac{1}{2\pi[\det(\boldsymbol{R})]^{1/2}} \times \exp\left\{ -\frac{1}{2} \left( \boldsymbol{z}_{i,j_{i,m},m} - \boldsymbol{\zeta}(\boldsymbol{x}_m, \boldsymbol{r}_i, \boldsymbol{s}_{n_{sr}}) \right)^T \boldsymbol{R}^{-1} \left( \boldsymbol{z}_{i,j_{i,m},m} - \boldsymbol{\zeta}(\boldsymbol{x}_m, \boldsymbol{r}_i, \boldsymbol{s}_{n_{sr}}) \right) \right\}, \tag{60}$$

where

$$\boldsymbol{R} = \begin{bmatrix} \sigma_d^2 & 0 \\ 0 & \sigma_\theta^2 \end{bmatrix}, \quad \boldsymbol{r}_i = \begin{bmatrix} r_{x,i} \\ r_{y,i} \end{bmatrix} \in \mathbb{R}^{2\times 1}, \quad \text{and} \quad \boldsymbol{s}_{n_{sr}} = \begin{bmatrix} s_{x,n_{sr}} \\ s_{y,n_{sr}} \end{bmatrix} \in \mathbb{R}^{2\times 1} \tag{61}$$

are the covariance matrix of $\boldsymbol{z}_{i,j_{i,m},m}$, the position of the $i$th receiver, and the position of the $n_{sr}$th source[10], respectively, and $\boldsymbol{\zeta}(\cdot,\cdot,\cdot)$ is a vector valued function such that

$$\boldsymbol{\zeta}(\boldsymbol{x}_m, \boldsymbol{r}_i, \boldsymbol{s}_{n_{sr}}) = \begin{bmatrix} \frac{1}{c}\left( \|\boldsymbol{s}_{n_{sr}} - \boldsymbol{x}_m\| + \|\boldsymbol{x}_m - \boldsymbol{r}_i\| - \|\boldsymbol{r}_i - \boldsymbol{s}_{n_{sr}}\| \right) \\ \frac{\pi}{2} + \pi(1 - u(x)) - \tan^{-1} \frac{(y_m - r_{y,i})}{(x_m - r_{x,i})} \end{bmatrix}, \tag{62}$$

where[11] $u(\cdot)$ is a step function taking value 1 over the interval $[0 \ +\infty[$. In other words, $\boldsymbol{\zeta}(\cdot,\cdot,\cdot)$ converts coordinates $(x_m, y_m)$ into the corresponding values of TDOA and bearing with respect to the $i$th receiver and the $n_{sr}$th source.

[8] $N_x, N_y \in \mathbb{N}$
[9] The $P_{i,m}$'s are assumed known.
[10] $n_{sr}$ is given by equation (34).
[11] Recall that $\|\cdot\|$ denotes the Euclidean norm of a vector and $c$ is the sound propagation velocity.

Dividing equation (57) by the pdf of $Z$ under the noise-only hypothesis, given by

$$f_0(Z) = \prod_{m=1}^{M} \prod_{i=1}^{N_r} \frac{1}{U_{i,m}^{N_{i,m}}}, \tag{63}$$

we obtain the likelihood ratio

$$\Lambda(Z; X, J) = \prod_{m=1}^{M} \prod_{i=1}^{N_r} \left\{ (1 - P_{i,m}) + P_{i,m} f(z_{i,j_{i,m},m}; \boldsymbol{x}_m) U_{i,m} \right\}, \tag{64}$$

which, by taking the logarithm, becomes

$$\Lambda(Z; X, J) = \sum_{m=1}^{M} \sum_{i=1}^{N_r} \log \left\{ (1 - P_{i,m}) + P_{i,m} f(z_{i,j_{i,m},m}; \boldsymbol{x}_m) U_{i,m} \right\}. \tag{65}$$

Observe that the likelihood ratio (65) is a function of the target state sequence $X$ and of the target originated measurement indexes $J$, which are unknown and, consequently, have to be estimated. To this end, the likelihood ratio (65) is maximized with respect to $X$ and $J$. The optimization with respect to $j_m$, $m = 1, \ldots, M$, is straightforward and solves the data association problem; in fact, given the ping index $m$ and the target state $\boldsymbol{x}_m$, for each receiver it is possible to select that measurement index[12], $\widehat{j}_{i,m}$ say, which minimizes the following Mahalanobis distance

$$\widehat{j}_{i,m} = \arg \min_{j_{i,m} \in J_m} \left( z_{i,j_{i,m},m} - \zeta(\boldsymbol{x}_m, \boldsymbol{r}_i, \boldsymbol{s}_{n_{sr}}) \right)^T \boldsymbol{R}^{-1} \left( z_{i,j_{i,m},m} - \zeta(\boldsymbol{x}_m, \boldsymbol{r}_i, \boldsymbol{s}_{n_{sr}}) \right). \tag{66}$$

It follows that (65) can be recast as

$$\Lambda(Z; X, \widehat{J}) = \sum_{m=1}^{M} \sum_{i=1}^{N_r} \log \left\{ (1 - P_{i,m}) + P_{i,m} f(z_{i,\widehat{j}_{i,m},m}; \boldsymbol{x}_m) U_{i,m} \right\}, \tag{67}$$

where

$$\widehat{J} = \left\{ \widehat{J}_1, \ldots, \widehat{J}_M \right\} \quad \text{with} \quad \widehat{J}_m = \left\{ \widehat{j}_{1,m}, \ldots, \widehat{j}_{N_r,m} \right\}. \tag{68}$$

Finally, in order to estimate the sequence of target positions, it is worth taking into consideration physical constraints on the target trajectory, that limit the maximum number of point transitions throughout $G$ between two consecutive pings. With this assumption in mind, an estimate of $X$ can be obtained as follows

$$\widehat{X} = \arg \max_{\substack{X \\ \boldsymbol{x}_m \in \mathcal{P}(\boldsymbol{x}_{m-1})}} \Lambda(Z; X, \widehat{J}), \tag{69}$$

where $\mathcal{P}(\boldsymbol{x}_{m-1})$ denotes the set of target states that can be reached from $\boldsymbol{x}_{m-1}$ under an upper bound on the target velocity. The above optimization problem can be solved by resorting to a Viterbi-like procedure to find the best path metrics in an expanded trellis diagram of depth $M$ whose states are the points of $G$ (see Figures 3). In Section 5 some illustrative examples show the effectiveness of the above approach.

[12] Recall that the $J_m$'s are finite discrete sets.

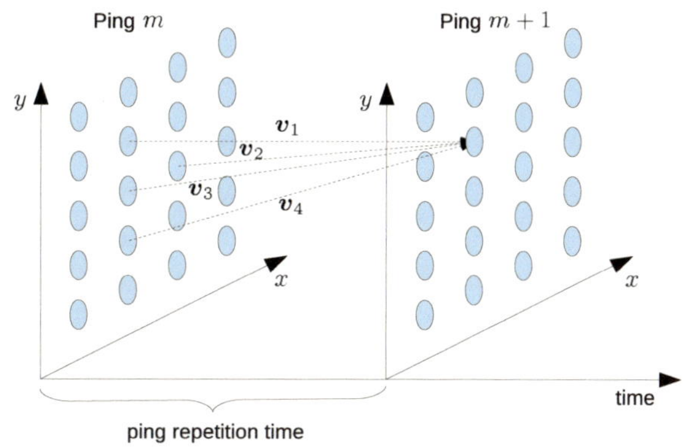


Fig. 3. The expanded trellis whose states are the points of $G$.

## 4. Sensor selection strategies for multistatic sonars

This section describes on a conceptual level a method to select those sensors whose measurements will be fed to the batch algorithm presented in Subsection 3.2. Consider the sensor network described in Subsection 2.2 and assume that SNR measurements, $\text{SNR}_{i,j,m}$ say, are also available at the fusion center. The statistical distribution of the SNR measurements is assigned as follows

$$\text{SNR}_{i,j,m} \sim \mathcal{N}_1(\mu_{i,m}, \sigma^2_{\text{SNR}}), \tag{70}$$

$$i = 1, \ldots, N_r, \quad j = 1, \ldots, N_{i,m}, \quad m = 1, \ldots, M, \tag{71}$$

where $\sigma_{\text{SNR}} > 0$ is the standard deviation and (Urick, McGraw-Hill 1983)

$$\mu_{i,m} = \begin{cases} \text{SL} + \text{TL}^{\text{ST}}_m + \text{TL}^{\text{TR}}_{i,m} + \text{TS}(\psi_{i,m}), & \text{if the contact is originated from the target,} \\ \gamma_{i,m}, & \text{otherwise.} \end{cases} \tag{72}$$

In (72), $\gamma_{i,m} \in \mathbb{R}$, SL is the source level, $\text{TL}^{\text{ST}}_m$ is the transmission loss between the $n_{sr}$th source, with $n_{sr}$ given by (34), and the target, $\text{TL}^{\text{TR}}_{i,m}$ is the transmission loss between the target and the $i$th receiver, and $\text{TS}(\psi_{i,m})$ is the target strength which depends on the aspect angle of the target, $\psi_{i,m}$ say, with respect to the $i$th receiver. The transmission loss $\text{TL}^{\text{AB}}$ between two points A and B spaced $D$ meters apart from one another is defined as follows (Urick, McGraw-Hill 1983)

$$\text{TL}^{\text{AB}} = \begin{cases} 10 \log_{10}\left(\dfrac{1}{D^2}\right), & \text{in case of spherical spreading of the sound wave,} \\ \\ 10 \log_{10}\left(\dfrac{1}{D}\right), & \text{in case of cylindrical spreading of the sound wave.} \end{cases} \tag{73}$$

The likelihood function of the SNR for the $i$th receiver at the $m$th ping can be written as follows

$$f^{i,m}_{\text{SNR}}(\alpha_{i,m}; j_{i,m}, \psi_{i,m}) = \frac{(1-P_{i,m})}{(\sqrt{2\pi}\sigma_{\text{SNR}})^{N_{i,m}}} \exp\left\{-\frac{1}{2\sigma^2_{\text{SNR}}} \sum_{k=1}^{N_{i,m}} (\text{SNR}_{i,k,m} - \gamma_{i,m})^2\right\}$$
$$+ P_{i,m} \frac{\exp\left\{-\dfrac{1}{2\sigma^2_{\text{SNR}}}\left[\displaystyle\sum_{k=1, k\neq j_{i,m}}^{N_{i,m}} (\text{SNR}_{i,k,m} - \gamma_{i,m})^2 + (\text{SNR}_{i,j_{i,m},m} - \text{C} - \text{TS}(\psi_{i,m}))^2\right]\right\}}{(\sqrt{2\pi}\sigma_{\text{SNR}})^{N_{i,m}}}, \tag{74}$$

where

- $\alpha_{i,m} = \left\{ \text{SNR}_{i,1,m}, \text{SNR}_{i,2,m}, \ldots, \text{SNR}_{i,N_{i,m},m} \right\}$;
- $j_{i,m} \in \{1, \ldots, N_{i,m}\}$ is the index of the target-originated measurement;
- $C = \text{SL} + \text{TL}_m^{\text{ST}} + \text{TL}_{i,m}^{\text{TR}}$.

In (74), $j_{i,m}$ and $\psi_{i,m}$ are not known and have to be estimated from the observables. To this end $f_{\text{SNR}}^{i,m}(\cdot;\cdot,\cdot)$ is maximized with respect to such parameters. More specifically, observe that

$$\widehat{\psi}_{i,m} = \arg \max_{\psi_{i,m} \in [0,\, 360)} f_{\text{SNR}}^{i,m}(\alpha_{i,m}; j_{i,m}, \psi_{i,m}) = \arg \max_{\psi_{i,m} \in [0,\, 360)} \text{TS}(\psi_{i,m}) \tag{75}$$

and that maximization over $j_{i,m}$ can be performed in the same manner as pointed out for (66). Next, we arrange the likelihood functions at the $m$th ping in ascending order, namely

$$f_{\text{SNR}}^{i_{1,m},m} < f_{\text{SNR}}^{i_{2,m},m} < \ldots < f_{\text{SNR}}^{i_{N_r,m},m} \tag{76}$$

and select the last $K$ to form the following set of indexes[13]

$$S_m = \{ i_{N_r-K+1,m}, i_{N_r-K+2,m}, \ldots, i_{N_r,m} \}. \tag{77}$$

Once sensors have been selected, the target trajectory can be estimated as follows

$$\widehat{X} = \underset{\substack{X \\ \boldsymbol{x}_m \in \mathcal{P}(\boldsymbol{x}_{m-1})}}{\arg\max} \left[ \max_J \Lambda_1(Z; X, J) \right], \tag{78}$$

where

$$\Lambda_1(Z; X, J) = \sum_{m=1}^{M} \sum_{i \in S_m} \log \left\{ (1 - P_{i,m}) + P_{i,m} f(\boldsymbol{z}_{i,j_{i,m},m}; \boldsymbol{x}_m) U_{i,m} \right\}. \tag{79}$$

Observe that (78) is less time-demanding than (69) since the maximizations over $J$ and $X$ are performed processing data from $K < N_r$ receivers.

## 5. Illustrative examples and discussion

### 5.1 Track-before-detect approach for raw data

The TBD approach described in Subsection 3.1 leads to a class of algorithms capable of handling raw hydrophone data. This subsection is aimed at proving the feasibility and the effectiveness of such algorithms using experimental bistatic sonar data collected by the NURC's DEMUS sensor array in the course of PreDEMUS'06 sea trial.

The setup of PreDEMUS'06 is shown in Figure 4 and involves a transmitter, three receivers and as target a towed sound source, called echo-repeater (E-R). The E-R, towed by a surface vessel at a depth of 80 m, retransmits received source signals with a specified amplification and a certain delay. The source transmits a frequency-modulated pulse waveform of duration $T_p = 1$ sec, bandwidth $B = 100$ Hz, and carrier frequency $f_c = 2500$ Hz. The resulting range resolution is about 7 m (we are assuming that $c = 1500$ m/sec). The receiver is a planar array (as shown in Figure 1) with $N_a = 9$ arms, $N_s = 8$ sensors, $\theta_a = 40$ degrees, and $d = 0.289$ m. The total number of sensors is $N = 64$ (remember that the arms share the central sensor).

[13] $K$ may be chosen according to a preassigned criterion.

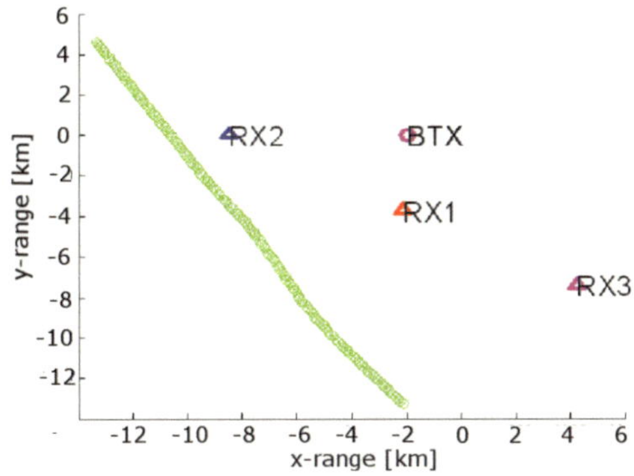


Fig. 4. PreDEMUS'06 setup. The fuchsia circle represents the transmitter, the red triangle represents the receiver 1, the blue triangle represents the receiver 2, finally the fuchsia triangle represents the receiver 3. The sequence of green circles is the target trajectory. [Courtesy M. Daun]

In the sequel, we examine the behavior of the receiver (51) only coupled with (48), where $\Omega_T$ is appropriately selected. Moreover, we consider a scenario where the useful signal echoes are strong enough to guarantee a good number of contacts for traditional tracking systems (strong-target scenario) and a scenario where the detection performance of the envelope detector is poor (weak-target scenario). The algorithm processes data corresponding to a little patch of the surveillance area and the target is moving within such a patch throughout the entire observation time. The processed patch takes up a region of about 2 seconds in range and at most 35 degrees in azimuth[14]. As to the threshold, it is set in order to have no detection when the algorithm is fed by noise-only data.

In Figures 5a-5c, we show the performance of the proposed algorithm in terms of target position estimate when it operates in the first scenario (for more details on the data set see Figure 4 of (Daun & Ehlers, JASP 2010)); more precisely, we process data from receiver 1 and pings 91-99 (scenario A1) in Fig. 5a, from receiver 2 and pings 76-84 (scenario A2) in Fig. 5b, and from receiver 3 and pings 100-108 (scenario A3) in Fig. 5c. The figures also report the trajectory of the E-R which is taken as reference. Observe that in this case the estimated target locations, represented by the line markers, closely follow the reference track. Observe that the estimated tracks exhibit some "fluctuations". This is partly due to the fact that estimation errors also depend on the undercurrent which modifies the receiver orientation (heading) and, consequently, introduces an additional uncertainty in the azimuth estimation. Otherwise stated, the position of the reference arm of each receiver is time-varying. Such heading errors can be partly compensated by using data recorded from an on-board compass. In Figure 5d, we plot the estimated target positions returned by the algorithm in the case of a low observable target; in particular, we consider data from receiver 1 and pings 16-26 (scenario B1). The figure highlights two important aspects of the TBD approach. First, the proposed algorithm can detect a low SNR target, whereas the envelope detectors, used by conventional tracking systems, are completely blind (Daun & Ehlers, JASP 2010). Second, it provides, as a by product, a rough estimate of the target locations. Notice that also in this figure, the heading errors make the curve representing the estimated trajectory "convoluted".

[14] A preliminary statistical analysis highlights that such data exhibit a "good" level of homogeneity.

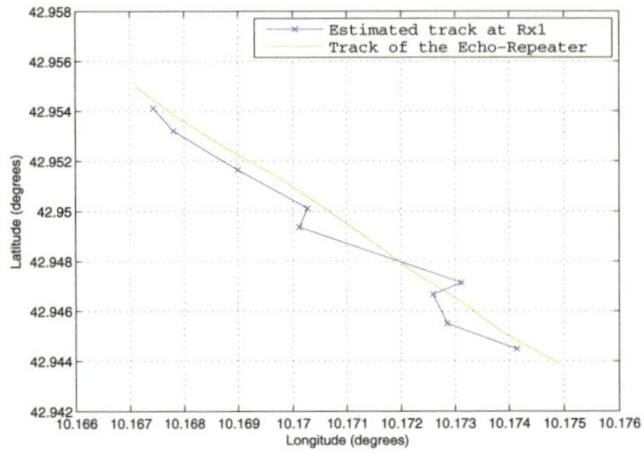


(a) Scenario A1 assuming $M = 9$ and $K > 2N$.

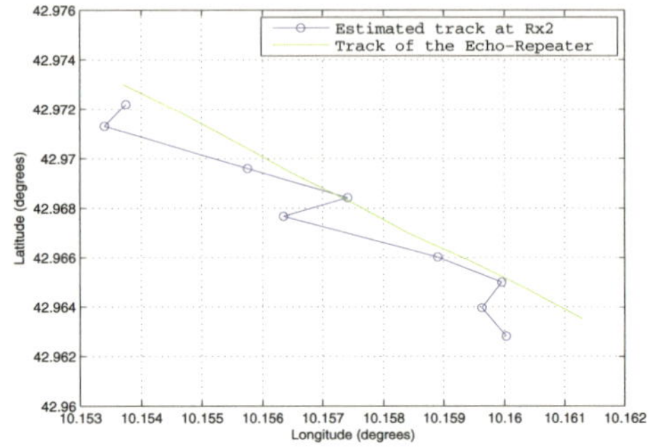


(b) Scenario A2 assuming $M = 9$ and $K > 2N$.

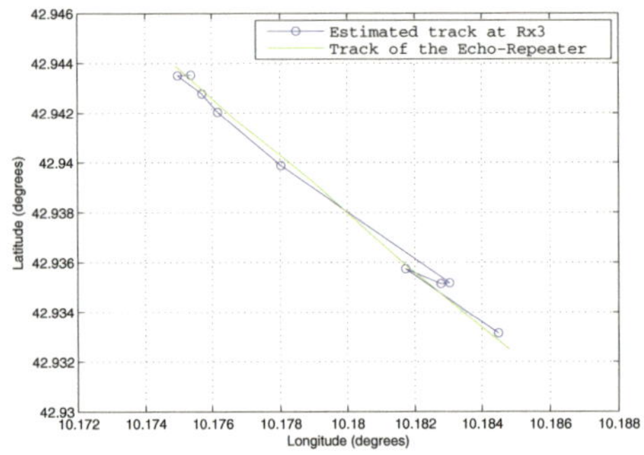


(c) Scenario A3 assuming $M = 9$ and $K > 2N$.

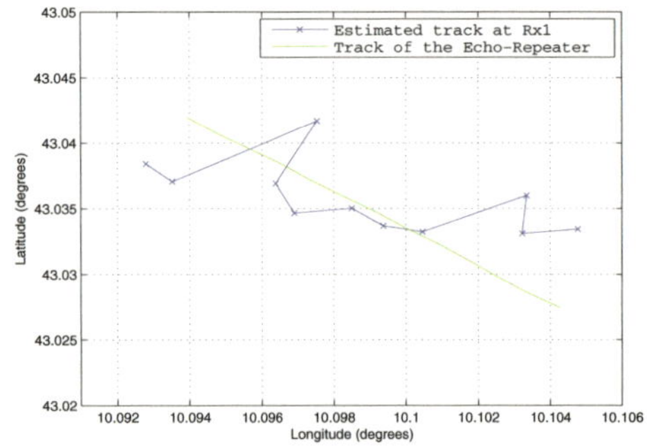


(d) Scenario B1 assuming $M = 11$ and $K > 2N$.

Fig. 5. Estimated target locations for scenarios A1, A2, A3, and B1.

As a final remark, the estimation of the $\boldsymbol{R}_m$'s can be automated modifying detector (51) as follows (Bandiera et al., IEEE-SPL 2006)

$$\max_{\Omega_T} \max_{\substack{\mathcal{D} \in (\Omega_R \times \Omega_A)^M \\ (l_m, \theta_{r,m}) \in \mathcal{P}((l_{m-1}, \theta_{r,m-1}))}} \sum_{m=1}^{M} \frac{|\boldsymbol{v}_m(\theta_{r,m})^\dagger \widehat{\boldsymbol{R}}_m^{-1} \boldsymbol{z}_{l_m,m}|^2}{\boldsymbol{v}_m(\theta_{r,m})^\dagger \widehat{\boldsymbol{R}}_m^{-1} \boldsymbol{v}_m(\theta_{r,m})} \underset{H_0}{\overset{H_1}{\gtrless}} \gamma, \tag{80}$$

where again $\widehat{\boldsymbol{R}}_m$ is given by (48).

### 5.2 Contact-based maximum-likelihood tracker

The subsection contains some illustrative examples to show the tracking performance of the contact-based tracker introduced in Subsection 3.2. Two different scenarios are considered.

#### 5.2.1 Test scenario 1

Consider $N_r = 12$ hydrophones (Rx) and $N_{sr} = 4$ sources (Sr) placed in the positions listed in Table 1. Moreover, we assume $P_{i,m} = P_d, \forall i, m$. The surveillance region is a $16740 \times 16740$ m$^2$ square with the upper left corner at $(37500, 34500)$ m and it is covered by a grid of $31 \times 31$ points 540 m spaced apart. Such a separation corresponds to a nominal target velocity of 6 knots. The total number of processed ping is $M = 20$ and the time interval between consecutive pings is $T_{pg} = 180$ seconds. As to the synthetic target, we assume that it moves

within the surveillance area with a constant velocity of about 6 knots and that after 10 pings its direction of motion changes and becomes orthogonal to the previous direction.
The standard deviations of target originated measurements are set as follows

$$\sigma_d = 0.4 \text{ sec} \quad \text{and} \quad \sigma_\phi = 8 \text{ degrees,} \tag{81}$$

while for each couple source-receiver the false contacts are distributed uniformly in the bistatic ellipse that contains the surveillance region.
Finally, the number of false contacts, namely those not originated from the target, obeys to the Poisson distribution with parameter

$$\lambda = P_{fa} N_{\text{cell}}, \tag{82}$$

where $P_{fa}$ is the probability of false alarm in any cell and $N_{\text{cell}}$ is the total number of cells. All simulation results assume $P_{fa} = 0.01$ and $\lambda = 9.6$.
In Figures 6a-6c we show the estimated tracks from 4 independent trials. Each figure refers to different values of $P_d$ and reports the actual trajectory of the target (ground truth). Observe that for low values of $P_d$ ($P_d = 0.3$) the proposed algorithm is still capable of providing an estimate of the target positions, even though the estimation error is significant. On the other hand, for greater values of $P_d$ the estimated target locations closely follow the ground truth. Such results are confirmed by Figure 6d, where we plot the root mean square (RMS) values based on 300 independent trials versus the $P_d$.

### 5.2.2 Test scenario 2 (from the multistatic tracking working group)

The intent of the multistatic tracking working group (MSTWG) is to foster the exchange of scientific and technical ideas, problems, and solutions related to multistatic tracking for sonar and radar. This includes the collaborative analysis of common data sets and culminate in workshops or special sessions disseminating the results. The MSTWG was founded as an ad hoc working group in December 2004, with a three year charter. In July 2007, based on mutual agreement between the International Society of Information Fusion (ISIF) board of directors and the current group membership, it was agreed to formalize the MSTWG as a working group under the auspices of ISIF. The objective of the ISIF MSTWG is to promote collaboration among its members in multisensor fusion and tracking, with a current focus on multistatic sonar and radar. This collaboration is achieved through regular meetings, participation in special sessions at conferences, and the analysis of common data sets. One of these data sets is

| Rx number | X (meters) | Y (meters) | Rx number | X (meters) | Y (meters) |
|---|---|---|---|---|---|
| 1 | 26752,91 | 8219,99 | 10 | 36033,99 | 36000 |
| 2 | 26752,91 | 26740 | 11 | 54581,52 | 17479,99 |
| 3 | 26752,91 | 45259,99 | 12 | 54581,52 | 36000 |
| 4 | 45285,83 | 8219,99 | **Sr number** | **X (meters)** | **Y (meters)** |
| 5 | 45285,83 | 45259,99 | 1 | 17486.46 | 54520 |
| 6 | 63818,75 | 8219,99 | 2 | 54581,52 | 17480 |
| 7 | 63818,75 | 26740 | 3 | 54581,52 | 54520 |
| 8 | 63818,75 | 45259,99 | 4 | 17486.46 | 17480 |
| 9 | 36033,99 | 17479,99 | | | |

Table 1. Sensor coordinates.

the Metron data set (Orlov, Metron Data set 2009). In the following a brief description of the data set is given (further details can be found in (Orlov, Metron Data set 2009)).
The operational field is a $72000 \times 72000$ m$^2$ square with the origin at $(0,0)$ m. The sensors are laid out as two concentric square grids centered about $(x,y) = (36000,36000)$ m. All sensors are stationary. The first grid consists of 16 sensors, laid out $4 \times 4$ beginning at $(x,y) = (8220,8220)$ m, with four receivers within a row 18520 m apart horizontally and four rows 18520 m apart vertically. All of these sensors are receivers. The second grid consists of 9 sensors, laid out $3 \times 3$ beginning at $(x,y) = (17480,17480)$ m, with three receivers within a row 18520 m apart horizontally and three rows 18520 m apart vertically. The four sensors at the corners of this grid are co-located source/receiver units, and the remaining five sensors are receivers. Figures 6e and 6f contain the results obtained by applying the proposed algorithm to the first MSTWG Metron data set described above.

## 6. Conclusions and hints for future works

This work has addressed new trends in the field of multistatic sonar systems. Specifically, the focus is on the design and analysis of tracking algorithms capable of operating in highly noisy environments. The TBD approach is investigated and it is shown that TBD (or TBD-based) algorithms can overcome the limitations exhibited by the conventional trackers in case of low SNR targets or low-quality sensors. In fact, the class of decision schemes described in Subsection 3.1 guarantees good detection performance of weak moving targets and, as a byproduct, provides a rough estimate of the target locations, while the batch algorithm introduced in Subsection 3.2 ensures acceptable performance also when the probability of detection per sensor is low.
Several important issues would deserve further attention. First, observe that since the spectral properties of the noise may change with the distance from the sensor, the homogeneity assumption does not generally hold. In such a case, it would be important to assess the performance of the TBD algorithm proposed in Subsection 3.1 coupled with other possible estimates of the noise covariance matrix, like the so-called normalized sample covariance matrix, introduced in (Conte et al., EUSIPCO 1994), or the one introduced in (Conte et al., IEEE-TSP 2002). In addition, the Gaussian assumption is not always met in realistic scenarios and, hence, it would also be of interest a performance analysis in non-Gaussian disturbance.
Another key point that has not been sufficiently investigated is the design and the analysis of TBD (or TBD-based) algorithms for extended and/or multiple targets; a first work on the design of TBD algorithms for multiple targets has been presented in (S. Buzzi et al., IEEE-TAES 2008).

### 6.1 Acknowledgment

This work was made possible through collaboration between NURC, a NATO Research Centre, and Prof. Giuseppe Ricci, Dipartimento di Ingegneria dell'Innovazione, Università del Salento, Italy.

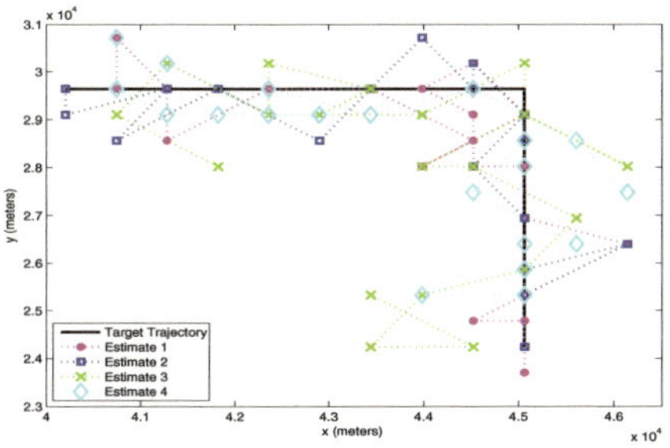


(a) Estimated tracks assuming $P_d = 0.3$.

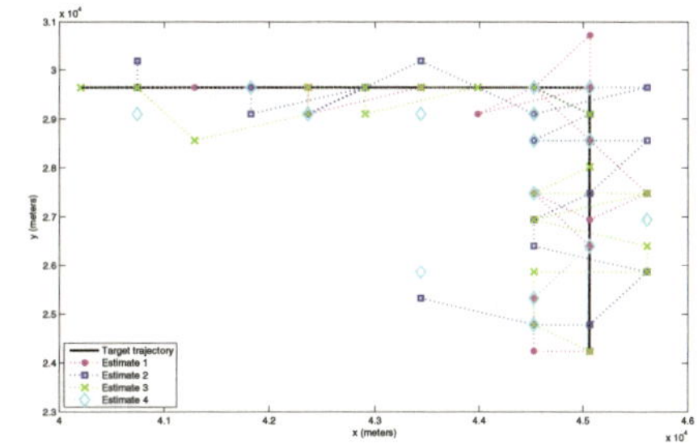

(b) Estimated tracks assuming $P_d = 0.5$.

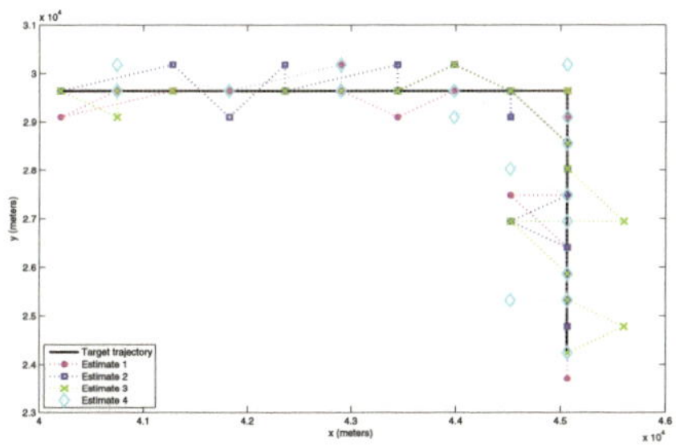

(c) Estimated tracks assuming $P_d = 0.8$.

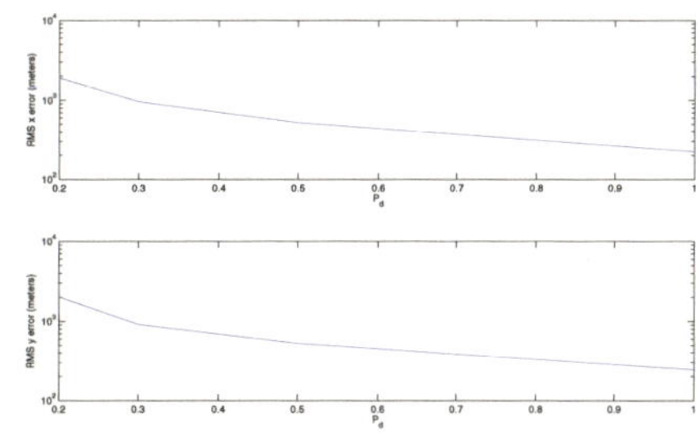

(d) RMS errors versus $P_d$.

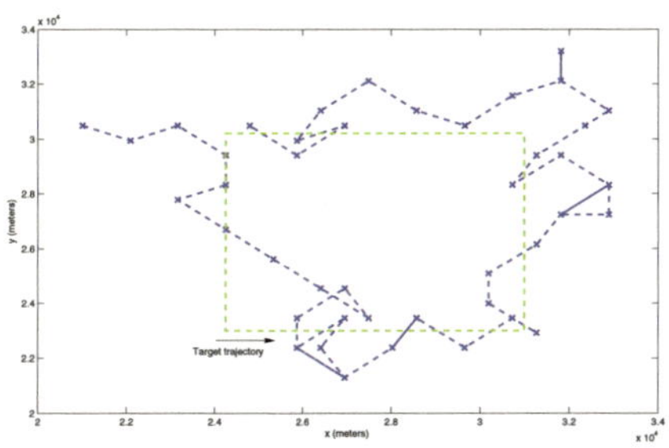

(e) Third target.

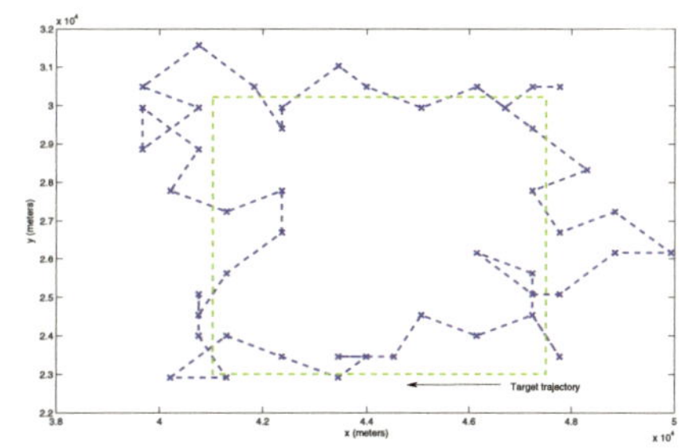

(f) Fourth target.

Fig. 6. Estimated tracks from 4 independent Monte Carlo trials, RMS errors, and estimated tracks from the first MSTWG Metron data set.

## 7. References


Daun, M. & Ehlers, F. (2010). Tracking Algorithms for Multistatic Sonar Systems. *EURASIP Journal on Advances in Signal Processing*, Vol. 2010, 2010.

Orlando, D.; Venturino, L.; Lops, M. & Ricci, G. (2010). Track-Before-Detect Strategies for STAP Radars. *IEEE Transactions on Signal Processing*, Vol. 58, No. 2, pp. 933-938, February 2010.

Orlando, D.; Ehlers, F. & Ricci, G. (2010). Track-before-detect Algorithms for Bistatic Sonars, *Proceedings of the 2nd International Workshop on Cognitive Information Processing*, Elba Island (Italy), June 2010.

Orlando, D.; Ehlers, F. & Ricci, G. (2010). A Maximum Likelihood Tracker for Multistatic Sonars, *Proceedings of the 13th International Conference on Information Fusion*, Edinburgh (UK), July 2010.

Orlov, K. (2009) *Description of the "MetronSimulation" data set for MSTWG*, Metron, Inc. Reston, Virginia, United States of America.

Willis, N. J. (2005) *Bistatic Radar*, Scitech Publishing Inc., 2005.

Bandiera, F; Orlando, D. & Ricci, G. (2009) *Advanced radar detection schemes under mismatched signal models*, Synthesis Lectures on Signal Processing No. 8, Morgan & Claypool Publishers, March 2009.

Varshney, P. R. (2005) *Distributed Detection and Data Fusion*, Springer-Verlag, New York, 1997.

Bar-Shalom, Y. & Fortmann, T. E. *Tracking and Data Association*, Academic Press, 1988.

Buzzi, S.; Lops, M. & Venturino, L. (2005). Track-before-detect procedures for early detection of moving target from airborne radars, *IEEE Transactions on Aerospace and Electronic Systems*, Vol. 41, No. 3, pp. 937-954, July 2005.

Buzzi, S.; Lops, M.; Venturino, L. & Ferri, M. (2008). Track-before-detect procedures in multi-targets environments, *IEEE Transactions on Aerospace and Electronic Systems*, Vol. 44, No. 3, pp. 1135-1150, July 2008.

Kramer, J. D. R. & Reid, J. W. S. (1990). Track-before-detect processing for an airborne type radar, *Proceedings of IEEE 1990 International Radar Conference*, Arlington, VA, USA, pp. 422-427, May 1990.

Wallace, W. R. (2002). The use of track-before-detect in pulse-doppler radar, *Proceedings of IEEE 2002 International Radar Conference*, Edinburgh, UK, pp. 315-319, October 2002.

Robey, F. C.; Fuhrman, D. L.; Kelly, E. J. & Nitzberg, R. (1992). A CFAR Adaptive Matched Filter Detector, *IEEE Transactions on Aerospace and Electronic Systems*, Vol. 29, No. 1, pp. 208-216, January 1992.

Forney, G. D. J. (1973). The Viterbi algorithm, *Proceedings of IEEE*, Vol. 29, No. 3, pp. 268-277, March 1973.

Bandiera, F.; Orlando, D. & Ricci, G. (2006). CFAR Detection of Extended and Multiple Point-like Targets without Assignment of Secondary Data, *IEEE Signal Processing Letters*, Vol. 13, No. 4, pp. 240-243, April 2006.

Urick, R. J. (1983) *Principles of underwater sound / 3d edition*, McGraw-Hill Inc., 1983.

Conte, E.; Lops, M. & Ricci, G. (2002). Adaptive Radar Detection in Compound-Gaussian Clutter. *Proceedings of Eusipco-94*, Edinburgh, UK, September 1994.

Conte, E.; De Maio, A. & Ricci, G. (2002). Recursive Estimation of the Covariance Matrix of a Compound-Gaussian Process and Its Application to Adaptive CFAR Detection, *IEEE Transactions on Signal Processing*, Vol. 50, No. 8, pp. 1908-1915, August 2002.

# 3

# Sonar Data Simulation

Didier Guériot and Christophe Sintes
*Institut Télécom ; Télécom Bretagne ; UMR CNRS 3192 Lab-STICC*
*Technopôle Brest-Iroise, CS 83818, 29238 Brest Cedex 3*
*Université européenne de Bretagne*
*France*

## 1. Introduction

### 1.1 What is the need for sonar data simulation ?

At-sea daily costs for performing data collection are high and acquiring specific data to either validate hypotheses or try new algorithms is often too expensive to be achieved. Thus, due to operational constraints for underwater data acquisition, simulating realistic sonar data, like images and swath bathymetry profiles, is crucial for designing and tuning detection and classification algorithms according to sensors settings, sea-bottom nature and topography. To sum up, an accurate and realistic simulation tool is a very good asset to enhance system & algorithm performances, thus appearing as a complement to real data collections.

### 1.2 Requirements for simulating realistic sonar data

The robustness of any performances estimation or prediction algorithms can be greatly improved as soon as such an accurate simulation tool provides a way to introduce within the simulation: a modular and flexible underwater world representation, multiple kinds of sensors, multiple kinds of environments (propagation, ...) and multiple kinds of acquisition conditions (including sensor trajectories). Indeed, controlling the full data acquisition process through a highly modular sea-world representation, allows to produce many datasets with various dedicated configurations.
However, another axis of modularity is required as different computational engines may be used depending on simulation cases (ray theory, normal mode or parabolic equation), all solving the Helmholtz equation with specific approximations (Burdic, 1984).
Starting with this equation and assuming high frequency signals, the first approximation leads to the eikonal equation that allows acoustic wave to be modeled as rays following a trajectory bent according to sound celerity gradients. The engines that will be described in this chapter derive from this model. Relying on wave guide propagation mechanisms, the second approximations open into modes solutions. Assuming extra constraints (specific stratification), the modes equation may become parabolic (Brekhovskikh & Lysanov, 1982) or using Perkins model, admit trigonometric function based solutions (Perkins & Baer, 1982). Green functions may also be introduced leading to Fourier transforms (Harrison, 1989). However all these methods (based on differential equations) are computationally expensive as numerical approaches (FDTD, TLM techniques) are required to compute their final solutions.

### 1.3 Chapter objectives

Therefore, responding to the presented needs, we have developed a framework for simulators allowing both underwater scene design and computational simulation engine choices. More precisely, this generic framework provides the reader with a common and simple software system in which various sensors, environments and computational engines can be plugged in.

The first section of this chapter presents the framework architecture pointing at the interactions between all its components. Subsequent sections of the chapter focus on specific computational engines describing all the phenomena to be considered and the problems to be answered in order to produce realistic simulated sonar data. Section 3 details how ray and tube engines collect exhaustively all the backscattered acoustic waves resulting from scene interactions with the transmitting acoustic wave.

Using these two engines and considering acoustic waves and scene interactions as energy based only, Section 4 presents realistic simulated results for several imaging systems (sidescan and front-looking sonars).

However, several phenomena such as speckle effect or diffraction, have a great impact on collected sonar data and they cannot be simulated using energy only: simulating full sonar signals (including the phase component) is absolutely required. Section 5 describes the difficulties such a complete simulation involves, and proposes solutions to model and implement the needed local signal/scene interactions. Finally, two standard phase-based sonar processing (beamforming an interferometry) fed with simulated full sonar signals, successfully produces output exhibiting very good behaviors compared to real measurement.

## 2. A multi-sensor simulator framework with engine plugins

The main goal of this simulation framework is to propose a common environment for developing modular simulation tools. The keyword "modular" stands for the simulation objects describing the underwater scene but also for all the algorithms developed to deal with acoustic propagation or object interactions (reflection, refraction ...). Thus, a user can pick among several families of components to tune its simulation environment. Some of these components are relative to static properties of the environment while others are meant to represent dynamic phenomena. Figure 1 offers a synthetic view of the software components of this framework.

### 2.1 Scene components

The scene components constitute the basic and common blocks for describing the simulation underwater environment; all the processing engines will operate on these blocks. Several families of scene components can be listed:

- *Objects*: All the geometric features of the scene are described through objects either using mathematical definitions (sphere ...) or lists of facets which are a more flexible way of description. Objects mean real objects like wrecks or containers but also stand for sea bottom elevation maps (or Digital Terrain Model). Vehicles supporting sensors are also considered as objects, as their geometric descriptions may impact the simulation.
- *Media*: Such medium elements gather all the intrinsic properties of the water (sound velocity profiles ...) because of the impact on the propagation (Ziomek & Polnicky, 1993).
- *Sensors*: Different kinds of sensor are defined like sidescan sonar or multibeam echosounders. The purpose of these sensor objects is to model their specific behaviors

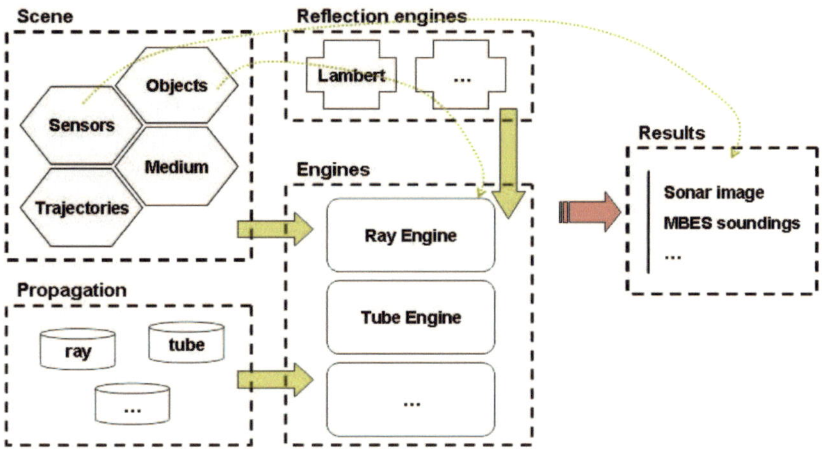


Fig. 1. Framework main software components

especially when aggregating the incoming acoustic waves and reconstructing the output "signal" (Bell & Linnett, 1997).

- *Trajectories*: Objects may be associated to trajectories (typically, vehicles supporting sensors during a survey). These trajectories allow a time varying environment to be taken into account.

This framework design aims at proposing flexibility for developing new items within these component families, allowing reuse and specialization of existing components.

### 2.2 Propagation

The propagation package gathers all the models available for simulating acoustic wave propagation within a medium (like rays or tubes, for instance). The medium properties greatly affect the propagation behavior: sound velocity variations according to depth may bend the rays depending on the sign of the velocity gradient (Urick, 1983).

### 2.3 Engines

The engines part relies on a scene description and a propagation model to propose a full simulation of the acoustic propagation within the environment. Therefore, it main goal is to generate a propagation history from the transmitting to the receiving sensor, gathering along the propagation path, all the interactions between the acoustic wave and the scene elements. Preserving such a propagation history allows each receiving sensor to aggregate all the incoming contribution to produce specific sonar simulated data.
In fact, one key feature of this framework is to separate the rendering part (through the various sensors) from the acoustic propagation simulation, thus allowing more flexible updates (with new sensors or simply for testing several sensor settings).

## 3. A tube engine for propagation history building

Within the global framework presented in the previous section, two engines have been developed in order to validate the required computational modularity. These first two engines basically rely on a ray approach to model the acoustic field propagation: the first considers

an acoustic wave as a series of rays while the second uses these rays as support for a volume description of the insonified area.
Previous works on simulating sonar images through acoustic ray tracing techniques have already shown interesting results (Bell, 1997). From these developments, the core engine dealing with ray propagation has been extracted and redesigned to fit in a framework compatible *ray engine*. The results obtained with this new implementation concur with simulations from the previous version (Guériot et al., 1998).

### 3.1 Why a volume approach ?

The main challenge of such an engine lays in the volume approach of the acoustic wave to be simulated. The previous engine represents the wave propagation as a series of rays always orthogonal to the current local wave front. The involved sampling of the studied volume, for both the propagation itself and the wave interactions with the scene elements, introduces artifacts as each ray has only a very localized interaction with the environment. Thus, tuning this sampling has a great impact on the quality of the simulated data (missing objects falling between two neighboring rays, missing object characteristics like refraction or border effects, artificial backscattering density due to the chosen sampling ...). It is then obvious that the key feature for this new engine is to implement a volume model of the acoustic wave along with propagation and scene interaction algorithms in order to guarantee an exhaustive energy transport.

### 3.2 Describing tube propagation

Figure 2 presents the geometrical sensor-centered reference, used for all the computations. Within this reference, $\wp$ corresponds to the selected acoustic propagation model used to build the rays supporting each tube.

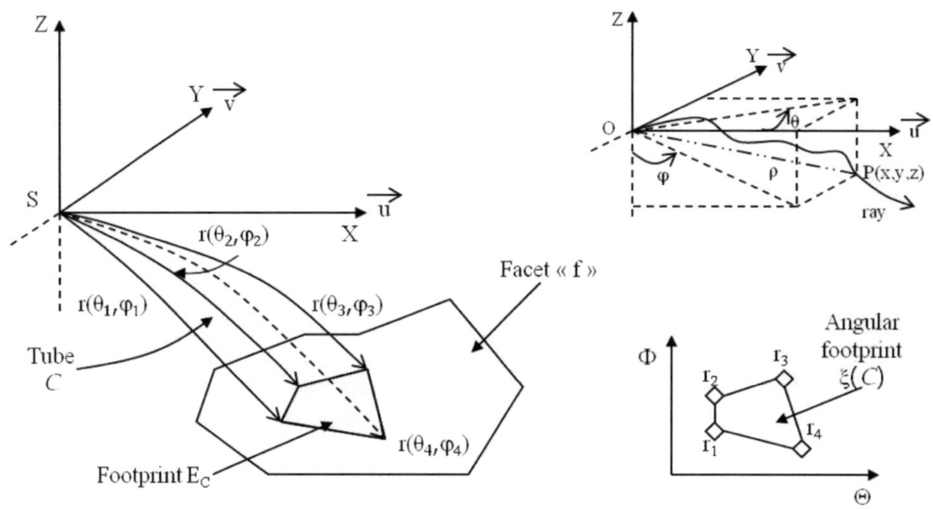


Fig. 2. From rays to tube definition

All the following definitions will be useful to clearly define the basic principles of the tube engine algorithm.

**Definition 1** (Ray). *A ray $r(A, \theta, \varphi)$ is defined as an acoustical trajectory $f_{\wp}(t)$ along the time t and according to the selected propagation model $\wp$. $\theta$ and $\varphi$ specify the launching angles of the ray r, from the point A. Thus, the tuple $(\theta, \varphi, t)$ corresponds to the propagation coordinates of a geographical point $(x, y, z)$ within the scene, according to the ray r. It is important to note that a geographical point may have several propagation coordinates.*

**Definition 2** (Tube). *A tube $C(A, r_i)$ is a 3D region of the scene bounded by a series of n rays $r_i(A, \theta_i, \varphi_i)$ such as the polygon $\xi(C)$ defined in $\Theta \times \Phi$, by the vertices list $(\theta_i, \varphi_i)$ is simple. $\xi(C)$ is the tube C angular footprint. A sector is a 4-rays tube; its goal is to easily split the scene volume space.*

**Definition 3** (Tube membership). *A ray $r(S, \theta, \varphi)$ belongs to a tube $C(S, r_i(S, \theta_i, \varphi_i))$ if the point $(\theta, \varphi)$ defined in $\Theta \times \Phi$, belongs to its angular footprint $\xi(C)$.*

**Definition 4** (Elementary tube). *A facet $\mathcal{F}$ in the scene is intersected by a tube C if it exists a ray r belonging to this tube such as r intersects this facet. In general, a tube may intersect several facets but an elementary tube intersects only one facet of the scene.*

**Definition 5** (Tube footprint). *As shown in Figure 2, the footprint $E_C$ of an elementary tube C on a facet, is the unique polygon which vertices are the intersections between the rays defining the tube and this facet.*

**Definition 6** (Footprint intersection). *The footprint intersection $E_1 \cap E_2$ between two elementary tubes $C_1$ and $C_2$ is either empty if the footprints $E_1$ and $E_2$ do not belong to the same facet, or a set of polygons (general case of intersection between two non-convex polygons).*

**Definition 7** (Tube intersection). *The intersection $C_1 \cap C_2$ between two tubes $C_1$ and $C_2$ is equal to a set of elementary tubes $\Im$ included in $C_1$ such as it exists no ray belonging to $C_1$ that has an intersection with the footprint $E_1 \cap E_2$ and does not belong to an elementary tube in $\Im$.*

### 3.3 Acoustic tube tracing algorithm

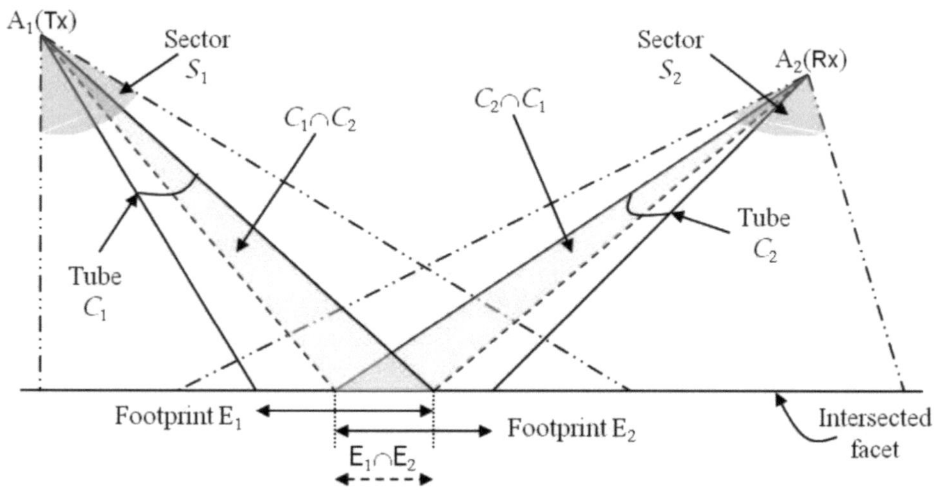


Fig. 3. Geometrical configuration for an elementary tube intersection from Tx to Rx sensors

With all these definitions, the tube engine basic computation algorithm is a 4-step processing (see Figure 3):

- The transmitting sector, defined by the Tx sensor properties, is split as long as it contains non-elementary tubes.
- The same splitting process is performed for the receiving Rx sensor.
- For each intersection between an elementary Rx tube and an elementary Tx tube, the corresponding footprint is computed allowing for the receiver to properly aggregate the transmitter contribution after its interaction with the intersected facet.
- In case of multiple paths simulations, all the footprints belonging to a facet are gathered in order for this facet to become a new transmitter. The amount of levels in this recursive loop defines the number of multiple paths to consider.

### 3.4 Tube engine generic output

Computing the full 3D acoustic wave propagation history is the major key feature of this algorithm. Each sensor uses it then to output data according to their specific appropriate *aggregation* algorithm (complete signal or amplitude for images...). Next section will use this propagation history in order for several sensors to output simulated data.

## 4. Multiple sensors simulation data

Based on the previously computed propagation history, this section presents simulated data from three different sensors: sidescan sonar, multibeam echosounder and acoustic camera. Following their specific aggregation algorithms, results are shown with focuses on specific simulator features.

### 4.1 Simulating sidescan sonar

#### 4.1.1 Aggregation principles

According to Figure 3, the geometrical configuration and properties of the facet supporting an elementary tube, characterize the energy backscattered through that tube. When arriving on the Rx sensor, this temporal backscattered energy is sampled according to the sensor sampling frequency to fill up the corresponding pixels in the output image. Processing all the Rx elementary tubes with respect to their intersection with Tx elementary tubes, completes the full image and involves a full energy preservation from the Rx sensor point of view.

#### 4.1.2 Imaging complex objects

High resolution simulated sonar images have been produced by our simulator when using this tube engine and a *virtual* sidescan sonar to perform the acoustic data collection. An oil well head model has been read from a "Wavefront" formatted file in order to extract the geometrical description of the object (mainly its vertices and facets). Using this facet-based description, a tube engine based simulation was launched assuming the oil well head lies on a flat sandy bottom. Several rotations (none, then 45° along the sonar path, then 30° more across the sonar path) have been applied to the oil well head in order to produce the images shown in Figure 4. The resulting images clearly agree with the 3D model (also presented).

#### 4.1.3 Imaging real sea-bottom

For this test, a Digital Terrain Model built with bathymetric measurements on a small area near the Brittany shores, has been used as sea bottom on which rocky patches appear along with a small ridge and rippled sandy regions. Figure 5 (upper part) shows the obtained simulated image on such a sea bottom and displays the links between the real sea bed features (rocky

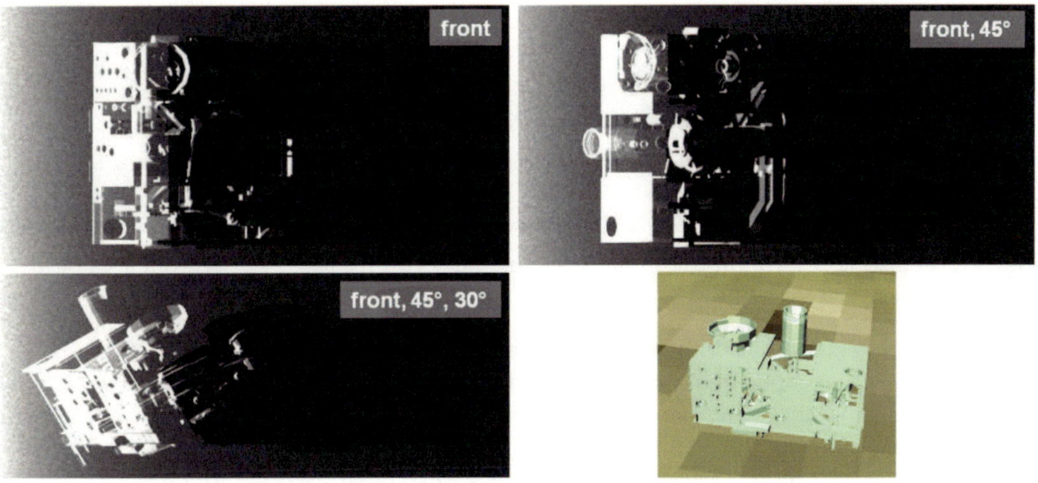


Fig. 4. Oil well head simulated sonar images (according to its 3D model)

patches, ridges, sandy areas limits) and how they are rendered on the simulated image. The main goal of this sketch is to show the consistency between the simulator inputs and the produced data.

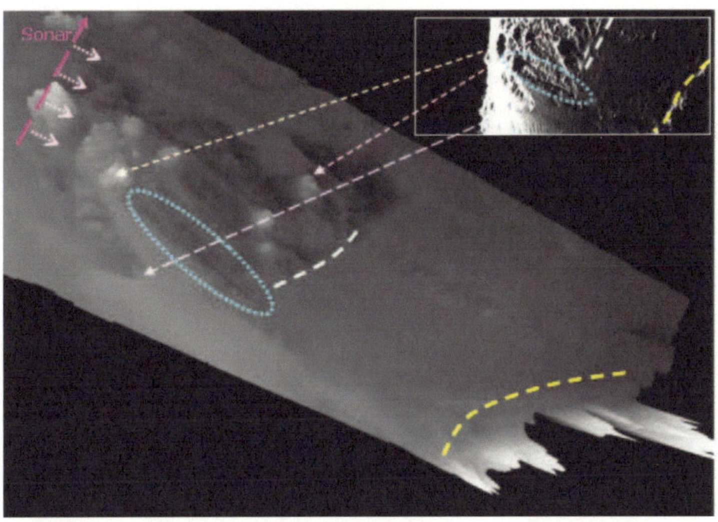


Fig. 5. From sea bottom topography to sonar image

#### 4.1.4 Ray vs. tube engines

As explained in section 3.3, the tube engine makes a complete energy summation over the whole solid angles. Even if tube borders are defined using rays, the inside volume is taken into account for all scene intersections, removing undersampling artifacts observed with the pure ray engine like missing sea-bottom features as shown in Figure 6. Moreover the medium absorption is directly taken into account through the volume approach involving a more realistic output (before a Time Varying Gain correction).

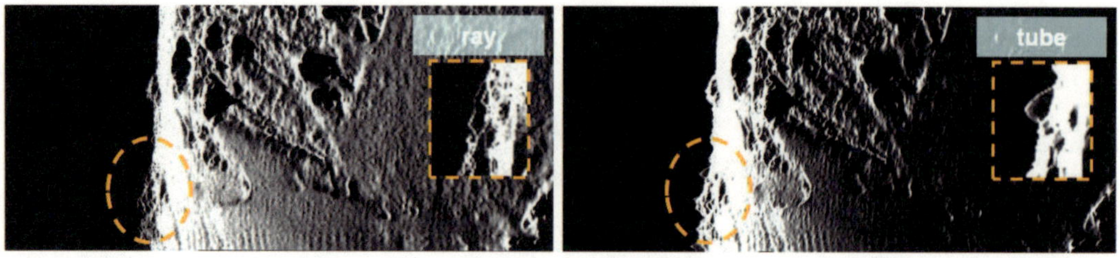


Fig. 6. Comparing ray vs. tube simulated sonar images on the same sea-bottom area (with subsampling effects).

#### 4.1.5 Comparing simulated with real sonar images

In order to check the quality of the simulator outputs, we compared real sonar images with simulated ones using the same acquisition conditions (sensor trajectory & attitude). Data collected for the *Shallow Survey '99* conference (Australia Defense Science & Technology Organization, 1999) in Sydney harbor have been used for this test according to the methodology given in Figure 7. During this survey, several kinds of sensors were used: a RESON 8101 multibeam echosounder for bathymetry data and a Klein 5400 for high resolution sonar images. Extra information such as sound velocity profiles and tide charts were also provided to correct bathymetry measurements in order to build a 2km by 3km DTM of the observed region with a 1.5m resolution cell.

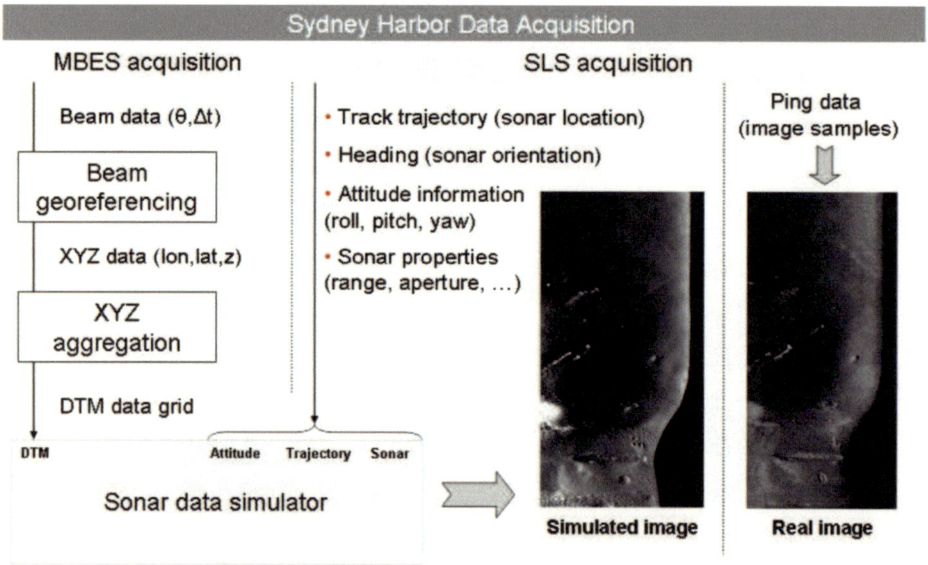


Fig. 7. Comparing simulated vs. real data procedure

The purpose of using such a survey is to compare real sonar images with simulated ones produced by the simulator fed with either raw data (sound velocity profile, trajectories ...) or preprocessed information (DTM), but excluding the real sonar images. The expectation is to find on the simulated images, the features appearing on the real ones.

A sidescan sonar antenna matching the KLEIN 5400 specifications has also been defined to produce images as output. Figure 8 presents both simulated and real sonar images. A quick

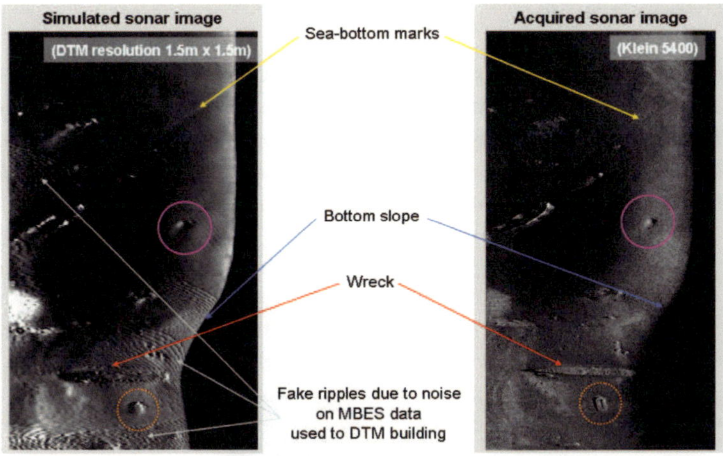


(a) Klein 5400 vs. simulated sonar image (DTM built with 1999 MBES data collection)

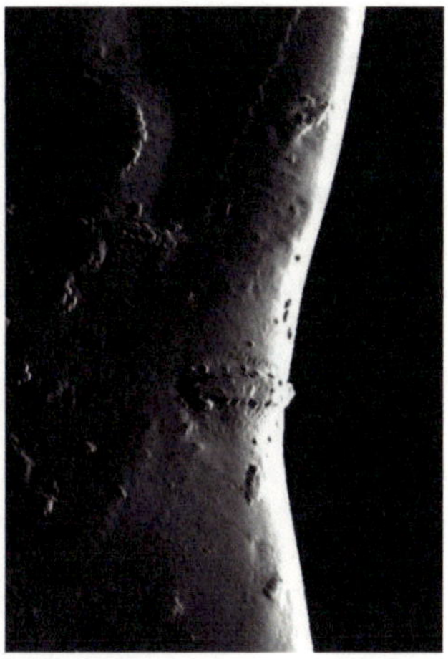

(b) Simulated sonar image using a DTM built from MBES data acquired in 2003

Fig. 8. Comparing Klein 5400 sonar images with simulated one on the same sea-bottom area (Sydney bay).

look shows that all the features (wrecks or bottom slopes) even the smaller ones, existing in reality (and shown on the real sonar image), also appear in the simulated data:

- *Objects*: both wrecks (main one on the bottom part, smaller one on the central part, end of range) can be identified. Two other spots corresponding to impacts (purple plain circle) or objects (orange dotted circle) also match.
- *Marks*: even fine sea-bottom marks appear on the simulated image.
- *Backscattering strength*: with recorded sidescan images only, determining the reason for having sandy lighter texture areas is not easy; their appearance in the simulated data involves they come from an elevation change (with a slope facing the sensor), besides sediment differences. This aspect is very interesting from a scene interpretation point of view.

The smoothing effect appearing on features of the simulated image can be explained by the DTM grid resolution (1.5m by 1.5m cell) compared to the resolution of the real or simulated sonar images (a few centimeters).
Some artifacts are also observed on the simulated data. These ripples come from the DTM building phase with not so well-calibrated MBES data. This lack of calibration introduced a few centimeters height difference between XYZ points coming from end-of-range overlapping tracks. As with this survey, only one or two points participated to the height of a DTM cell located at the end of the sonar range, for these cells, the poor amount of data leads to alternating heights between neighboring cells.
Therefore, having an accurate world representation is crucial as the simulation accuracy puts into relief all its details. Thus, when using a DTM built on the same area but with different MBES data (2003 instead of 1999 survey), these artifacts do not show off. Figure 8(b) presents such a simulated image with a sensor trajectory slightly shifted to the left: the sonar path flies just over the main wreck, as revealed by the water column limit.

### 4.2 Simulating multibeam echosounders

#### 4.2.1 Aggregation principles

Each Rx beam from the multibeam echosounder is described by a specific Rx sector which is split into Rx elementary tubes according to the same process as previously. A series of energy contribution along time is then computed for each Rx beam. Based on amplitude detection, the first temporal return among these contributions, provides the beam sounding characterizing the return time travel between the sensor and the intersected scene.

#### 4.2.2 Results

A RESON 8101 survey has been simulated on the Sydney area using a previously built DTM. On the simulated data, the two wrecks observed on the simulated sidescan images are logically present along with some small rocks over the slopes of the basin.

Fig. 9. Multibeam echosounder simulation on the Sydney harbor area

### 4.3 Simulating front-looking imaging sonar : DIDSON front looking camera

Based on high frequency sonar, DIDSON acoustic cameras provide short range, high resolution images of underwater environments when optical systems get blind mainly due to water high turbidity. Moreover, their image refresh rate can be rather high providing several views of quite the same area, allowing various applications like inspection (Belcher et al., 2001), navigation or surveillance missions (Asada et al., 2007).

#### 4.3.1 Aggregation principles

Besides the generic propagation history building, some DIDSON specificities need to be taken into account to create realistic images. DIDSON system uses acoustic lenses to perform direct and energy saving beamforming. The antenna length is split into 48 or 96 cells or beams, each beam being 0.3° width with spacing ranging from 0.3° to 0.4°, for a full 29° azimuth coverage. Sampling along time the energy focused inside each cell provides per-beam backscattered signal according to range (Belcher et al., 2002). A DIDSON image gathers all these collected backscattered signal samples per beam and per range.
However, according to the system frequency (high or low resolution), a full DIDSON image (48 or 96 beams) is built from the reverberation received by 4 or 8 sets of 12 non-adjacent beams fired simultaneously. This interlacing technique requires several Tx/Rx cycles to produce a complete image but aims at preventing beams cross-talk. Indeed, the acoustic beam pattern width for one beam is almost 2°, showing that two neighboring (0.3° spaced) beams cannot be used within a single cycle. Nevertheless, depending on the shape and nature of the observed object, cross-talk phenomena are still observed on real images (see Figure 10(a)) as side lobes

may gather enough energy contributions from round-shaped objects that reflect transmitted energy to all directions.
All the built images are stored within a DDF file (DIDSON format for sequences of recorded images) as frames.

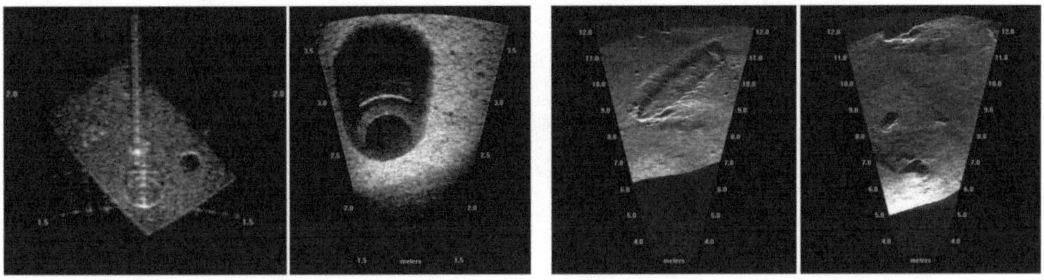

(a) Real DIDSON images (metal plate and tire)

(b) Simulated images extracted from a simulated sequence on Sydney bay

Fig. 10. Examples of real and simulated DIDSON images.

### 4.3.2 Results

The component in charge of building DIDSON images from the propagation history gathers the backscattered energy described through the lists of Rx elementary tubes according to sensor Tx and Rx patterns. Moreover, depending on the considered cycle, only sectors corresponding to the fired beams are used for data collection.

#### 4.3.2.1 Image simulations

Figure 10(b) exhibits two frames extracted from a simulated DDF file using Sydney harbor survey data (DTM). These images clearly show the big wreck and the rocky patches already displayed by both the previous sensors.

#### 4.3.2.2 Observing cross-talk phenomenon

When performing simulation on round-shaped objects, cross-talk phenomena are observed the same way as with images coming from a real DIDSON camera. Figure 11 presents both real and simulated images on a scene containing a sphere lying on a flat bottom. Cross-talks are logically observed on both real and simulated images as round objects reflect the transmitted acoustic wave to all directions. This realistic result clearly shows the ability of the simulator to take into account the actual sensor beam pattern and its specific aggregation algorithm, without questioning the propagation history principle.

### 4.4 Why using georeferenced perturbations ?

Using facets or DTM cells for describing items inside an underwater scene defines the resolution of the geometrical features seen by the sensors during any simulation. However, many phenomena come from higher resolution scene features. For instance, on real DIDSON images collected in a flat bottom pool, some patterns between consecutive images are observed, besides the fluctuating noise. Being clearly georeferenced, these patterns show that modeling a flat bottom by a flat DTM only, is not enough. Indeed, they reveal the bottom response occurs at a finer resolution than the one used for DTM description.
Next sections will present two approaches to allow a more detailed scene description, producing either *macro-textures* or *micro-textures*.

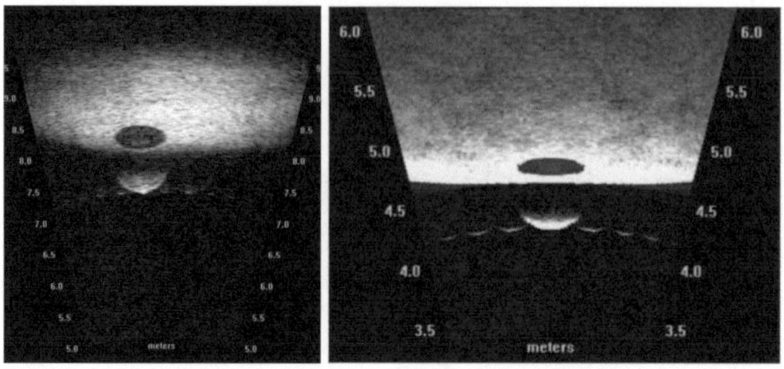


(a) Real (b) Simulated

Fig. 11. Real & simulated DIDSON images with crosstalk phenomenon on a sphere lying on a flat bottom

#### 4.4.1 Macro-textures

This first approach introduces georeferenced perturbations to the elevation surface modeling the observed scene, by decomposing each facet into smaller ones (sub-facets). Adding the global elevation coming from the underlying facet (DTM or object) to a small variation produced by a local and finer elevation map, gives each sub-facet elevation. Thus, each sub-facet produces a specific response giving a more realistic characterization of the encompassing facet.

Various techniques may produce adequate local elevation maps depending on the textures to be simulated. These textures clearly depend on the scene sea-bottom sedimentology. Rocky sea-bottoms may be obtained using standard fractal techniques like fractional Brownian motion (Pentland, 1984). Moreover, shifting fractional Brownian motion spectral components allows producing oriented textures; the direction and amplitude of the produced ripples depend on the direction and length of the spectral translation (Guillaudeux & Maillard, 1998).

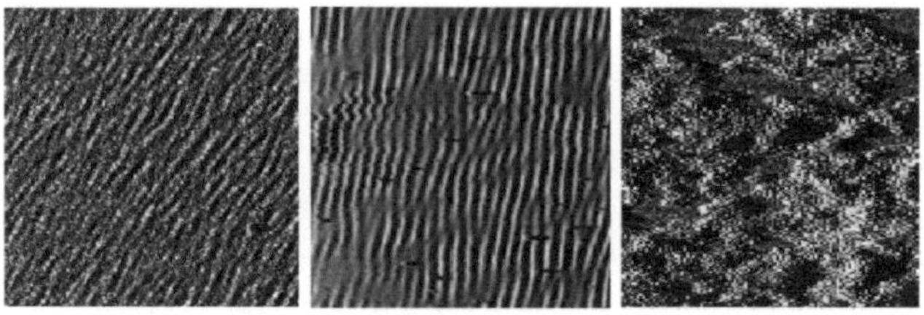

(a) Soft ripples (b) Sharp ripples (c) Rocky area

Fig. 12. Using macro-textures to simulate sea-bottom elevation features

Figure 12 shows examples of simulated texture images using these techniques to produce local elevation maps. Thus, without any modification of the simulator core, introducing local georeferenced perturbations to the global elevation map, allows producing more realistic sea-bottoms textures (with structured shadows) reflecting local seabed altitude variations.

#### 4.4.2 Micro-textures

The second approach that produces *micro-textures* does not involve elevation perturbations but introduces georeferenced perturbations to the facet reflection properties. In that case, the

standard backscattered energy transmitted by each Rx elementary tube is modulated by a local contribution linked to both the statistical properties of the facet (sediment nature, material. . .) and the location on this facet.
Figure 13 illustrates the impact of such georeferenced backscattering modifications on simulated images with various micro-textures corresponding to several sandy sea-bottoms. Indeed, modifying the backscattered energy at the elementary tube resolution allows tuning image granularity inside facets, but unlike the macro-texture approach, this micro-texture approach cannot produce textures with shadows.

(a) Mud (b) Sand (c) Gravel

Fig. 13. Using micro-textures to simulate various kind of sandy sea-bottoms

As for macro-textures, micro-textures can be generated using existing maps or georeferenced perturbation functions computed during the simulation. The underlying micro-texture used for the next simulation results has been generated online using Perlin noise (Perlin, 2002) and counted for 25% of the diffuse reflection component.

#### 4.5 Simulation results

As previously mentioned, consecutive DIDSON frames exhibit a meaningful level of correlation between images. Micro-textures, being georeferenced by definition, induce such cross-correlations between simulated images observing same scene areas as illustrated by Figure 14 which shows several consecutive images extracted from a sequence simulated with a flat sea-bottom with a straight constant speed trajectory. The yellow dotted circle outlines a visible pattern between the three frames. As the sensor moves along its trajectory, the pattern slowly drifts towards the sensor water column (first return recorded by the sensor).

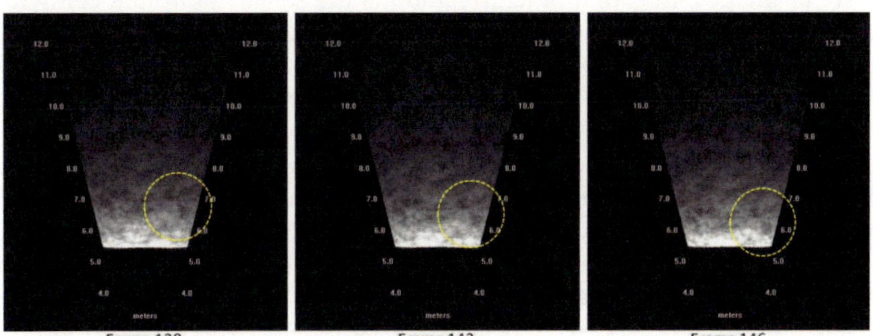

Frame 138 Frame 142 Frame 146

Fig. 14. Neighboring frames (with consistent micro-textures) extracted from a simulated DIDSON sequence with a straight constant speed sensor trajectory

Finally, Figure 15 compares two simulations performed with the oil well head scene. As the sea-bottom from the left image is only modeled by a flat DTM with no micro-textures, no structured pattern appears: the backscattered energy is only affected by random and uncorrelated noise. On the contrary, the right image integrates georeferenced perturbations into the reflection: the underlying flat bottom produces uneven intensities across the image. Moreover, as micro-textures adds spatially structured contributions to the basic backscattering, some smoothing effect is also observed.

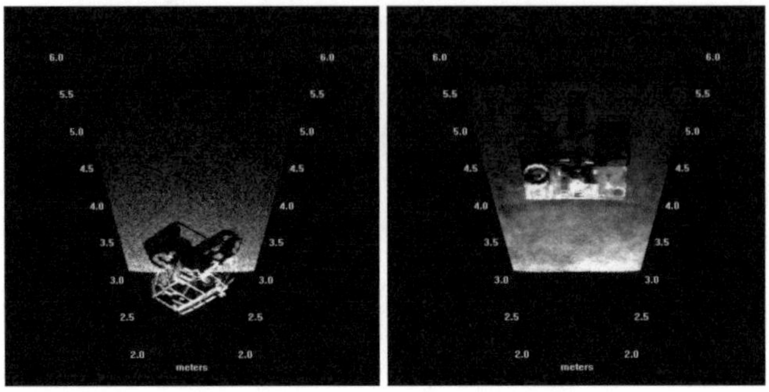


(a) No cross-talk, no micro-texture (b) Cross-talk, micro-texture

Fig. 15. DIDSON simulated images and georeferenced perturbations using Perlin noise

## 5. Simulating sonar full signal

Previous sections have explained how realistic images can be produced by simply gathering incoherent contributions from the observed sea-bottom. No speckle effect although very obvious on sidescan sonar image, can be produced with this approach. However, extra contributions looking like speckle noise may be added to the simulated images, after simulation. Partial diffraction effects for specific configurations may also be integrated to simulated images, using the same kind of shortcuts. However, simulating phase signal is absolutely required in the general case and clearly to feed standard sonar processing like interferometry or beamforming. Simulating images only is not enough.

Simulating the phase of a sonar signal is a matter of aggregation resolution. Indeed, when simulating sonar images, the scale of aggregation is the sensor sampling cell which defines the scale used to aggregate the backscattered contributions. However, coherent phenomena involving the signal phase component, occur at the signal wavelength scale. Simulating the phase signal thus requires the simulator to process the propagation history at this scale. Hopefully, this propagation knowledge is already available.

The notion of *coherence* is the key feature for any realistic signal simulation. Indeed, the sampled backscattered signal is the result of coherent summations of contributions from sea-bottom tiny parts. These tiny reflectors behaving according to many parameters such as local sea-bottom properties, the summation is strongly linked to the observed sea-bottom small area. Indeed, estimating the correlation between two distinct sea-bottom patches, returns 0 while the correlation of two partially overlapping areas is only proportional to their common surface: the specific identity of every sea-bottom part affected by the transmitted signal, controls the coherent summation.

Thus, any simulation engine looking forward to being used as input for sensor array processing (beam processing or interferometry) has to deliver a georeferenced phase associated to every sampling cell using the underlying coherent summation. The challenge is to find a sea-bottom sampling model which is computationally simple enough to be used for full track simulations. Next paragraphs will describe a proposed attempt to get realistic phase simulation. First, the required formalism is introduced to fully explain the approach. Then, a Brownian-based optimization is proposed to lighten the computational burden needed to get reflectors behaving according to wished statistics properties.

### 5.1 Requirements for building a coherent sea-bottom simulation

The first engine plugged in this simulator uses ray tracing and only involves geometrical interactions between the acoustic wave and the scene. These local interactions are closely related to punctual scatterers but how to use these rays to simulate coherent backscattering still remains a big issue since two major parameters of the acoustic wave are not taken into account: the signal wavelength and the signal bandwidth which quantifies the quantity of information transported by the carrier frequency and thus, the sensor resolution.

#### 5.1.1 Sonar resolution cell

Each time sample of the backscattered signal (after sonar demodulation) corresponds to a footprint on the sea-bottom, also called resolution cell. The energy level contained in such a sample results from a very intricate combination of several contributing parts within the resolution cell. Figure 16(a) illustrates such a resolution cell according to the sampling interval $\tau$ and the sound speed $c$. Thus, the challenge of this simulator is to use the acoustic propagation history in order to create a summation process, representative of the coherent aggregation.

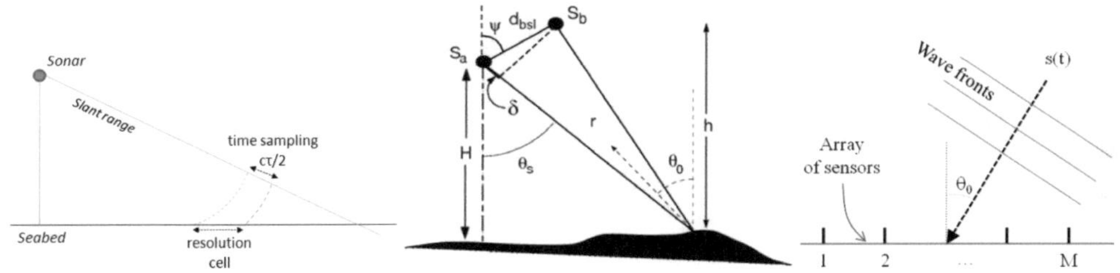

(a) Resolution cell definition (b) Interferometry (c) Beamforming

Fig. 16. Basic principles for interferometry and beamforming

#### 5.1.2 Sonar processing involving phase signal: interferometry and beamforming principles

Simulating full sonar signal including the phase component is absolutely required to feed standard sonar processing like interferometry or beamforming with signal generated by the simulator. This section quickly describes how the phase component is crucial for such processing and clearly shows that phase simulation is intrinsically linked to the sea-bottom coherence.

5.1.2.1 Interferometry principles

Interferometry is a widely-used method that provides accurate bathymetric information by measuring the phase difference between two sensors (Lurton, 2003)(Denbigh, 1989). Figure 16(b) depicts the standard geometrical configuration for interferometric sonar systems where $\theta_0$ is the direction of arrival (DOA) of the backscattered signal and $\psi$, the steering angle. Assuming a non-dispersive medium hypothesis (i.e. the phase and group velocity have a linear connection), Equation 1 gives the searched phase difference $\Delta\varphi$ which represents indeed a backscattered signal delay between the sensors.

$$\Delta\varphi = \frac{2\pi}{\lambda}\delta = \frac{2\pi}{\lambda} d_{bsl} \cos(\theta_0 + \psi) + 2\pi k \quad (1)$$

where $\lambda$ is the acoustic wavelength, $d_{bsl}$ measures the spacing (or baseline) between the sub-arrays centers and $k$ corresponds to the number of $2\pi$ phase ambiguities as $\Delta\varphi$ can go beyond $2\pi$ for high angles and large baselines, while the measured phase difference (the cosine term) remains trapped within the $[0, 2\pi]$ interval.

5.1.2.2 Beamforming principles

Beamforming is also widely used to build the backscattered signal from the receiving antenna considered as an array of several elementary sensors with wide aperture. Basic beamforming thus consists in an in-phase summation of the backscattered signal collected by all the $M$ elementary sensors. As shown in Figure 16(c), when assuming the incoming signal $s(t)$ comes from a far field source with a direction of arrival $\theta_0$, Equation 2 gives the signal $\vec{y(t)}$ arriving on each sensor of a $d$-spaced line array.

$$\vec{y(t)} = a(\vec{\theta_0})s(t) + \vec{b(t)} \quad (2)$$

where $\vec{b(t)}$ represents noise contribution on each sensor and $a(\vec{\theta_0})$, the goniometric vector

$$a(\vec{\theta_0})^T = \left[e^{2\pi j \frac{0}{\lambda}\sin(\theta_0)}, ..., e^{2\pi j \frac{d.(M-1)}{\lambda}\sin(\theta_0)}\right] \quad (3)$$

Building the final backscattered signal $y_{BF}(t)$ through beamforming consists in introducing a phase correction on each sensor to compensate for the $a(\vec{\theta_0})$ effect before summing all the elementary contributions $y_k(t)$ from $\vec{y(t)}$. Equation 4 gives this in-phase summation.

$$y_{BF}(t) = \sum_{k=0}^{M-1} e^{-2\pi j \frac{d.k}{\lambda}\sin(\theta_0)} y_k(t) \quad (4)$$

### 5.1.3 How to model the phase brought by resolution cells ?

Many works model various properties of these resolution cells and how they contribute to the backscattered signal in terms of phase behavior (Mourad, 1989)(Kuo, 1964).

5.1.3.1 Exact and continuous models

Two main approaches try to exactly describe resolution cells backscattering behavior using continuous models. The first one takes benefit from the rough surface theory with Kirchoff condition while the second one deals with diffraction phenomena through *micro-roughness* with Bragg wavelength (Jackson et al., 1986). These models heavily rely on finding Green function solutions to solve the propagation equation making the computation of one resolution cell backscattered signal very complex and time consuming. Moreover, all these models make strong assumptions about isotropy or angle of incidence.

5.1.3.2 Single scatterers approach

A third approach considers a resolution cell as a set of point scatterers. Then the backscattered signal $F$ of a resolution cell is equal to the coherent summation of the incoming signal $s(t)$ transmitted back by the $N$ single scatterers belonging to this resolution cell. As shown in Equation 5, this atomistic model is very appealing when the amount of scatterers remains small as each scatterer $i$ receives the incident signal $s(t)$ and transmits it back with modified power $a_i$ and phase $\phi_i$ corresponding to the scatterer properties.

$$F = \sum_{i=1}^{N} a_i e^{j\phi_i} s(t) \tag{5}$$

As within a resolution cell, the amount $N$ of scatterers, the amplitude $a_i$ and phase $\phi_i$ of every scatterer are random, it easily explains the reverberation diversity such a model is able to produce. However, finding basic single scatterer properties fitting both experimental and theoretical models is the main drawback of this model as the involved mathematical developments may become rapidly complicated due to the manipulation of complex random variable summation.

Moreover, when performing interferometry, several sensors observe almost the same sea-bottom from different points of view. To each sensor corresponds a series of resolution cells. A small patch of sea-bottom is then covered by one resolution cell per sensor, each one having slightly different boundaries on the sea-bottom due to the various points of view. Spatial and angular decorrelations are then introduced between these resolution cells as they do not include exactly the same set of single scatterers (Sintes et al., 2010). The spatial distribution of the scatterers is thus a crucial point to consider in order to perform consistent correlation between the resolution cells and compute for instance, their phase difference.

Thus controlling both the spatial correlation and random properties of single scatterers are fundamental to obtain a realistic full sonar signal simulation.

## 5.2 Simulating phase for sensor resolution cells

Generally speaking, the key feature to create a simulated signal (both intensity and phase), is to make interfere different parts of the resolution cells.

### 5.2.1 An atomistic approach : single scatterers and ray tracing

This paragraph gives a short introduction on how to simulate the response of a resolution cell through a single scatterer model and a ray tracing approach within a simulation context.

5.2.1.1 Building a resolution cell response from single scatterers

Equation 6 gives the temporal response $h(t)$ of a single scatterer which distance from the receiver is $r/2$, $\delta$ standing for the Dirac function.

$$h(t) = \delta(t - r/c) \tag{6}$$

As shown in Equation 7, we consider the sonar transmits a simple truncated sinusoidal signal $s(t)$ where $T$ is the signal duration and $f$, the carrier frequency. $\mathbb{1}_A(u)$ stands for the indicator function and is equal to 1 when $u \in A$, 0 anywhere else.

$$s(t) = \mathbb{1}_T(t - r/c).e^{j2\pi ft} \tag{7}$$

The received signal $x(t)$ depends on $s(t)$ with a phase shift depending on the total traveled distance $r$.

$$x(t) = \mathbb{1}_T(t - r/c).e^{j2\pi f(t-r/c)} \tag{8}$$

Finally, demodulating the received signal $x(t)$ gives $y(t)$ with

$$y(t) = x(t)e^{-j2\pi ft} = \mathbb{1}_T(t - r/c).e^{-j2\pi fr/c} \tag{9}$$

This signal $y(t)$ is sampled in a $z(k)$ vector according to the sampling step (or resolution cell duration) $T_e$. Assuming the sampling step is close to the signal duration ($T \approx T_e$), Equation 10 computes the contribution of one $r_i$ distant single scatterer $i$ to the sampling cell $z(k+1)$.

$$z(k+1) = \mathbb{1}_T\left((k+1)T_e - r_i/c\right).e^{-j2\pi fr_i/c} = e^{-j2\pi fr_i/c} \tag{10}$$

The sampling cell $z(k+1)$ gathers the contributions of all the scatterers $i$ within the corresponding resolution cell: the scatterers $i$ must check $kT_e < r_i/c < (k+1)T_e$ to be summed into $z(k+1)$ as shown in Equation 11.

$$z(k+1) = \sum_i a_i.e^{-j2\pi fr_i/c} \tag{11}$$

5.2.1.2 Simulation through ray tracing

Performing the sensor aggregation step according to Equation 11 is straightforward when using a ray engine. However, the amount of rays hitting each resolution cell has to be dense enough to compute the contribution of every single scatterer within this resolution cell.
Relying on this model, Figure 17 proposes two simulated interferometric signals on a flat sea bottom. In this example, the amplitude of every single scatterer is kept constant. The difference between the two signals lies on the average amount of scatterers per resolution cell, at the end of the swath: 3 for Figure 17(a) and 40 for Figure 17(b). The second simulated signal seems more realistic because of a higher noise level, coming from a bigger amount of contributions per resolution cell. Indeed, as the resolution cell has a finer description (more single scatterers), the spatial decorrelation effect is better described.

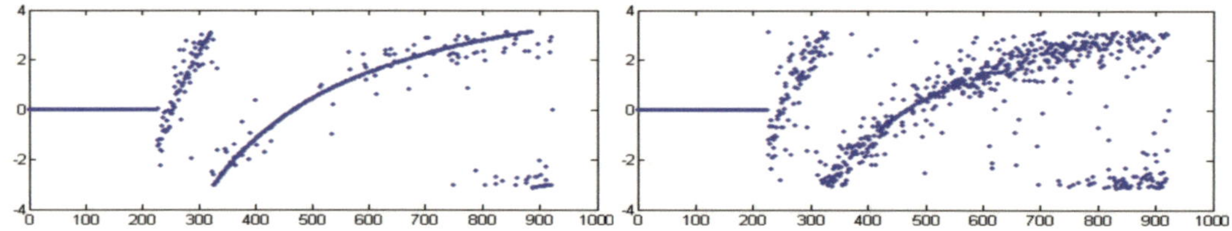


(a) 3 single scatterers (b) 40 single scatterers

Fig. 17. Simulated interferometric signal using ray tracing engine

This first attempt is very interesting as it clearly shows the simulator (using the ray engine) can be used to simulate interferometric signals. Nevertheless, some key features remain to be addressed: for instance, choosing a unique constant coefficient of reflection for all single scatterers involves important constraints on the random properties of the simulated backscattered signal. As shown, the random properties of the resolution cell are also affected by the amount of scatterers per resolution cell (i.e. the amount of rays hitting this cell).
A more continuous approach may be used to solve this acoustic wave discretization effect: this atomistic model may be extended for being used by a tube engine. It mainly requires to rewrite the summation of the incoming contributions within a continuous context.

### 5.2.2 Towards a continuous approach : extended single scatterers and tube tracing

Going from a discrete to a continuous model, Equation 12 gives the contribution $z$ of the observed sea-bottom within a time interval $[t_1, t_2]$: all the sea-bottom parts $u = r/c$, backscattering signal within the interval $[t_1, t_2]$ participate in $z$.

$$z = \int_{t_1}^{t_2} e^{-j2\pi f u} du = e^{-j2\pi f \frac{t_1+t_2}{2}} . \frac{\sin(\pi f (t_2 - t_1))}{\pi f} \tag{12}$$

Based on the linearity of the integral, a Chasles relation can be derived from Equation 12 between neighboring surfaces sharing the same properties. Equation 13 shows such a relation when considering three instants $t_1$, $t$ and $t_2$.

$$\begin{aligned} z &= e^{-j2\pi f \frac{t_1+t}{2}} . \frac{\sin(\pi f (t - t_1))}{\pi f} + e^{-j2\pi f \frac{t+t_2}{2}} . \frac{\sin(\pi f (t_2 - t))}{\pi f} \\ &= e^{-j2\pi f \frac{t_1+t_2}{2}} . \frac{\sin(\pi f (t_2 - t_1))}{\pi f} \end{aligned} \tag{13}$$

#### 5.2.2.1 Building a continuous resolution cell response

The sea-bottom contribution to the $k+1$ sampling cell is easily computed with $t_1 = kT_e$ and $t_2 = (k+1)T_e$ as shown in Equation 14.

$$z(k+1) = e^{-j2\pi f (kT_e + \frac{T_e}{2})} . \frac{\sin(\pi f T_e)}{\pi f} \tag{14}$$

A major benefit from this model is that as the tube engine splits a resolution cell into angular intervals which are different for one sensor to another, Equation 13 reveals it is possible to choose any splitting of one resolution cell, for a subsequent and adequate merging with an other overlapping resolution cell.

However, Equation 14 also points at one major drawback of this approach: no phase signature is associated to each radiating sub-surface composing a resolution cell as $z(k+1)$ only depends on $k$ and $T_e$. It involves any interferometric processing using such modeled signals on every resolution cell will give no informative correlation (1 everywhere) which is useless. This behavior is logically not observed using the ray engine with a non-uniform distribution of scatterers as each single scatterer returns a phase linked to its distance from the transmitter.

#### 5.2.2.2 Extended single scatterers

The proposed solution is then to give each contribution an unique identity. Associated with this ID, a single scatterer becomes an *extended single scatterer*. Indeed, in order to make backscattered simulated signals interfere, each scatterer should be coherent with itself but not with its neighbors.

Thus, the key idea is to divide the sea-bottom into elementary patches or *extended single scatterers*, with specific properties. Through this local identification, the continuous splitting of the sea-bottom preserves both the conservation of energy when integrating on surfaces and the decorrelation phenomenon between slightly different areas of the sea-bottom.

The flexibility of this approach comes from Equation 13 that allows an extended scatterer overlapping two resolution cells, to be split into two extended scatterers while preserving both phase information and energy. Figure 18 illustrates such a configuration with two sensors slightly shifted in relation to the sea-bottom. In this example, all the extended scatterers have a $T_e/2$ time contribution within the $T_e$ duration resolution cells. Then the integration duration $T_e$ is divided into 4 parts, to take into account all the combination between the overlapping scatterers from sensor #1 and sensor #2 within the two resolution cells but also the time delay between the cells. Each part has its own phase properties. Sensor #1 and sensor #2 signals can then be computed according to Equation 12, over the appropriate integration area. Using the simulated signals from sensor #1 and #2, the interferometric signal (and its phase) can be obtained through standard interferometry processing as given in Equation 1.

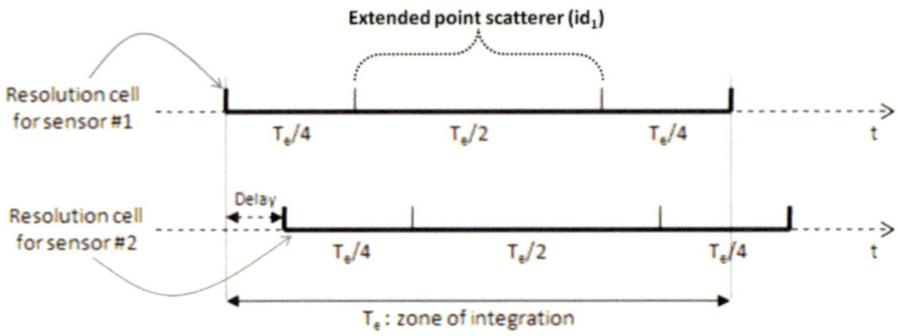


Fig. 18. How the continuous approach with extended scatterers allows taking into account spatial decorrelation effect when integrating backscattered contributions within the final resolution cell

### 5.2.3 Results

Next paragraphs present some results of full signal simulation (including phase) using the proposed extended scatterer approach. Such signals have been used as inputs feeding two typical sonar processing relying on phase information: interferometry and beamforming.

5.2.3.1 Computing phase differences for interferometry

Figure 19 presents simulated interferometric signals using the extended single scatterers technique. Figure 19(a) is obtained using 3 extended single scatterers, for each resolution cell located at end of the swath. Figure 19(b) presents the same simulation with 40 extended single scatterers, for the same conditions.

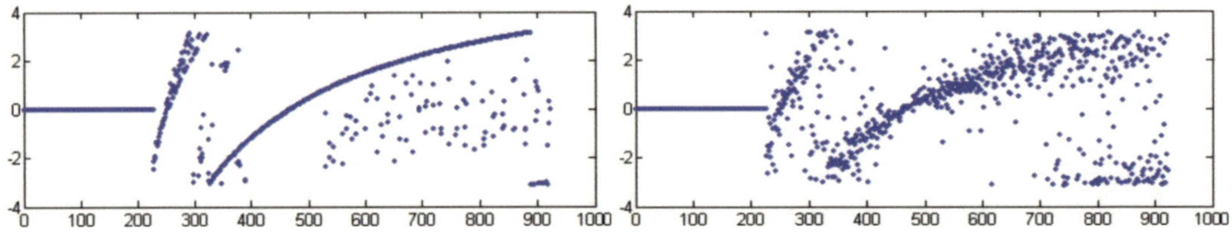

(a) 3 single scatterers (b) 40 single scatterers

Fig. 19. Simulated interferometric signal using tube tracing engine on a flat sea-bottom

The interferometric signal computed from the simulated full signal for two sensors observing the same sea-bottom clearly indicates the proposed approach embedding extended single scatterers succeeds.

5.2.3.2 Application to beamforming

The extended scatterer approach can also be used prior to beamforming. Figure 20 shows simulated raw signal for a multibeam echosounder consisting of 60 sensors uniformly spaced on a horizontal line. The frequency carrier $f$ is 300 kHz and the transmitted signal is a pulsed wave with $f/\Delta f = 10$. The simulated multibeam echosounder observes a flat sea-bottom with a cube located in the middle of the swath. Both amplitude and phase results are very realistic: for instance, on the simulated phase image, interferometric fringes are clearly observed around the object.

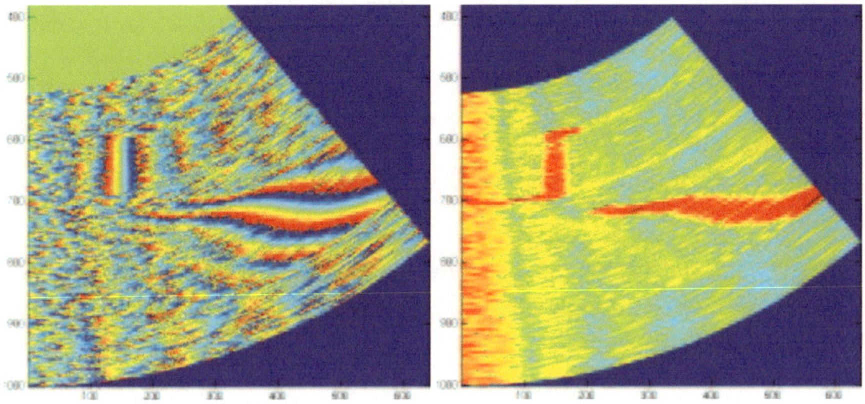

(a) Phase image (b) Amplitude image

Fig. 20. Simulated full signal (both phase and amplitude) using tube tracing engine for a multibeam echosounder observing a small object lying on flat sea-bottom

#### 5.2.4 Discussion

The proposed approach clearly allows the simulation of realistic full sonar signal. However, two features still remain to be studied: what is the best strategy for representing a sea-bottom with unique extended scatterers? what is the influence of extended scatterers distribution in the resolution cell response properties.

It has been shown that the sea-bottom needs to be divided into extended scatterers with unique identifications (like beacons). A global strategy of splitting is required in order for simulated signals over same sea-bottoms to have meaningful correlations. This strategy is independent of sensor trajectory; however, the resolution of the splitting depends on the wavelength of the signal to be simulated. Indeed, at least three extended scatterers are required within a sensor resolution cell, to simulate a realistic signal.

The second question deals with the simulation signal statistics evolution in relation to the amount of extended scatterers per resolution cell. To estimate this impact on a final interferometric signal, a simulation has been performed, considering a constant delay between the two sensors with an amount of extended scatterers varying from 3 to 100. The experience shows the amount of extended scatterers roughly does not affect globally the simulated amplitude. However, having more extended scatterers greatly reduces the variance of the simulated signals from 3 to 20 scatterers per resolution cell. After 20, the variance remains almost stable. This simple analysis clearly shows that to simulate specific random response of a sea-bottom, the statistics nature, distribution and amount of extended scatterers gathered within a resolution cell should be carefully taken into account.

### 5.3 Optimizing signal simulation through a Brownian sheet technique

As previously mentioned, the extended scatterers approach has two main drawbacks. The first one concerns the statistical properties of the generated signals because they do not correspond to actual phase statistics (Tough et al., 1995). The second drawback deals with the identity assigned to each extended scatterer: although assigning along simulation, a random phase to each single extended scatterer grants good interferences (Sintes & Guériot, 2008), this identification strategy is not robust enough to be used from track to track or survey to survey. Thus, an extended scatterer identity (or individual phase) clearly needs to be linked with the scatterer geographic location in order for several overlapping tracks to correlate when observing the same scatterers, the level of correlation depending on the amount of common observed scatterers. For this two-dimensional problem, the idea is to create a map where each position (or pixel) is a random variable, holding phase statistics that allows such correlation properties.

To produce such a georeferenced map, different techniques may be implemented such as Perlin (gradient) noise (Perlin, 2002) widely used in computer texture graphic technology, or fractional Brownian motion (Pesquet-Popescu & Vehel, 2002) bringing interesting Gaussian properties. Perlin noise main advantage is to compute the value of a given pixel independently of its neighborhood; however, the generated signal has no real statistical sense. Conversely, a Brownian-based solution does permit the generation of uncorrelated Gaussian signals from which it is possible to derive various kinds of statistics.

Thus, the next section discusses the proposed Brownian-based solution and how it has been adapted to generate both amplitude and phase information. Interferometry is then used to check the correctness of such signal simulations.

#### 5.3.1 Assigning specific phase to extended scatterers using Brownian-based model

The effort described in this paragraph consists in assigning each extended scatterer a random phase with specific properties, instead of using a uniform random phase.

Starting from Equation 12, the sea-bottom contribution $z(k+1)$ to the $k+1$ resolution cell can be written in a more general way. Equation 15 proposes such an expression and introduces the function $g(u)$ that gives the random signal associated to any extended scatterer $u$.

$$z(k+1) = \int_{t_1}^{t_2} g(u) \cdot e^{-j2\pi f u} du \tag{15}$$

Using Equation 15 for a system supporting 2 antennas (#1 and #2), the sea-bottom contribution $z_1(k+1)$ (resp. $z_2(k+1)$) to sensor #1 (resp. sensor #2) $k+1$ resolution cell can be written for each receiver as:

$$\begin{aligned} z_1(k+1) &= H(t_b) - H(t_a) \\ z_2(k+1) &= H(t_d) - H(t_c) \end{aligned} \tag{16}$$

where $[t_a, t_b]$ (resp. $[t_c, t_d]$) defines temporal boundaries of the $k+1$ resolution cell from receiver #2 (resp. receiver #1) as illustrated in Figure 21 and $H(t_x)$ denotes the antiderivative or cumulative function at $t_x$, of the integrand in Equation 15.

$$H(t_x) = \int_{-\infty}^{t_x} g(u) \cdot e^{-j2\pi f u} du \tag{17}$$

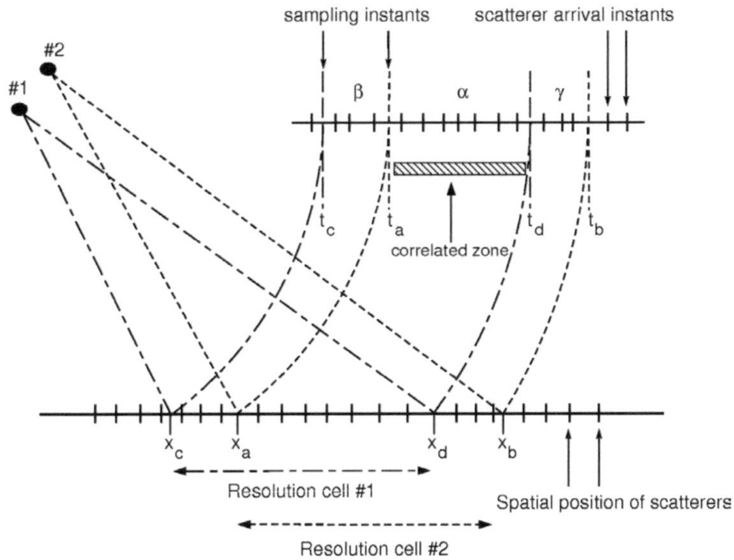


Fig. 21. Splitting integration range due to decorrelation between receivers.

In order to model $H(x)$, let us evaluate the cross-correlation function between the received signals:

$$z_1(k+1).z_2^*(k+1) = [H(t_d) - H(t_c)][H(t_b) - H(t_a)] \quad (18)$$

$$= [\underbrace{H(t_d) - H(t_a)}_{\alpha} + \underbrace{H(t_a) - H(t_c)}_{\beta}].[\underbrace{H(t_b) - H(t_d)}_{\gamma} + \underbrace{H(t_d) - H(t_a)}_{\alpha}]$$

$$= \alpha^2 + \alpha\gamma + \beta\alpha + \beta\gamma \quad (19)$$

with $\alpha$, $\beta$ and $\gamma$ being defined in Figure 21 as the increments between antiderivatives. Thus, $\alpha$ gives the correlated contribution while $\beta$ and $\gamma$ represents the uncorrelated contributions between the two resolution cells.
Equation 19 permits the modeling of the received signals. Indeed, in order to ensure both a correct cross-correlation between signals and independent increments, Equations 20 must be checked.

$$\mathbb{E}\{z_1(k+1).z_2^*(k+1)\} = \mathbb{E}\{\alpha^2\} \Longrightarrow \mathbb{E}\{\alpha\gamma\} = \mathbb{E}\{\alpha\beta\} = \mathbb{E}\{\beta\gamma\} = 0 \quad (20)$$

Now, assuming the signal $g(u)$ associated to extended scatterers, is a white Gaussian-distributed noise, its integration in Equation 17 leads the antiderivatives $H(t_x)$ to be a Wiener process also known as Brownian motion (Saupe, 1988). As a result, the increments given in Equation 16 are also Gaussian distributed, producing Gaussian signals at reception.

#### 5.3.2 Brownian sheet simulation

A 1-D Brownian motion can be simulated by creating a zero-mean Gaussian random vector and computing the cumulative sum function. In order to check the properties of cross-correlation between resolution cells, a 1024-length Brownian vector composed of 1024

observations was created. Then, the cross-correlation function was computed between two ten-units length resolution cells. In the case the resolution cells were 3 units apart (resp. 6), the resulting correlations were 0.7 (resp. 0.4) on average, as expected.
For a two-dimensional case, the simulation of a Brownian sheet is carried out by creating a zero-mean Gaussian random matrix, applying the cumulative sum function along one axis direction and then, applying this function again to the result along the other axis direction. As in the previous case, the measured cross-correlations behave as expected.
The main drawback of this model is that computing every pixel of the Brownian sheet requires the whole map to be processed. Moreover, as already said, as the map resolution depends on sonar characteristics (signal wavelength, baselines...), this map may become huge. Thus, one solution is to find a way for computing Equation 15 without requiring each time, a full map processing. Using a series expansion of the integral in Equation 15, it has been possible to simulate coherent signals, bypassing such a costly processing (Llort-Pujol et al., 2010).

### 5.3.3 Results

Using the Brownian optimization approach (and the proposed extended scatterers), Figure 22 presents simulated signals for a two-sensors interferometer along with the produced interferogram. Fringes of interference are correctly produced showing adequate correlations properties have been taken into account when simulating the two signals over the sea-bottom.

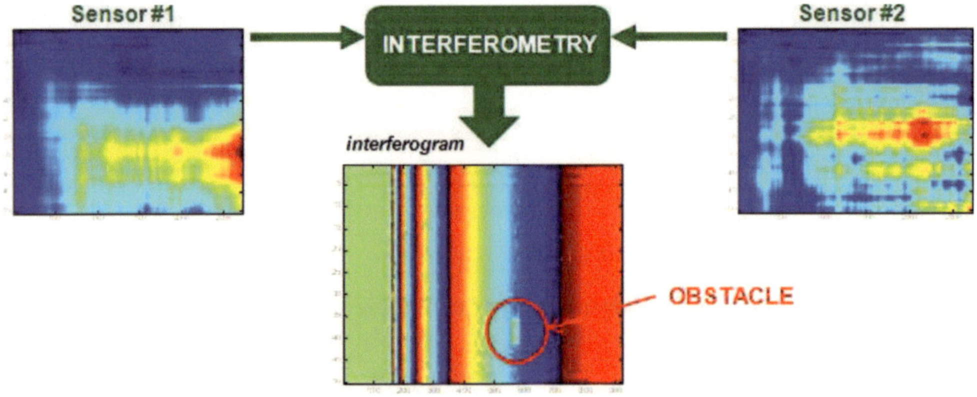


Fig. 22. Simulating sonar signals coming from two different tracks and resulting interferogram.

Nevertheless, the georeferenced grid size remains a major drawback for this method, when computing (once) the Brownian sheet. Indeed, for instance, to generate a 100 meters by 100 meters map for simulating a sidescan sonar which transmits 0.15 ms duration signals, the simulator engine should store the position of about seven million scatterers in order for each resolution cell to encompass only three scatterers. Generating complex signals from the resulting Brownian sheet thus requires a huge amount of both memory and pre-computation processing power.
A more attractive approach, compatible with polygonal ray tubes footprint, would consist in using extensions of the Brownian bridge (Khoshnevisan & Xiao, 2007) to enable on-demand refinement of an already simulated rough grid. The idea is to develop techniques close to Perlin noise generation which can deliver a random number with appropriate properties from its 2D coordinates only.

## 6. Conclusion

Using tubes for modeling acoustic waves underwater propagation allows the framework we presented in this chapter, to simulate very realistic sonar images. Indeed, the whole energy transmitted by one sensor, can be tracked down through scene interactions, till its aggregation in the receiving sensor. Moreover, this energy-based method has been successfully enhanced to simulate full sonar signal (intensity + phase) developing a coherent aggregation processing relying on the tubes propagation. Indeed, using a sea-bottom representation through extended single scatterers, a coherent summation of their backscattered contributions has allowed a complete backscattered signal to be simulated. The realism of such simulated signals have been proven as beamforming and interferometry have successfully worked on these signals.
Some improvements remains to be integrated within this simulation framework such as more complex signals transmission, Doppler effect checking and diffraction propagation through the tube history. To achieve these new goals, other methods (modes simulation or FDTD techniques) mostly based on analytical calculations, are required to finely model both propagation and interaction of the acoustic wave with the observed scene (sea-bottom, objects...). However, their computational burden is very high and to produce realistic output within an acceptable computation time, they should be used only locally in the scene where specific and impacting phenomena (like diffraction) occur. Such an hybrid behavior has already been successfully implemented for radar simulation, outside this simulator (Lurton, 2010). While electromagnetic wave propagation is performed through rays, interactions with the sea surface rely on FDTD computation. Next developments around this simulator will then consist in integrating this promising hybrid scheme, as the current framework already support multiple engines.

## 7. References

Asada, A., Maeda, F., Kuramoto, K., Kawashima, Y., Nanri, M. & Hantani, K. (2007). Advanced surveillance technology for underwater security sonar systems, *OCEANS 2007 - Europe* pp. 1–5.

Australia Defense Science & Technology Organization, D. (1999). Common data set, *Shallow Survey 99 - International Conference on High Resolution Surveys in Shallow Water*, Sydney, Australia.

Belcher, E., Hanot, W. & Burch, J. (2002). Dual-frequency identification sonar DIDSON, *Underwater Technology, 2002. Proceedings of the 2002 International Symposium on* pp. 187–192.

Belcher, E., Matsuyama, B. & Trimble, G. (2001). Object identification with acoustic lenses, *OCEANS, 2001. MTS/IEEE Conference and Exhibition* 1: 6–11 vol.1.

Bell, J. & Linnett, L. (1997). Simulation and analysis of synthetic sidescan sonar images, *IEE Proceedings—Radar, Sonar and Navigation* 144(4): 219–226.

Bell, J. M. (1997). Application of optical ray tracing techniques to the simulation of sonar images, *Optical Engineering* 36(6): 1806–1813.

Brekhovskikh, L. & Lysanov, Y. (1982). *Fundamentals of ocean acoustics*, Springer-Verlag Heidelberg, New York.

Burdic, W. S. (1984). *Underwater Sound System Analysis*, Prentice-Hall, Englewood Cliff.

Denbigh, P. N. (1989). Swath bathymetry: Principles of operation and an analysis of errors, *IEEE Journal of Oceanic Engineering* 14(4): 289–298.

Guériot, D., Daniel, S. & Maillard, E. P. (1998). Sonar images and swath bathymetry simulations for environment restitution through acoustical ray tracing techniques, *Fourth European Conference on Underwater Acoustics*, Vol. 1, Rome, Italy, pp. 425–431.

Guillaudeux, S. & Maillard, E. (1998). Seabed simulations with fractal techniques, *OCEANS '98 Conference Proceedings*, Vol. 1, pp. 433 –437 vol.1.

Harrison, C. (1989). Ocean propagation models, *Applied Acoustics* 27(3): 163 – 201.

Jackson, D. R., Winebrenner, D. P. & Ishimaru, A. (1986). Application of the composite roughness model to high-frequency bottom backscattering, *The Journal of the Acoustical Society of America* 79(5): 1410–1422.

Khoshnevisan, D. & Xiao, Y. (2007). Images of the brownian sheet, *Trans. Amer. Math. Soc.* 359: 3125–3151.

Kuo, E. Y. T. (1964). Wave scattering and transmission at irregular surfaces, *The Journal of the Acoustical Society of America* 36(11): 2135–2142.

Llort-Pujol, G., Sintes, C., Chonavel, T., Gueriot, D. & Garello, R. (2010). Simulation on large scale of acoustic signals for array processing, *OCEANS 2010 IEEE - Sydney*, pp. 1 –7.

Lurton, T. (2010). *Modélisation et simulation de l'interaction entre onde électromagnétique et surfaces de mer*, PhD thesis, Dépt. ITI (Télécom Bretagne), France.

Lurton, X. (2003). Theoretical modelling of acoustical measurement accuracy for swath bathymetric sonars, *International Hydrographic Review* 4(2): 17–30.

Mourad, P.D.; Jackson, D. (1989). High frequency sonar equation models for bottom backscatter and forward loss, *OCEANS '89. Proceedings* 4: 1168–1175.

Pentland, A. P. (1984). Fractal-based description of natural scenes, *Pattern Analysis and Machine Intelligence, IEEE Transactions on* PAMI-6(6): 661 –674.

Perkins, J. S. & Baer, R. N. (1982). An approximation to the three-dimensional parabolic-equation method for acoustic propagation, *The Journal of the Acoustical Society of America* 72(2): 515–522.

Perlin, K. (2002). Improving noise, *SIGGRAPH '02: Proceedings of the 29th annual conference on Computer graphics and interactive techniques*, ACM, New York, NY, USA, pp. 681–682.

Pesquet-Popescu, B. & Vehel, J. L. (2002). Stochastic fractal models for image processing, *IEEE Signal Processing Magazine* 19(5): 48–62.

Saupe, D. (1988). *Algorithms for random fractals*, The science of fractal images, Springer-Verlag, New York, chapter 2, pp. 71–136.

Sintes, C. & Guériot, D. (2008). Interferometric sonar signal simulation through tube tracing, *Oceans 2008 MTS/IEEE Conf. Proc.*, Kobe, Japan, pp. 1–8.

Sintes, C., Llort-Pujol, G. & Guériot, D. (2010). Coherent probabilistic error model for interferometric sidescan sonars, *Oceanic Engineering, IEEE Journal of* 35(2): 412–423.

Tough, R. J. A., Blacknell, D. & Quegan, S. (1995). A statistical description of polarimetric and interferometric synthetic aperture radar data, *Proc. Royal Society* 449: 567–589.

Urick, R. J. (ed.) (1983). *Principles of underwater sound*, third edn, Peninsula Publishing, Los Altos, California.

Ziomek, L. J. & Polnicky, F. W. (1993). The RRA algorithm: Recursive ray acoustics for three-dimensional speeds of sound, *IEEE Journal of Oceanic Engineering* 18(1): 25–30.

# Part 2

## Sonar Signal Processing

# 4

# Beamforming Narrowband and Broadband Signals

John E. Piper
*NSWC*
*USA*

## 1. Introduction

The history of beamforming in sonar applications goes back many years. Perhaps Collodon and Sturm's use of a horn receiver in their 1826 measurement of the speed of sound in water is a first example. Their 13-km range certainly required a receiver with a good beam pattern to increase the signal to a measureable level. Although receivers have changed over the years, beamforming is still an active area of interest and research.

Modern sonars often utilize multiple sensors in an array configuration. More sensors yield more signal information and can help suppress noise and interference. The key to extracting information from the array is to exploit the fact that the signals arrive at the different sensors at different times due to the array geometry and signal angle. An example of the time delay for a signal incident on two sensors is shown below.

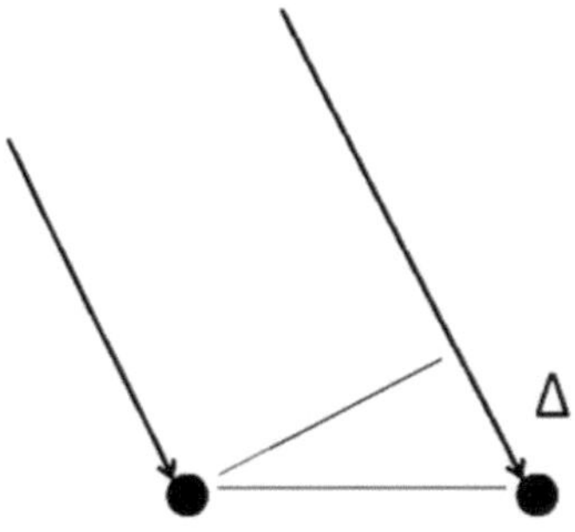


Fig. 1. Time delay, Δ, associated with a signal incident on a two-sensor array.

Modelling the signal interaction with the array can be conveniently done using matrices. The measured data, $y(t)$, over an array of $n$ sensors is represented as a mapping of the signal, $s(t)$, onto the array plus noise.

$$\begin{pmatrix} y_1(t) \\ y_2(t) \\ \vdots \\ y_n(t) \end{pmatrix} = \begin{pmatrix} 1 \\ \Delta_2 \\ \vdots \\ \Delta_n \end{pmatrix} s(t) + noise \tag{1}$$

The mapping involves a series of time delays, which are functionally represented by Δ. The signal is referenced to the first array sensor, and the time delays are also referenced to the first array sensor. It is convenient to call this mapping vector, $D$.

*The essence of beamforming* involves inverting (1) to yield the best estimate of the signal from the measured data. The optimal solution in a least-squares sense can be constructed using the Moore-Penrose inverse. This inverse is well known and widely used in a variety of applications. It is implemented by first left multiplying (1) by the complex transpose of the mapping vector, $D^{\dagger}$. Then, since $D^{\dagger}D$ is a square matrix (in this case a scalar), left multiplying by $(D^{\dagger}D)^{-1}$. This allows the best estimate of the signal, $s(t)$, to be expressed in terms of the measured sensor data, $y(t)$, and the mapping vector, $D$.

$$s(t) = (D^{\dagger}D)^{-1}D^{\dagger}\begin{pmatrix} y_1(t) \\ y_2(t) \\ \vdots \\ y_n(t) \end{pmatrix} \tag{2}$$

or

$$s(t) = \frac{1}{n}\begin{pmatrix} 1 & \Delta_2^{\dagger} & \cdots & \Delta_n^{\dagger} \end{pmatrix}\begin{pmatrix} y_1(t) \\ y_2(t) \\ \vdots \\ y_n(t) \end{pmatrix} \tag{3}$$

since,

$$D^{\dagger}D = n \tag{4}$$

The mathematical interpretation of (3) corresponds to what is commonly called delay-and-sum beamforming. This equation forms the foundation for beamforming methods.

An important observation needs to be noted. The model in (1) was developed for only one signal. It is the optimal solution for the case that includes only one signal. However, the one-signal model is often used in multiple-signal environments. Much work has gone into mitigating the effects of other signals on (3) by introducing weights on the sensors that reduce the leakage from other directions or can sometimes put a null in the direction of the interfering signal. However, these methods have their limitations. It is best to use a multiple-signal model in a multiple-signal environment.

## 2. Narrowband signals

The required time shifts in the mapping matrix have a particularly simple representation for narrowband signals. This is based on the observation that shifting the phase of a narrowband signal approximates a time shift. The resulting equations allow for high resolution beamforming and direction-of-arrival estimation.

### 2.1 Phase shift approximation

The extra distance, Δ, that the signal, $s$, has to travel to the second array element in Fig. 1 is geometrically determined by the distance between the sensors, $a$, and the angle of incidence, $\vartheta$. This extra distance is simply

$$\Delta = a \sin(\vartheta) \tag{5}$$

For narrowband signals it is convenient to express this distance as the radian measure of the fraction of a wavelength extra distance that the signal travels.

$$\varphi = \frac{2\pi}{\lambda} a \sin(\vartheta) \tag{6}$$

This phase angle can then be used to simulate advancing or delaying a narrowband signal by simple multiplication of the phase term.

$$e^{i\varphi} e^{i\omega t} = e^{i(\omega t + \varphi)} \tag{7}$$

It should be noted that this is only an approximation in the sense that it does not actually shift the signal in time. Instead it only changes the phase to match a signal shifted in time. This phase-shifted signal still starts and stops at the same time samples as the original signal. So, it is not actually time shifted, but it does approximate a time-shifted signal over part of its interval. This can be problematic for short signal pulses and large time shifts.

### 2.2 Narrowband beamforming

The general sonar problem involves multiple sensors and multiple signals. This can be expressed as a mapping, $D$, of the $m$ signals onto the $n$ sensors in the array.

$$\begin{pmatrix} y_1(t) \\ y_2(t) \\ \vdots \\ y_n(t) \end{pmatrix} = D \begin{pmatrix} s_1(t) \\ s_2(t) \\ \vdots \\ s_m(t) \end{pmatrix} + noise \tag{8}$$

The narrowband mapping or steering matrix, $D$, can be expressed in terms of the phase shifts, $\varphi$, associated with the various directions of arrival. Here, each column in the matrix corresponds to a diffent signal. For a uniform linear array, which is often the case, the steering matrix has the following Vandermonde stucture.

$$D = \begin{pmatrix} 1 & 1 & \cdots & 1 \\ e^{i\varphi_1} & e^{i\varphi_2} & \cdots & e^{i\varphi_m} \\ \vdots & \vdots & \ddots & \vdots \\ e^{i(n-1)\varphi_1} & e^{i(n-1)\varphi_2} & \cdots & e^{i(n-1)\varphi_m} \end{pmatrix} \tag{9}$$

Other array geometries can be easily accommodated with this approach by correctly modeling the various phase delays associated with the various time delays.

Beamforming requires inverting (8). As before, the Moore-Penrose inverse should be used. This yields the following representation for the signals.

$$\begin{pmatrix} s_1(t) \\ s_2(t) \\ \vdots \\ s_m(t) \end{pmatrix} = (D^{\dagger} D)^{-1} D^{\dagger} \begin{pmatrix} y_1(t) \\ y_2(t) \\ \vdots \\ y_n(t) \end{pmatrix} \tag{10}$$

It is important to note that in this representation the $m$ signals are decoupled or separated. This is generally not the case when conventional beamforming techniques are applied in a multiple signal environment, since conventional beamforming uses a one-signal model that cannot correctly solve the multiple signal case.

It is worthwhile to understand the mathematical details in (10). The term $D^{\dagger}y$ can be thought of as forming $m$ conventional beams from the array data. These beams are then multiplied by the matrix $(D^{\dagger}D)^{-1}$ as prescribed by Moore-Penrose. The $(D^{\dagger}D)^{-1}$ term allows the beams to be be decoupled. This matrix is more fully described in the next section where the importance of the off-diagonal terms in this matrix becomes clear. There is no analog in the conventional approach since the $(D^{\dagger}D)^{-1}$ term reduces to a simple scalar in the one-signal model.

A simulation example is shown in Fig. 2 for two overlapping narrowband +20 dB signals arriving on a two-sensor array with half-wavelength spacing at incident angles of ±3°. This example simulates a direct signal plus a multipath signal that reflects off the water's surface and arrives at a slightly later time. The top plot shows the output using simple beam steering. The two signals are not separated since the beamwidth of this approach cannot differentiate signals with small angular separation. The next two plots show the outputs of the Moore-Penrose beamformer directed at +3° and −3°. The two signals are seen to be clearly separated with this approach.

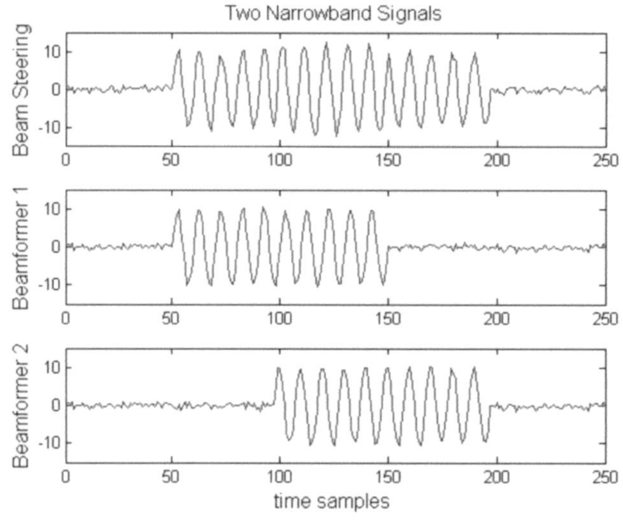


Fig. 2. Simple beam steering and Moore-Penrose beamformer outputs for a two-sensor array with two overlapping narrowband +20 dB signals at ±3°.

Sonar data was collected with a synthetic aperture rail system at the NSWC test pool. A short 20-kHz narrow beam signal was projected and two vertical wide beam elements received the signal. Fig. 3 shows the results from simple beam steering and from the Moore-Penrose beamformer output. The vertical axis corresponds to the ping number along the rail length, and the horizontal axis corresponds to the relative time sample. The bottom target is clearly seen between time samples 1100 and 1500 in both images. A multipath signal can also be seen between time samples 2500 and 2900 using simple beam steering, since it is within the array's mainlobe. However, the output from the two-signal Moore-Penrose beamformer steered towards the bottom target effectively eliminates or cancels the multipath signal.

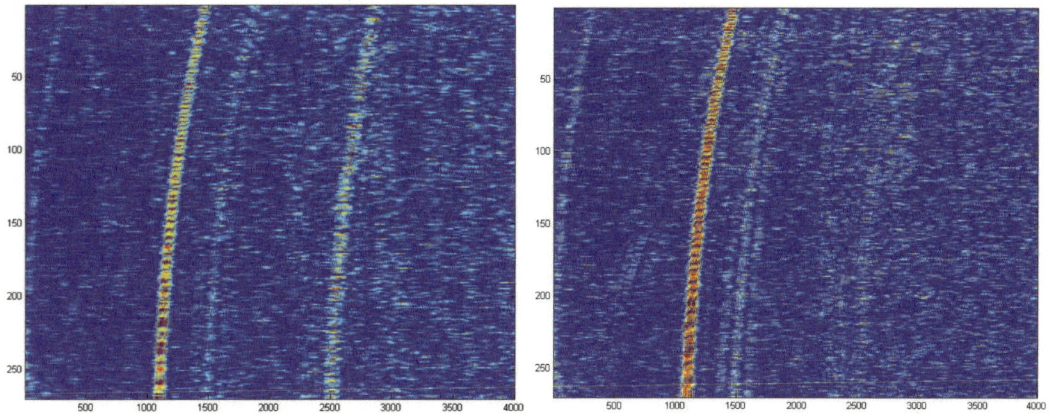

Fig. 3. Direct and multipath signals with beam steering (left) and only a direct signal with the Moore-Penrose beamformer (right).

The separation of the direct path and the multipath signals in Fig. 2 and 3 is interesting since both signals exist deep within the classical mainlobe of the steered beam. The ability to separate the signals comes from using a multiple-signal model. Null beam steering is somewhat similar in the sense that by placing a null at a second signal effectively leads to decoupling. However, null beam steering is limited by its fundamental dependency on a one-signal model, which has no analog to the important $(D^{\dagger}D)^{-1}$ term.

### 2.3 Likelihood function

The least-squares method offers a robust high resolution approach for direction-of-arrival estimation. The likelihood function, $L$, can be constructed as the least-squares difference between the measured array data, $y$, and the parametric model, $Ds$.

$$L = \left\| y - Ds \right\|^2 \tag{11}$$

It is useful to insert the Moore-Penrose representation for $s$ into (11). After multiplying terms, this yields the following representation for $L$.

$$L = y^{\dagger}y - y^{\dagger}D(D^{\dagger}D)^{-1}D^{\dagger}y \tag{12}$$

It is convenient to drop the constant term $y^{\dagger}y$ and flip the sign to yield a maximization problem

$$L = y^{\dagger}D(D^{\dagger}D)^{-1}D^{\dagger}y \tag{13}$$

This representation can be interpreted as an *inner product* of the vector $D^{\dagger}y$ and its Hermitian conjugate, $y^{\dagger}D$. The $(D^{\dagger}D)^{-1}$ term should be interpreted as the inner product metric tensor. Inner product spaces are of considerable interest in physics and mathematics. This array processing example is likewise interesting.

It is worthwhile to investigate the properties of this metric tensor for the two-signal and $n$-element case. The metric tensor is constructed from the steering matrix, $D$.

$$D^{\dagger}D = \begin{pmatrix} 1 & e^{-i\phi_1} \cdots & e^{-i(n-1)\phi_1} \\ 1 & e^{-i\phi_2} \cdots & e^{-i(n-1)\phi_2} \end{pmatrix} \begin{pmatrix} 1 & 1 \\ e^{i\phi_1} & e^{i\phi_2} \\ \vdots & \vdots \\ e^{i(n-1)\phi_1} & e^{i(n-1)\phi_2} \end{pmatrix} = \begin{pmatrix} n & \dfrac{1-e^{in\delta}}{1-e^{i\delta}} \\ \dfrac{1-e^{-in\delta}}{1-e^{-i\delta}} & n \end{pmatrix} \tag{14}$$

where $\delta = \phi_2 - \phi_1$ . The inverse is then

$$(D^{\dagger}D)^{-1} = \frac{1}{n^2 - \dfrac{(1-\cos(n\delta))}{(1-\cos(\delta))}} \begin{pmatrix} n & -\dfrac{1-e^{in\delta}}{1-e^{i\delta}} \\ -\dfrac{1-e^{-in\delta}}{1-e^{-i\delta}} & n \end{pmatrix} \tag{15}$$

It can be seen that when $n$ is large, the inner product metric tensor tends to become diagonally dominant. In this case the off-diagonal terms tend to be relatively less important, and the conventional signal processing approach starts to assume some validity.

The off-diagonal terms in (15) are particularly interesting. These terms are a measure of the coupling between the signals. Interestingly, as the angle between the two signals becomes small, the off-diagonal terms approach the magnitude of the diagonal terms.

$$\lim_{\delta \to 0} \frac{1-e^{in\delta}}{1-e^{i\delta}} = n \tag{16}$$

Since the off-diagonal terms can grow to be nearly as large as the diagonal terms when the separation angle is small, they clearly cannot be simply ignored.

There are special cases when the off-diagonal terms can be safely ignored. These special cases occur when the off-diagonal terms are zero, which occur when

$$n\delta = \pm 2\pi, \pm 4\pi, \cdots . \tag{17}$$

Because the off-diagonal terms are zero at these values, the problem naturally decouples and the signals can be completely separated. These zeros correspond to the zeros commonly seen in conventional sidelobe structures. This is the goal of null steering, which adaptively adds weights to the steering matrix to produce nulls in the beam pattern in the direction of an unwanted signal by exploiting the condition in (17) where the off-diagonal terms go to zero.

The likelihood function is commonly expressed using a projection operator and a sample covariance matrix representation. This may be derived by taking the trace of (13) and rotating the vector $y^{\dagger}$ to the right side of the equation.

$$L = tr(D(D^{\dagger}D)^{-1}D^{\dagger}yy^{\dagger}) = tr(D(D^{\dagger}D)^{-1}D^{\dagger}R) \tag{18}$$

The least-squares function for the one-signal case reduces to the familiar periodogram representation, since $(D^{\dagger}D)^{-1}$ reduces to a simple scalar. The peak in this function corresponds to the best estimate of the parameters.

$$L = \left\| D^{\dagger} y \right\|^2 \tag{19}$$

### 2.4 Direction-of-arrival estimation

Direction-of-arrival estimation of signals is an important function of array processing. This information can be used for localization applications or used as input for high resolution beamforming applications. The likelihood function approach can be used to solve this problem by finding the parameters that maximize the function.

A simulation example is shown in Fig. 4 for the same two overlapping narrowband +20 dB signals arriving on an array at incident angles of ±3° as seen in Fig. 2. The likelihood function is calculated over the 250 time sample interval. Two cases are considered; the first one with an array of four sensors and the second with eight sensors with half-wavelength spacing.

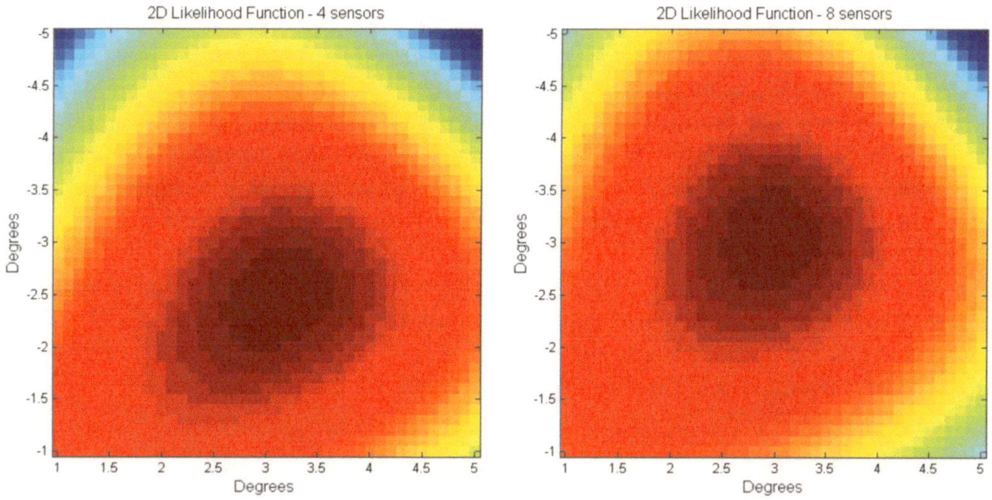


Fig. 4. Example likelihood functions for two narrowband signals with incident angles of ±3° on arrays with 4 and 8 sensors.

The peak in the likelihood function corresponds to the best estimate of the directions of arrival. This approach has been shown to approach the Cramer-Rao bound in applications [Stoica 1989]. The Cramer-Rao bound has been derived in the literature and is given as

$$CRB = (SNR)^{-1}\,(n(n^2 - 1)/6)^{-1} \tag{20}$$

So, the ability to estimate the directions of arrival depends on two terms. The bound is inversely proportional to the signal-to-noise ratio. The bound is also inversely related to the number of sensors in a nonlinear way. Four sensors yield ten-fold better estimates than two sensors. Likewise, eight sensors yield an eight-fold better estimate than four sensors. This increase in resolution with more sensors is seen in the above figure.

An alternative and popular approach to direction-of-arrival estimation is the signal subspace method [Schmidt 1986]. This alternative method depends on the eigenvalue structure of the sample covariance matrix. There are a couple of problems with this method. Noise in the sample covariance matrix can lead to poor estimation of the true eigenvalues. Additionally, coherent signals such as seen in multipath lead to problems in the structure of the sample covariance matrix. A number of signal subspace techniques have been developed to address these problems [Wang 1985], [Di Claudio 2001].

### 2.5 Number of signals resolvable by an array

The number of signals resolvable by an array is somewhat dependent on the array processing methods used. Signal subspace methods are strictly limited to $n-1$ signals with an array of $n$ sensors [Friedlander 1991]. This limit is imposed since of the $n$ eigenvalues of the sample covariance matrix at least one needs to be assigned to the noise subspace. Hence, that leaves at most $n-1$ eigenvalues to be assigned to the signal subspace. It may be possible to assign all the eigenvalues to the signal subspace, but then it would technically not be a subspace.

Both the Moore-Penrose inverse and the likelihood function require the inversion of the term $D^{\dagger}D$. This inverse may or may not exist. In order for it to exist, it is necessary for $D^{\dagger}D$ to be full rank. This requirement constrains the number of resolvable signals.

Consider the $n$ by $m$ general steering matrix for $n$ uniformly spaced array elements and $m$ signals. If the incident angles for the incoming signals are all different, then $D$ has $m$ independent columns. For uniformly spaced sensors and most other geometries there are $n$ independent rows. Since the rank of an $n$ by $m$ matrix is less than or equal to min($m$,$n$), D has full rank if and only if rank(D) = min($m$,$n$). In this case, rank($D^{\dagger}D$) = min($m$,$n$), so that when the incident angles for the incoming angles are distinct, the $m$ by $m$ matrix $D^{\dagger}D$ is invertible if and only if $n \geq m$.

This result that an $n$-sensor array can resolve $n$ signals differs from the signal subspace result of $n-1$ signals. Simulation and test pool examples of two signals resolved by two sensors are seen in Fig. 2 and 3. These signals are resolved in the sense that they can be fully decoupled in beamforming and direction-of-arrival methods.

## 3. Time-shift operators

Proper array processing requires shifting the signals in time. Time-shift operators, $\Delta$, are useful to translate a digitized signal by an arbitrary amount of time, $\tau$. These can be functionally expressed as

$$\Delta(\tau)\, s(t) = s(t+\tau) \tag{21}$$

Unitary matrix operators can be constructed for this task [Piper 2009]. Although the specific time-shift matrices developed in this chapter are very useful for solving the general broadband problem, they have numerous applications in other signal processing problems [Laakso 1996].

### 3.1 Unit time shifts

The simplest time-shift operator is the identity matrix, which shifts the time-series signal by a zero amount. Using matrices with ones along other diagonals, results in time shifts by a discrete number of time samples. It is worthwhile to consider the unit time-shift operator, $\Delta(1)$. This may be written in a Toeplitz matrix form as

$$\Delta(1)=\begin{pmatrix} 0 & 1 & 0 & 0 & \cdots \\ 0 & 0 & 1 & 0 & \cdots \\ 0 & 0 & 0 & 1 & \cdots \\ 0 & 0 & 0 & 0 & \cdots \\ \vdots & \vdots & \vdots & \vdots & \ddots \end{pmatrix} \tag{22}$$

The effect of this operator is to shift vector elements up or backwards in time by one unit. It should be noted that this simple operator is quite independent of the shape of the waveform it operates on. It is convenient to require these operators to be unitary so that inverses are simply complex transposes.

### 3.2 Fractional time shifts

Construction of the fractional time-shift operator, $\Delta(\tau)$, for an arbitrary time shift, $\tau$, is best done in the frequency domain where the fractional time shift can be effected by a phase shift. This operation can be accomplished by multiplying three matrices; the Fourier matrix, $\mathcal{F}$, the phase shift matrix, $\Delta(\theta)$, and the inverse Fourier matrix, $\mathcal{F}^{-1}$. The resultant time-shift operator can then be functionally written as

$$\Delta(\tau)=\mathcal{F}^{-1}\Delta(\theta)\mathcal{F} \tag{23}$$

The Fourier, phase shift, and inverse Fourier matrices are listed below

$$\mathcal{F}=\begin{pmatrix} 1 & 1 & 1 & \cdots & 1 \\ 1 & f & f^2 & \cdots & f^{N-1} \\ 1 & f^2 & f^4 & \cdots & f^{2(N-1)} \\ \vdots & \vdots & \vdots & \ddots & \vdots \\ 1 & f^{N-1} & f^{2(N-1)} & \cdots & f^{(N-1)^2} \end{pmatrix} \tag{24}$$

$$\Delta(\theta)=\begin{pmatrix} 1 & 0 & 0 & \cdots & 0 \\ 0 & p & 0 & \cdots & 0 \\ 0 & 0 & p^2 & \cdots & 0 \\ \vdots & \vdots & \vdots & \ddots & \vdots \\ 0 & 0 & 0 & \cdots & p^{N-1} \end{pmatrix} \tag{25}$$

$$\mathcal{F}^{-1}=\frac{1}{N}\begin{pmatrix} 1 & 1 & 1 & \cdots & 1 \\ 1 & f^{-1} & f^{-2} & \cdots & f^{-(N-1)} \\ 1 & f^{-2} & f^{-4} & \cdots & f^{-2(N-1)} \\ \vdots & \vdots & \vdots & \ddots & \vdots \\ 1 & f^{-(N-1)} & f^{-2(N-1)} & \cdots & f^{-(N-1)^2} \end{pmatrix} \tag{26}$$

where $f = e^{-2\pi i/N}$ and $p = e^{2\pi i\tau/N}$.

Multiplication of these matrices yields the following exact analytical expression for the fractional time-shift operator

$$\Delta(\tau) = \begin{pmatrix} \Delta_0 & \Delta_1 & \Delta_2 & \cdots & \Delta_{N-1} \\ \Delta_{-1} & \Delta_0 & \Delta_1 & \cdots & \Delta_{N-2} \\ \Delta_{-2} & \Delta_{-1} & \Delta_0 & \cdots & \Delta_{N-3} \\ \vdots & \vdots & \vdots & \ddots & \vdots \\ \Delta_{-(N-1)} & \Delta_{-(N-2)} & \Delta_{-(N-3)} & \cdots & \Delta_0 \end{pmatrix}. \tag{27}$$

This matrix has a Toeplitz structure. The diagonal terms, $\Delta_k$, are geometric series and are easily summed.

$$\Delta_k = \frac{1}{N}(1 + pf^k + p^2 f^{2k} + \cdots + p^{N-1} f^{(N-1)k}) = \frac{1}{N}\frac{1-p^N}{1-pf^k} \tag{28}$$

An example of the fractional time-shift operator advancing a unit impulse signal by 2.2 time samples is shown below in Fig. 5.

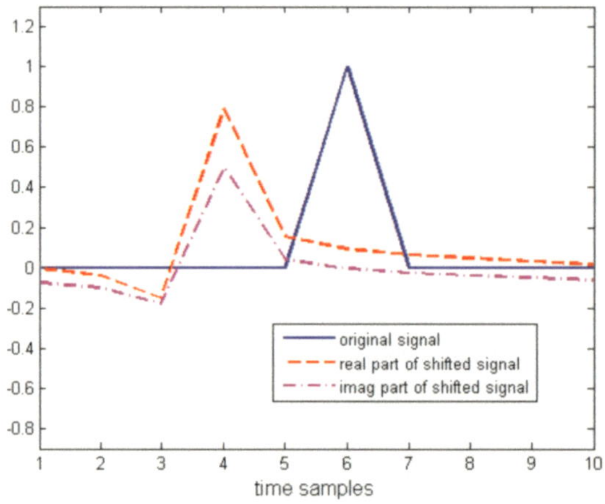


Fig. 5. Example of fractional shift of 2.2 samples applied to unit impulse signal.

## 4. Broadband signals

Broadband signals are much more useful in sonar systems than narrowband signals. The increased bandwidth leads to increased range resolution, and the longer signal duration can lead to higher signal-to-noise ratios. This type of signal is important, and considerable research effort has gone into array processing of broadband data. The key to solving this problem is to use the same functional approach used in the narrowband case but use time-shift operators to perform the necessary time shifting.

### 4.1 One-signal beamformer

The one-signal model for an $n$-sensor array can be written in matrix form using the time-shift operators, $\Delta$, as

$$\begin{pmatrix} y_1 \\ y_2 \\ \vdots \\ y_n \end{pmatrix} = \begin{pmatrix} 1 \\ \Delta_2 \\ \vdots \\ \Delta_n \end{pmatrix} s + noise \tag{29}$$

Inversion of this equation in a least-squares sense can be done with the Moore-Penrose inverse. This representation can be interpreted as a simple delay-and-sum approach.

$$s = \frac{1}{n}\begin{pmatrix} 1 & \Delta_2^{\dagger} & \cdots & \Delta_n^{\dagger} \end{pmatrix} \begin{pmatrix} y_1 \\ y_2 \\ \vdots \\ y_n \end{pmatrix} \tag{30}$$

It is possible and somewhat popular to solve the above problem in the frequency domain. This can be realized by inserting Fourier transforms and their inverses into (30). Performing the multiplications reveals that the signal can be constructed in the frequency domain using simple phase shift operators, $\Delta(\theta)$, defined in (25). Since the phase-shift operators only have non-zero elements along the main diagonal, some computational simplicity can be obtained with this approach.

$$s = \frac{1}{n}\mathcal{F}^{-1}\begin{pmatrix} 1 & \Delta^{\dagger}(\theta_2) & \cdots & \Delta^{\dagger}(\theta_n) \end{pmatrix} \begin{pmatrix} \mathcal{F}y_1 \\ \mathcal{F}y_2 \\ \vdots \\ \mathcal{F}y_n \end{pmatrix} \tag{31}$$

### 4.2 Two-signal beamformer

For two signals and $n$ sensors the sonar model can be written as

$$\begin{pmatrix} y_1 \\ y_2 \\ \vdots \\ y_n \end{pmatrix} = \begin{pmatrix} 1 & 1 \\ \Delta_{12} & \Delta_{22} \\ \vdots & \vdots \\ \Delta_{1n} & \Delta_{2n} \end{pmatrix} \begin{pmatrix} s_1 \\ s_2 \end{pmatrix} + noise \tag{32}$$

The resultant Moore-Penrose inverse solution for the two-signal broadband problem can be functionally written as

$$\begin{pmatrix} s_1 \\ s_2 \end{pmatrix} = \begin{pmatrix} n & 1+\Delta_{12}^{\dagger}\Delta_{22}+\cdots \\ 1+\Delta_{22}^{\dagger}\Delta_{12}+\cdots & n \end{pmatrix}^{-1} \begin{pmatrix} 1 & \Delta_{12}^{\dagger} & \cdots \\ 1 & \Delta_{22}^{\dagger} & \cdots \end{pmatrix} \begin{pmatrix} y_1 \\ y_2 \\ \vdots \\ y_n \end{pmatrix} \tag{33}$$

Using the above equation with time-shift operators developed in the previous section leads to the matrix to be inverted in (33) being rank deficient. This is a problem. It is therefore necessary to use an *approximate inverse*. The following equation uses a simple approximate inverse that allows the signals to be decoupled or separated. However, this approximate inverse is not well normalized.

$$\begin{pmatrix} s_1 \\ s_2 \end{pmatrix} \approx \frac{1}{n} \begin{pmatrix} 1 & -(1+\Delta_{12}^{\dagger}\Delta_{22}+\cdots)/n \\ -(1+\Delta_{22}^{\dagger}\Delta_{12}+\cdots)/n & 1 \end{pmatrix} \begin{pmatrix} 1 & \Delta_{12}^{\dagger} & \cdots \\ 1 & \Delta_{22}^{\dagger} & \cdots \end{pmatrix} \begin{pmatrix} y_1 \\ y_2 \\ \vdots \\ y_n \end{pmatrix} \tag{34}$$

A simulation example of this approach is shown in Fig. 6 for two overlapping +20 dB chirp signals arriving on a two-element array at incident angles of ±3°. This example is similar to the one seen in Fig. 2 in that it also simulates a direct signal plus a delayed multipath signal. The top plot shows the measured signal of sensor 1. The next two plots show the two outputs of the Moore-Penrose beamformer directed at +3° and −3°. The two chirp signals can be seen to be clearly separated with this approach.

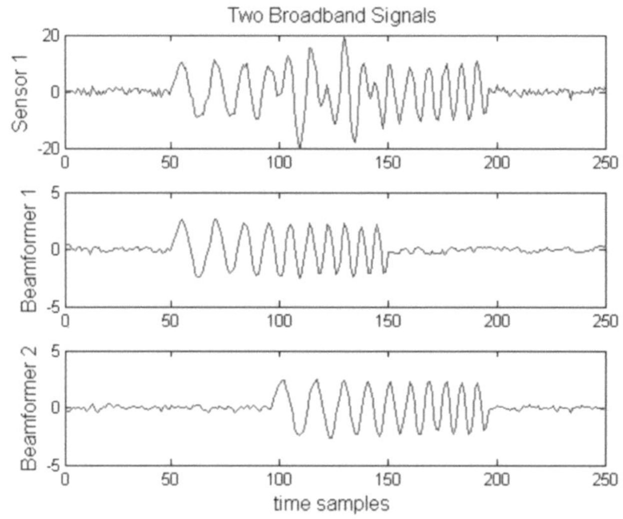


Fig. 6. Sensor 1 measured signal and Moore-Penrose beamformer outputs for a two-sensor array with two overlapping broadband +20 dB signals at ±3°.

The two-signal case may be generalized to the multiple-signal case by expanding the signal vector and steering matrix to account for more signals. The inner product metric term, $D^{\dagger}D$, will also need to be expanded and an approximate inverse found. For a large number of signals this could be challenging.

It may be computationally advantageous to solve this problem in the frequecy domain instead of the time domain. This can be realized by inserting Fourier transforms and their inverses into (34) and multiplying to yield the following frequency domain representation. The $\Delta(\theta)$ terms are the diagonal phase shift matrices defined in (25).

$$\begin{pmatrix} s_1 \\ s_2 \end{pmatrix} \approx \frac{1}{n} \begin{pmatrix} \mathcal{F}^{-1} & 0 \\ 0 & \mathcal{F}^{-1} \end{pmatrix} \begin{pmatrix} 1 & -(1+\Delta^{\dagger}(\theta_{12})\Delta(\theta_{22})+\cdots)/n \\ -(1+\Delta^{\dagger}(\theta_{22})\Delta(\theta_{12})+\cdots)/n & 1 \end{pmatrix} \times \begin{pmatrix} 1 & \Delta^{\dagger}(\theta_{12}) & \cdots \\ 1 & \Delta^{\dagger}(\theta_{22}) & \cdots \end{pmatrix} \begin{pmatrix} \mathcal{F}y_1 \\ \mathcal{F}y_2 \\ \vdots \\ \mathcal{F}y_n \end{pmatrix} \tag{35}$$

### 4.3 Broadband direction-of-arrival estimation

As in the narrowband case, high resolution direction-of-arrival estimates can be obtained using a likelihood function approach. The likelihood function is derived from the least-squares method and is given as

$$L = y^{\dagger} D(D^{\dagger} D)^{-1} D^{\dagger} y. \tag{36}$$

It is again necessary to use an approximate inverse in (36). The best estimate of the angles can be found by numerically evaluating (36) to find its maximum.
Two examples are shown in Fig. 7. These correspond to the two chirp signals seen in Fig. 6. These signals have incident angles of ±3° on an array with either 4 or 8 sensors.

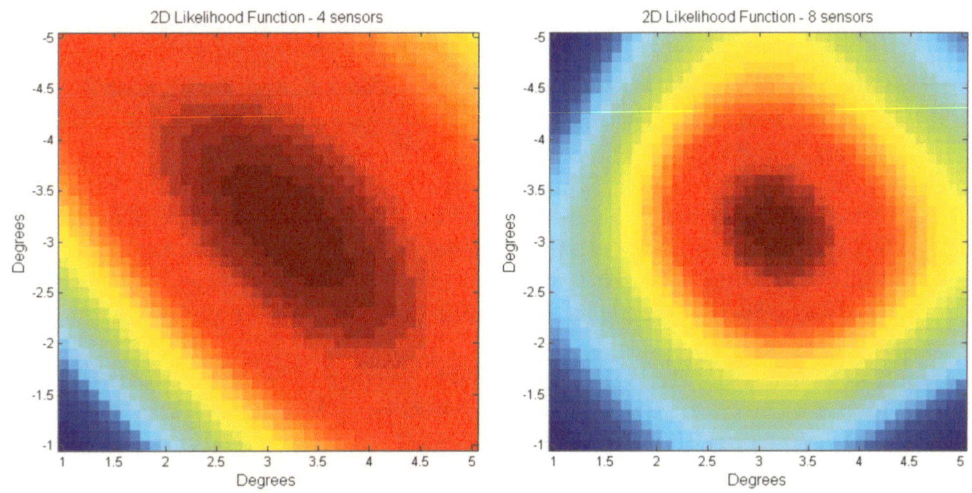


Fig. 7. Example likelihood functions for two broadband signals with incident angles of ±3° on arrays with 4 and 8 sensors.

## 5. Comments

*Array beamforming* has been largely dominated by one-dimensional thinking. One dimensional in the sense that one-signal models are applied and expected to work in multiple-signal environments. Optimal results can only be expected when the number of signals in the model equals the number of signals in the environment.
*The Moore-Penrose inverse* has been found to be absolutely ideal for beamforming applications in a multiple-signal environment. This inverse is essentially a two-step process. The first step, which is a multiplication by the complex transpose of the steering matrix, $D^{\dagger}$, can be thought of as simple beamforming. The second step, which is a multiplication by the inner product metric tensor, $(D^{\dagger}D)^{-1}$, decouples the signals.
*Time-shift operators* allow broadband signals to be beamformed nearly as easily as narrowband signals. A matrix representation of this operation has been given in this chapter. Other approaches are also possible.
The motivation for this chapter has been the *multipath cancellation problem*. This is a significant problem due to the interference of multipath signals that are often seen in sonar data as reflections off the surface and the bottom. Since traditional array processing is

limited in its abilities to mitigate this problem, it has been necessary to look at what other approaches can be used for this problem. This chapter is a result of that long-term interest and presents a way forward.

## 6. Acknowledgment

I would like to thank Prof. Henry Unruh for his advice to "make matrices do what you want them to do." I would also like to thank Dr. Quyen Huynh for his suggestion to use an approximate inverse. I am indebted to Prof. John Bird for pointing out an error in my work. In addition, I would like to acknowledge partial support for this work by the In-Laboratory Independent Research program funded by the Office of Naval Research.

## 7. References

Stoica, P. & Nehorai, A. (1989). MUSIC, Maximum Likelihood, and the Cramer-Rao Bound, *IEEE Transactions on Acoustics, Speech, and Signal Processing,* Vol.37, No.5, (May 1989) pp. 720-741, ISSN 0096-3518

Schmidt, R. (1986). Multiple Emitter Location and Signal Estimation, *IEEE Transactions on Antennas and Propagation,* Vol.34, No.3, (March 1986) pp. 276-280, ISSN 0018-926X

Wang, H. & Kaveh, M. (1985). Coherent Signal-Subspace Processing for the Detection and Estimation of Angles of Arrival of Multiple Wide-band Sources, *IEEE Transactions on Acoustics, Speech, and Signal Processing,* Vol.33, No.5, (August 1985) pp. 823-831, ISSN 0096-3518

Di Cladio, E. & Parisi, R. (2001). WAVES: Weighted Average of Signal Subspaces for Robust Wideband Direction Finding, *IEEE Transactions on Signal Processing,* Vol.49, No.10, (October 2001) pp. 2179-2191, ISSN 1053-587X

Friedlander, B. & Weiss, A. (1991). On the Number of Signals Whose Directions can be Estimated by an Array, *IEEE Transactions on Signal Processing,* Vol.39, No.7, (July 1991) pp. 1686-1689, ISSN 1053-587X

Piper, J. (2009). Exact and Approximate Time-Shift Operators, *Proceedings of SPIE Algorithms for Synthetic Aperture Radar Imagery,* ISSN: 0277-786X, Vol.7337, Orlando, Florida, USA, April 16-17, 2009

Laafso, T.; Valimaki, V.; Karjalainen, M. & Laine, U. (1996). Splitting the Unit Delay, *IEEE Signal Processing Magazine,* Vol.13, No.1, (January 1996) pp. 30-60, ISSN 1053-5888

# 5

# Passive Sonar Signal Detection and Classification Based on Independent Component Analysis

N. N. de Moura, J. M. de Seixas and Ricardo Ramos
*Federal University of Rio de Janeiro, COPPE/Poli, Signal Processing Laboratory, Rio de Janeiro Brazil*

## 1. Introduction

Sonar systems use the sound propagation in underwater environments for detection, communication and navigation. The main purpose of these systems is to analyse the underwater acoustic waves received from different directions by a sensor system and identify the type of target that has been detected in a given direction. Sonar systems may either be passive or active. Both passive and active sonar systems are mainly employed in military settings, although they are also used in commercial and scientific applications, e.g. detecting shoal fishes, performing tomography on sea to exploit a given area, to measure the depth of a region, and so on (Burdic, 1991).

In order to detect and classify signals against background noise, passive sonar systems (Waite, 2003) listen to the noise radiated by targets (ships or submarines) using an array of hydrophones. The background noise may be produced by the sea ambient noise or the self-noise of the sonar platform. From the acquired signals, the direction of arrival (DOA) is estimated, in order to inform the eventual presence of a target in a determined direction (bearing). After DOA estimation, relevant features of the target may be extracted from a given direction.

There are two types of analysis that can be implemented to obtain the signal relevant features: DEMON (Detection Envelope Modulation On Noise) (Nielsen, 1991) and LOFAR (Low Frequency Analysis and Recording) (Di Martino, 1993). The DEMON is a narrowband analysis that furnishes the propeller characteristic: number of shafts, shaft rotation frequency and blade rate of the target. On the other hand, LOFAR, which is a broadband analysis, estimates the noise vibration of the target machinery. Both analysis are based on spectral estimation and support detection and classification of targets.

Depending on the bearing resolution, signal interference may occur for neighbour directions, which contaminates the acquired signals and makes even more difficult the target detection and classification tasks. To minimize these interferences, algorithms using ICA (Independent Component Analysis) (Hyvarinen, 2001), ( Jutten, 2004) may be applied to recover the original sources of the resulting signal mixture and obtain optimal target detection and classification for each direction.

The detection is implemented using the classical signal demodulation to obtain the propeller characteristics. On the other hand, efficient classification is often obtained through neural

networks (Soares Filho, 2001). For both DEMON and LOFAR analysis, a signal preprocessing using ICA may be implemented to reduce the signal interference in neighbour directions.
On the other hand, underwater acoustic signals suffer fluctuations as a function of the sea features (salinity, temperature, and so on). In addition, noise radiated from targets may vary according to operational conditions. As a consequence, the stationarity of passive sonar signals may be affected. Therefore, it is necessary to monitor changes in the statistics of the passive sonar signals in real time. When changes are detected, the independent component extraction phase may be reloaded for updating the feature extraction procedure. An unsupervised clustering method using a modified ART (Adaptive Resonance Theory) neural network can perform such monitoring task.
The chapter is organised as it follows. In section 2, both DEMON and LOFAR analysis are detailed. Section 3 addresses signal interference and its removal in frequency-domain using ICA. Section 4 briefly presents the independent component analysis and the algorithm that was used to implement the blind source separation scheme. Section 5 gives some signal detection and classification results from experimental data, which were acquired from a passive sonar system that has been installed in a submarine. Finally, Section 6 presents conclusions and the perspectives for passive sonar signal processing.

## 2. Spectral analysis

In this section, both DEMON and LOFAR analysis are described. Their aim is to detect and classify the targets from a given DOA.

### 2.1 DEMON analysis

For signal detection, DEMON analysis is normally applied. DEMON is a narrowband analysis that operates over the cavitation noise of the target propeller (Nielsen, 1991). As resulting parameters provide a detailed knowledge of the target propellers (the propeller noise is characteristic for each target), often efficient detection is achieved. Figure 1 shows the block diagram of classical DEMON analysis.

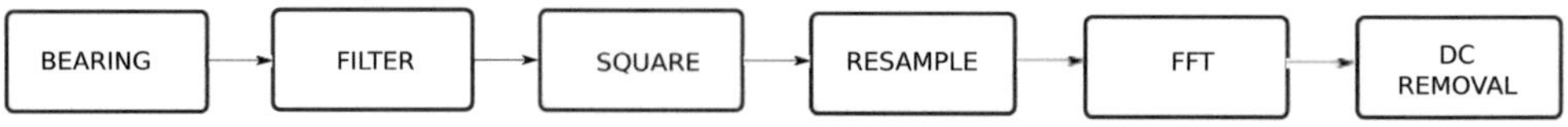


Fig. 1. DEMON analysis block diagram.

Given a direction of interest (bearing), a bandpass filter is implemented to limit the cavitation frequency range (Burdic, 1991). The cavitation frequency goes from hundreds until thousands of Hz. Therefore, it is important to select the cavitation band and obtain the maximum information for ship identification. Following, the signal is squared as in traditional demodulation (Van Trees, 2001), (Yang, 2007). In most cases, the signal sampling rate is relatively high, so that the band of interest is sampled with coarse resolution with respect to observation needs. Thus, it is necessary to decimate the signal for better observation in the range of interest (Rabner, 1983), as shown in Figure 2. The signal is sampled at a frequency

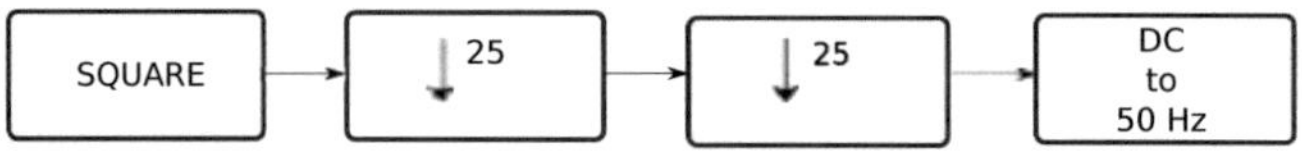


Fig. 2. Signal decimate.

of 31,250 Hz. When the first decimate by a factor of 25 is applied the down sampling goes to 1,250 Hz. In the sequence, a further decimate by a factor of 25 is realized and the range from DC to 50 Hz is searched for. This range contains the rotations of interest that go from 0 to 1,500 (rpm). Finally, a short-time Fast Fourier Transform algorithm (Diniz, 2010) is applied for performing signal analysis in frequency-domain and the TPSW (Two Pass Split Window) algorithm is used to reduce the background noise (Nielsen, 1991).

### 2.2 LOFAR analysis

The LOFAR (Waite, 2003) is a broadband analysis that provides the machinery noise to the sonar operator and goes from DC to 15,625 Hz. The block diagram of the LOFAR analysis is shown in Figure 3.

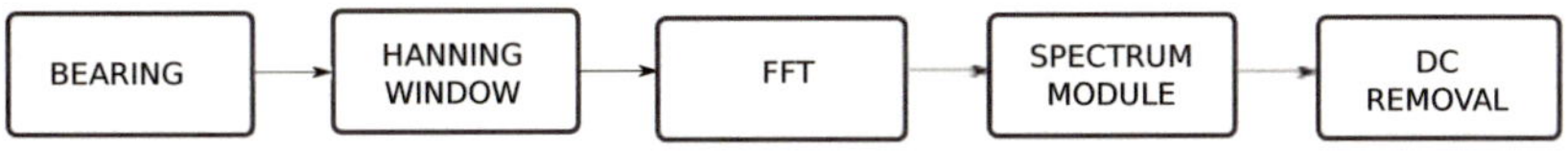


Fig. 3. LOFAR analysis block diagram.

After bearing, the signal is multiplied by a Hanning window to emphasize the frequency range of interest (Diniz, 2010). Then, the signal is separated in blocks of 1,024 samples, which are transformed into frequency-domain using a short time Fourier transform. A spectrum module is implemented and the spectra are normalized using the TPSW algorithm. The normalization may be implemented estimating the background noise that is present at each spectrum and computing a normalized frequency bin using this estimation as normalization factor. This estimation removes the spectrum bias and equalizes the spectrum amplitude (Soares Filho, 2011).

## 3. Signal interference

A passive sonar system is typically used by submarines to realize the surveillance in a determined operation area. The beamforming aims at estimating the direction of arrival (DOA) from a given target. The Figure 4 shows the beamforming display. The main purpose of the DOA is to estimate the target energy for a particular direction of interest. The horizontal axis represents the bearing position ($-180^o$ to $180^o$) and the vertical axis represents time (waterfall display). In this case, an acquisition window of one second was considered (Krim, 1996).

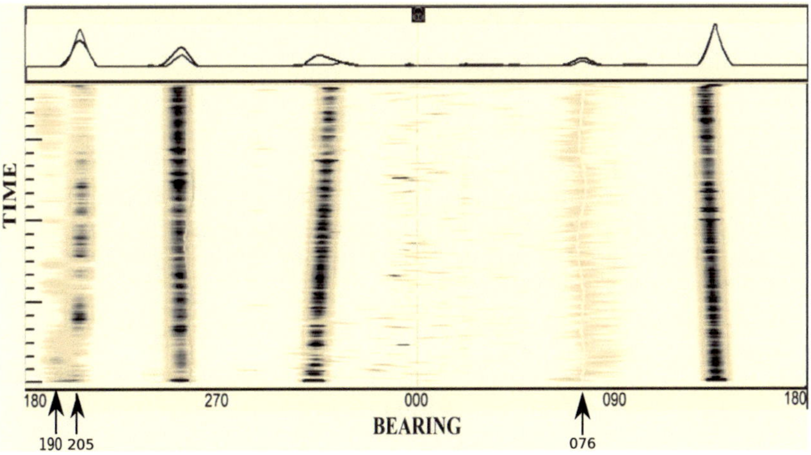


Fig. 4. Bearing time display of a passive sonar system.

When signal interference from neighbor bins occurs, as it may be the case for bearings 190° and 205° (see Figure 4), the original target features may be masked and both target detection and classification efficiencies may be affected. Beyond that, the self-noise, bearing 076°, may interfere on the detection of both targets by the sonar operator. Thus, a preprocessing scheme may be developed aiming at reducing signal interferences, facilitating target identification and classification. Here, signal preprocessing is addressed by independent component analysis (Hyvarinen, 2001), which is performed in frequency-domain (see Figure 5). After beamforming a spectral analysis is implemented at each direction of interest and an algorithm to extract the independent components is implemented. Then both DEMON and LOFAR analysis are performed over the independent sources (ICA space).

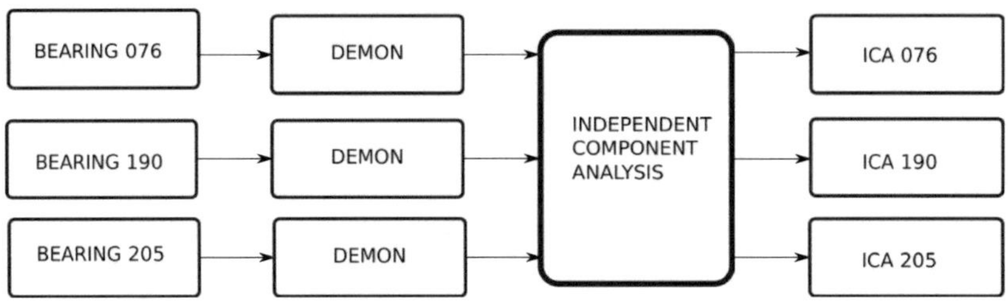


Fig. 5. Interference removal in frequency-domain.

## 4. Independent component analysis

The ICA provides a linear representation of nongaussian data, so that components are statistically independent, or as much independent as possible (Hyvarinen, 2001), (Yan, 2000). The basic ICA model assumes the existence of $n$ independent signals $s_1(t), ..., s_n(t)$ and the observation of as many linear and instantaneous mixtures $x_1(t), ..., x_n(t)$, as shown in equation 1:

$$\mathbf{x}(t) = \mathbf{As}(t) \tag{1}$$

where $\mathbf{s}(t) = [s_1(t), ..., s_n(t)]^T$ is a $n$x1 column vector corresponding to the source signals, and $\mathbf{x}(t)$ similarly collects the $n$ observed signals. $\mathbf{A}$ is the mixing matrix that contains the mixture coefficients. The ICA problem consists in recovering the source vector $\mathbf{s}(t)$ using only the observed data $\mathbf{x}(t)$, with the assumption of independence between the entries of the source vector $\mathbf{s}(t)$. The ICA problem may also formulated as the estimation of a $n$x$n$ demixing matrix $\mathbf{B}$, which allows original sources to be recovered:

$$\mathbf{y}(t) = \mathbf{Bx}(t) \tag{2}$$

where $\mathbf{y}(t)$ becomes the estimated source vector.
Due to the well known central limit theorem (Barany, 2007), the ICA estimates can be obtained from the determination of the directions for which the nongaussianity is maximum. Efficient algorithms are available today (Hyvarinen, 2001), among which Newton-like algorithm has been claimed to be very efficient (Akuzawa, 2001).

### 4.1 Multiplicative newton-like algorithm for ICA

This ICA algorithm was proposed by Akuzawa and Murata (Akuzawa, 2000). Using kurtosis as the cost function, this method applies second order optimization (through a Newton-like algorithm) in the search for independent components (instead of first-order gradient iterations used in most ICA algorithms). Kurtosis is a fourth-order cumulant that measures the shape of the distribution. For a zero-mean random variable $\mathbf{x}$, kurtosis is defined as (Kim, 2004):

$$K_4 = E\{\mathbf{x}^4\} - 3[E\{\mathbf{x}^2\}]^2 \tag{3}$$

This ICA algorithm does not require pre-whitening and thus operates directly over the data. Modifications on the method have also been proposed (Akuzawa, 2000) aiming at reducing the computational cost by substituting the pure-Newton optimization by quasi-Newton iterations (Akuzawa, 2001).

## 5. Experimental results

The raw data used here were acquired from a passive sonar system that is installed in a submarine of the Brazilian Navy. Raw data were processed according to Figure 5, in which the ICA block was performed by the Newton-like algorithm. Performance is evaluated for both signal detection using DEMON analysis and signal classification from LOFAR analysis.

### 5.1 Signal detection

From Figure 4, one may observe that bearings $190^0$ and $205^0$ suffer from mutual interference. Figure 6 displays the DEMON analysis for each of these directions. For bearing $190^o$ the largest peak is at 146.4 rpm, which corresponds to the target shaft rotation. The next harmonic represents the target blade rate. But interferences are also observed at 120.1 rpm and 304.7 rpm, which are relative to the $205^o$ and the self-noise ($076^o$), respectively. Similarly, interference is seen at 304.7 rpm for $205^o$, which is the peak of the self-noise.
Figure 7 shows the resulting DEMON analysis from ICA preprocessing. In general it is noted that the baseline noise is reduced for bearings $190^o$ and $205^o$, which reflects in a better signal-to-noise ratio. Additionally, interferences from others bearings and the self-noise are reduced.
A manner to measure the independence between signals is the mutual information, which may quantify how much independent are the components that were extracted by the Newton-like algorithm.

#### 5.1.1 Mutual information

Assuming we have two random variables $\mathbf{v}$ and $\mathbf{w}$ with marginal distribution $p(\mathbf{v})$, $p(\mathbf{w})$ and joint distribution $p(\mathbf{v},\mathbf{w})$, the mutual information (Hild, 2001), (Hoffmann, 2006) is a measure of the amount of information that the variable $\mathbf{v}$ has about the variable $\mathbf{w}$, as shown in Equation 4. The mutual information $I(\mathbf{v},\mathbf{w})$ must be non-negative. When $I(\mathbf{v},\mathbf{w}) = 0$, $\mathbf{v}$ and $\mathbf{w}$ are independent.

$$I(\mathbf{v},\mathbf{w}) = \sum_{v_i \in \mathbf{v}} \sum_{w_j \in \mathbf{w}} p(v_i,w_j) \frac{\ln p(v_i,w_j)}{p(v_i)p(w_j)} \tag{4}$$

As it can be seen from equation 4, it is necessary to compute the probability density function (pdf) of the variables. To estimate the variable pdf, a non-parametric method (Kernel estimation) may be applied (Peng, 2005). Suppose $v_1,\ldots,v_n$ independent identically distributed (i.i.d.) variables, then, the kernel density approximation of its probability density function may be given by:

$$\hat{p}_h(x) = \frac{1}{nh}\sum_{i=1}^{n} K\left(\frac{v-v_i}{h}\right) \tag{5}$$

where $K$ is some kernel function and $h$ is a smoothing parameter called the bandwidth. Here, $K$ is taken to be a standard zero-mean Gaussian function having unitary variance.
For comparison, the mutual information is computed before and after the interference removal process (see Figure 5). For controlling artificially the amount of signal interference, a weighting matrix is created:

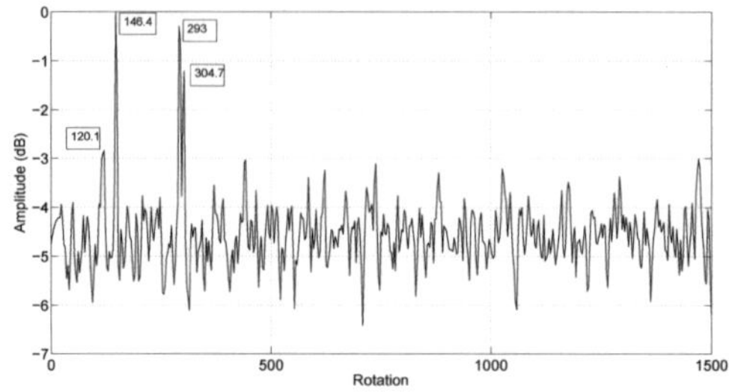


(a) Bearing 190.

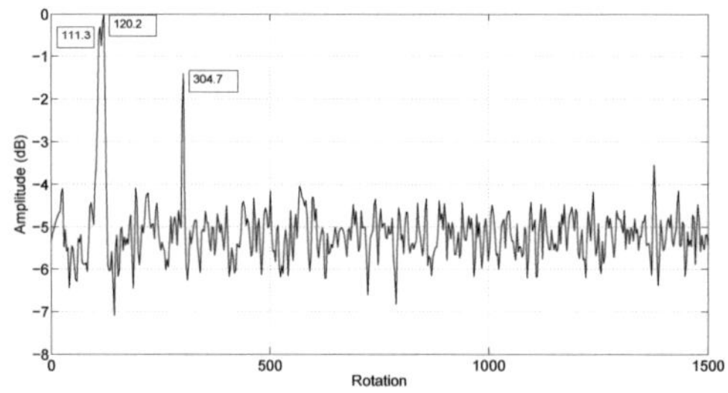


(b) Bearing 205.

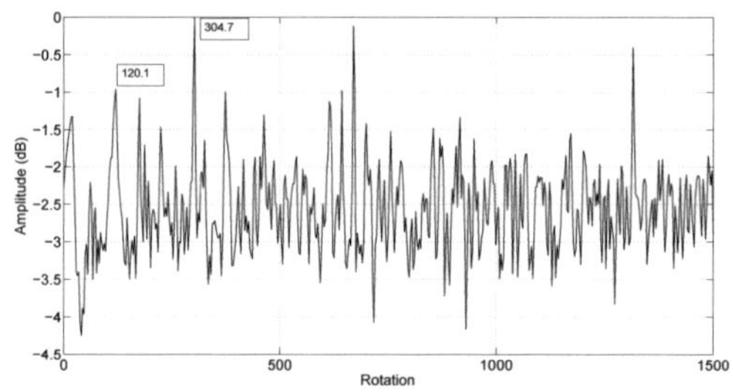


(c) Bearing 076.

Fig. 6. Standard DEMON analysis for: (a) $190^o$, (b) $205^o$ and (c) $076^o$ directions.

$$W = \begin{bmatrix} p & 1 \\ 1 & p \end{bmatrix} \tag{6}$$

The $p$ parameter varies from 0 to 1 with steps of 0.01. With such weighting process, it is possible to control the mixture on signals and components. This allows to calibrate the measure of the mutual information. Figure 8 shows the mutual information between bearings $190^o$ and $205^o$ as a function of weighting value. Although the components are multiplied by the weighting matrix, the mutual information is always smaller for extracted components than for the original mixing signals. Thus, despite increasing the interference through the weighting matrix, ICA outputs remains less susceptible to interferences.

### 5.2 Signal classification

For classification tests, data from two bearings $31^o$ and $146^o$ were chosen (see Figure 9). As it may be observed from Figure 9, during around the first three minutes, signals from these two

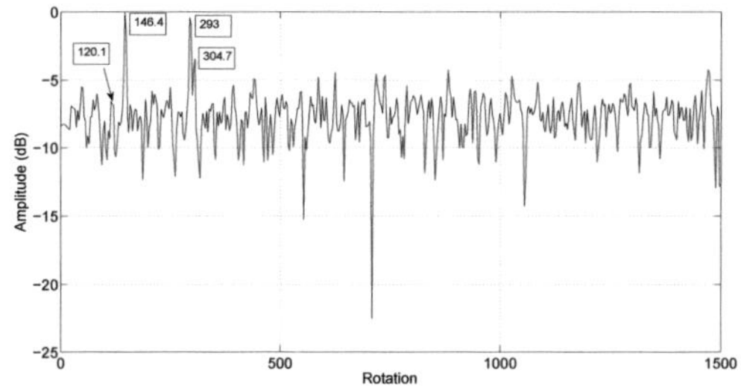


(a) Bearing 190.

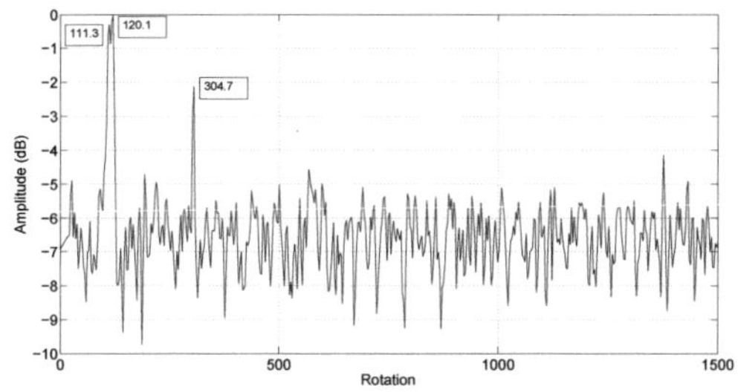


(b) Bearing 205

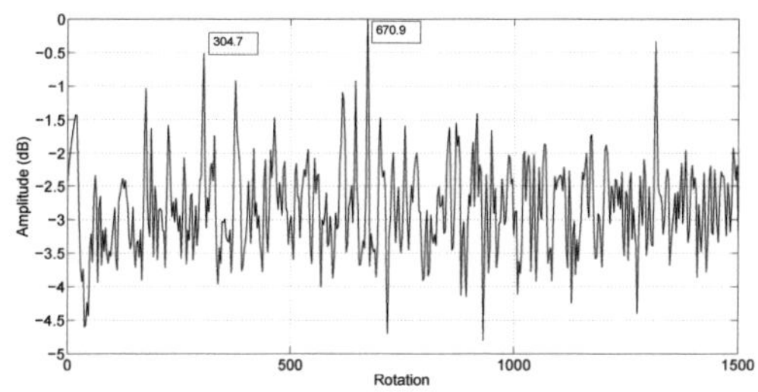


(c) Bearing 076

Fig. 7. Preprocessed DEMON analysis for: (a) $190^o$, (b) $205^o$ and (c) $076^o$.

bearings are well separated. This will make a good characterization of the contacts in their bearings for the neural training. In the sequence, the detected target at $31^o$ approaches the one in $146^o$, causing mutual interferences.
The signal classification was implemented through spectra acquired from the LOFAR analysis (see Figure 3). Figure 10 shows the lofargram for both bearings $31^o$ and $146^o$, which shows that target characteristics can be extracted through the this analysis.
For evaluating neural classification robustness based on independent component information signals were combined through a mixture matrix $A$ (Equation 7). The matrix $A$ corresponds to the beamforming obtained for these bearings.
After the mixture, a Gaussian noise at each bearing that correspond a signal-noise ratio of about 10 dB . The neural network was tested with these signals to verify the performance of the classify.

$$A = \begin{bmatrix} 0.2 & 0.8 \\ 0.8 & 0.2 \end{bmatrix} \tag{7}$$

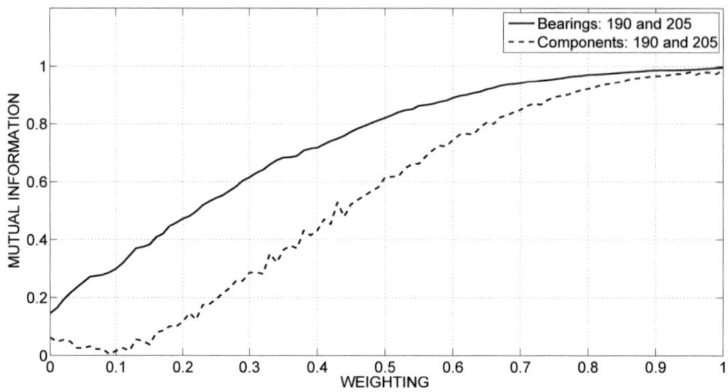


Fig. 8. Mutual information between the bearings 190 and 205 for original signal and extracted from ICA.

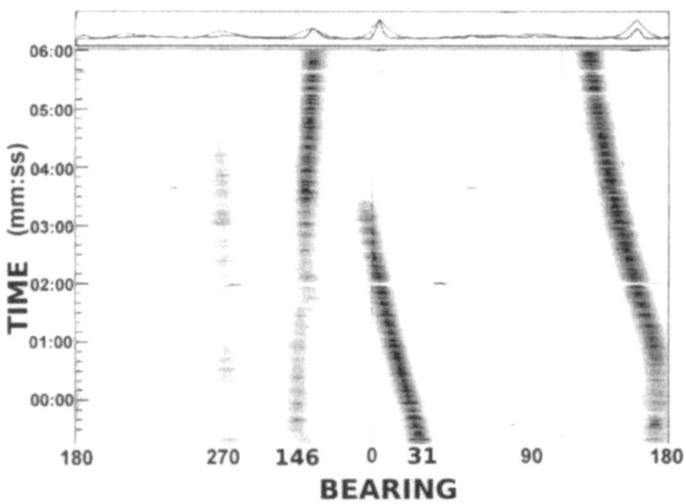


Fig. 9. Bearing time display of a passive sonar system (classification example).

- First, the classifier was tested with the separated signal;
- Next, noise was added to bearings;
- In sequence, bearing signals were mixed;
- Finally, the bearings were mixed, noise was added and the ICA algorithm, was applied, in order to recover from mixture.

#### 5.2.1 Neural classifier

The neural classifier is based on MLP - Multi-Layer Perceptron topology (Haykin, 2008), comprising three layers without feedback. This neural network was designed with one input layer having 256 nodes, a single hidden layer with 4 neurons and an output neuron, which was trained for assigning $+1$ for data from bearing $31^o$ and $-1$ for $146^o$. The neurons from both hidden and output layers have hyperbolic tangent as the activation function. The training

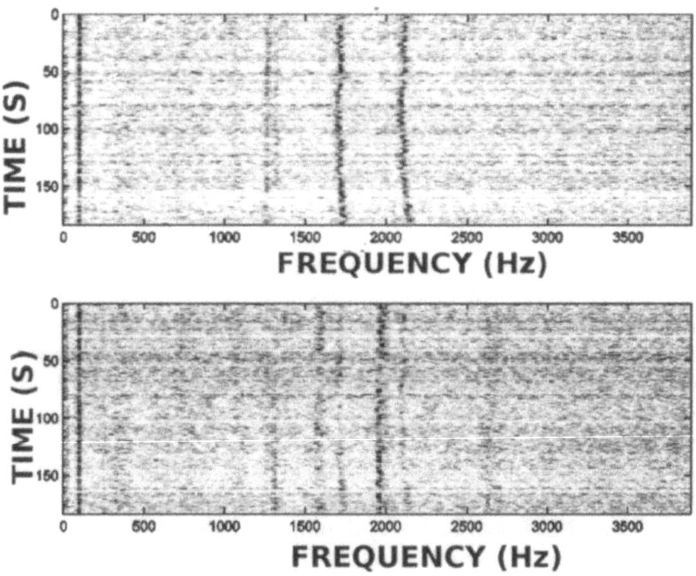


Fig. 10. LOFAR analysis for bearings $31^o$ (top) and $146^o$ (bottom), respectively.

phase has been performed in batch and used half of the spectra data from both targets. The training algorithm was backpropagation. After training, the classification performance is evaluated from the testing set (the other half of data that did not participate in the training step).

| | |
|---|---|
| Test signal | 94.3 |
| Without mixer and with noise | 89.0 |
| With mixer | **82.7** |
| With ICA | **88.0** |

Table 1. Classification performance (see text).

From original unmixed signals the classification efficiency was 94.3%. Adding noise reduces efficiency, which reaches 82.7%. Mixing signals provides a further reduction in classification efficiency and applying ICA, efficiency is recovered, setting back to 88%.

### 5.3 Stationarity analysis

The passive sonar signals are not stationary due to the ambient noise characteristics and external conditions. Then, signal stationarity monitoring is required and when the statistics changes an update of the independent component estimation may be requested for signal processing. A manner to implement this is through clustering using neural networks.

#### 5.3.1 Clustering

Clustering may be based on non-supervised training. The main idea is to receive spectra from the DEMON analysis and observe when a change may occur on data statistics. The spectra information is from the DEMON analysis. The input spectra comprises of 513 frequency bins. Principal component analysis (PCA) (Chen, 2000), (Soares Filho, 2001) is firstly applied for

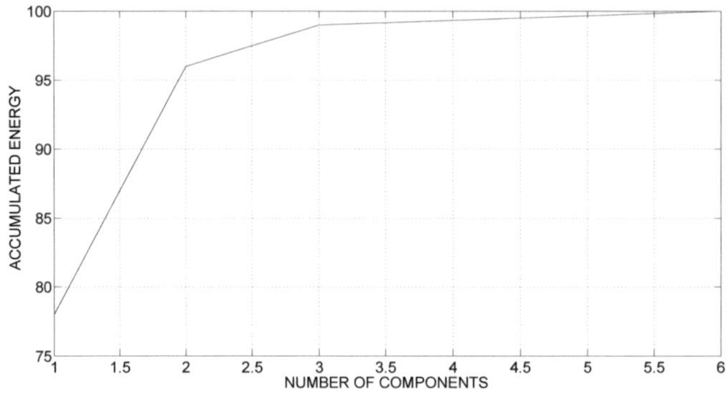


Fig. 11. Accumulated energy in principal components for PCA.

data dimensionality reduction. The Figure 11 displays the accumulated energy curve which shows that the clustering process may be realized projecting data onto only two components. For clustering, a modified ART (Adaptive Resonance Theory) ( Vassali, 2002). The training is competitive and based on a Kohonen layer, attaching a neuron at each new input pattern. The control parameters are the cluster vigilance radius and the learning rate. The vigilance radius of the cluster is defined by the most probable value obtained from measuring the distance of all input data patterns.
The conceptual model of the clustering consists in:

- Information obtained from the sonar operator points out that the passive sonar signal must be evaluated at each 10 seconds to verify a statistical change. How the DEMON analysis presents each line in 500ms then, it was acquire 20 FFT windows through that corresponds to these 10 seconds;
- As data dimension is very high, a PCA was applied to reduce data dimension;
- Clusters are formed from these 20 windows;
- Changes on number of clusters are evaluated, in conjunction with possible shifts of the center of these clusters; and
- The statistical changes are then identified.

A signal of the bearing $190^o$ with 200 seconds was recorded to system development. As mentioned early the spectra were obtained through the DEMON analysis. Specialists indicate that the passive sonar signal must be evaluated at each 10 seconds to verify if a statistical changes have been occurred. Then 20 spectra were chosen as a start point to network parametrization.
Because of the high dimension of the input data and having the knowledge the clustering presents better results in reduced dimensions (Duda, 2000), principal components analysis (Soares Filho, 2001) was implemented. The Figure 11 shows the accumulated energy in principal components for PCA. It may be noted that only with the first two components were obtained from 95% of the variance of the process. Therefore, the clustering was implemented using in the input net the data projections of the two principal components.
Clustering is to give one neuron to each cluster from the similarity between each input pattern, which represents an FFT window. Figure 12 shows the evolution of the number of clusters. This information states that there was a new cluster or more than one. May be observed that there was the forgotten one or more clusters and this is very important to indicate that there was a statistical change in the process. For example, one can notice that there was a great

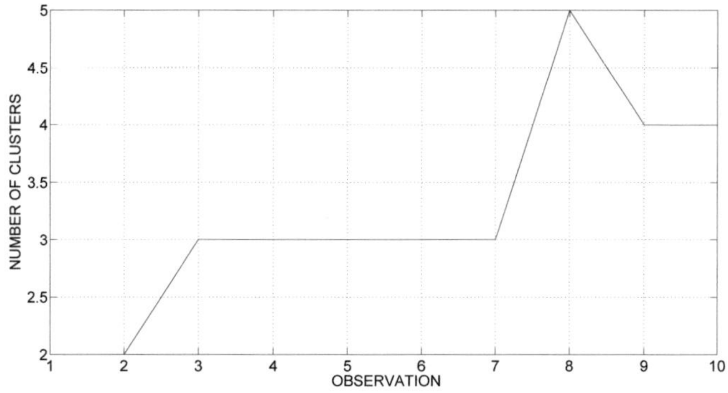


Fig. 12. Number of clusters after clustering at each observation.

increase in observation 7 to 8, since the number of clusters jumps from 3 to 5. This is a strong indication that the statistics of the signal changed significantly.

## 6. Conclusions

The passive sonar system aims at detecting and classifying targets from given directions. Additional difficulties arise when interferences occur in neighbour directions Therefore, a preprocessing scheme may be developed for reducing interferences and facilitating detection and classification tasks for the sonar operator. This preprocessing step may be implemented using the independent component analysis in frequency-domain and improvements in both detection and classification efficiencies have been observed from experimental data.

As original sources have to be estimated blindly, the stationarity of the passive sonar signals should be monitored. When statistics change, the preprocessing data chain may be reloaded for updating the source estimation.

Other algorithms of blind source separation may be implemented to improve the signal noise interference removal in passive sonar systems. Among them, the nonnegative matrix factorization (NMF) (Chichocki, 2009) which uses similarity measures to extract the original free of interference signals looks attractive. Additionally, algorithms of blind source separation using nonstationarity and convolutive mixture may be used.

## 7. References

Akuzawa T.: Extended Quasi-Newton Method for the ICA. *Proc. Int. Workshop Independent Component Analysis and Blind Signal Separation*, Helsinki, Finland, pp.521-525 (2000).

Akuzawa T. and Murata N., Multiplicative Nonholonomic Newton-like Algorithm. *Chaos, Solutions and Fractals*, Volume 12, Number 4, 3 January, pp. 785-793 (2001).

Barany Imre and Van Vu, Central limit theorems for Gaussian polytopes. *The Annals of Probability (Institute of Mathematical Statistics)*, 35 (4), pp. 1593-1621 (2007).

Buchner Herbert, Aichner Robert and Kellermann Walter, A generalization of blind source separation algorithms for convolutive mixtures based on second-order statistics, *IEEE Transactions on Speech and Audio Processing.*, Vol. 13, No. 1, pp. 120-134 (2005).

Burdic William S., *Underwater Acoustics System Analysis*, Peninsula Publishing, 2nd Ed. (1991).

Chen Chin-Hsing, Lee Jiann-Der and Lin Ming-Chi, Classification of Underwater Signals Using Neural Networks, *Tamkang Journal of Science and Engineering*, Vol. 3, No. 1, pp. 31-48 (2000).

Cichocki Andrzej, Zdunek Rafal, Huy Phan Anh and Amari Shun-ichi, *Nonnegative Matrix and Tensor Factorizations Applications to Exploratory Multi-way Data Analysis and Blind Source Separation*, John Wiley and Sons (2009).

Di Martino, J.C.; Haton, J.P.; Laporte, A., Lofargram line tracking by multistage decision process. *IEEE International Conference on Acoustics, Speech, and Signal Processing*, Minneapolis, USA, vol. 1, pp. 27-30 April (1993).

Diniz P. S. R. , Silva E. and Netto S. L., *Digital Signal Processing: System Analysis and Design.*, Cambridge University Press, 2nd. (2010).

Duda Richard O., Hart Peter E. and Stork David G., *Pattern Classification* , Pattern Classification , 2nd Ed. (2000).

Haykin, S., *Neural Networks and Learning Machines*, Prentice Hall, 3rd Ed. (2008).

Hild, K.E., II; Erdogmus, D.; Principe, J., Blind source separation using Renyi's mutual information, *IEEE, Signal Processing Letters*, June, 174-176 (2001).

Hoffmann G.M., Waslander S.L and Tomlin C.J., Mutual Information Methods with Particle Filters for Mobile Sensor Network Control, *45th IEEE Conference on Decision and Control*, San Diego, USA, December, 1019-1024 (2006).

Hyvarinen A., Karhunen J. and Oja E. , *Independent Component Analysis*, Wiley (2001).

Jutten C. and Karhunen J., Advances in blind source separation (BSS) and independent component analysis (ICA) for nonlinear mixtures, *International Journal of Neural Systems.*, Vol. 14, No. 5, pp. 267-292 (2004).

Kim T. and White H., On more robust estimation of skewness and kurtosis. *Finance Research Letters*, vol. 1, pp. 56-73 (2004).

Krim H. and Viberg M., Two decades of array signal processing research: The parametric approach. *IEEE Signal Processing Magazine*, vol. 13, Issue 4, pp. 67-94 (1996).

Nielsen R. O., *Sonar Signal Processing*, Artech House Inc., Northwood, MA (1991).

Peng H.C., Long F., and Ding C., *Feature selection based on mutual information: criteria of max-dependency, max-relevance, and min-redundancy*, IEEE Transactions on Pattern Analysis and Machine Intelligence, Vol. 27, No. 8, pp. 1226-1238 (2005).

Rabiner Lawrence R. and Crochiere Ronald E., *Multirate Signal Processing*, Prentice Hall (1983).

Xueyao Li ; Fuping Zhu; Harbin Eng. and Harbin Univ., Application of the zero-crossing rate, LOFAR spectrum and wavelet to the feature extraction of passive sonar signals, *Proceedings of the 3rd World Congress on Intelligent Control and Automation, 2000.*, vol.4 , pp. 2461-2463 (2000).

Soares Filho W., Seixas J. M. and de Moura N. N., Preprocessing passive sonar signals for neural classification. *IET Radar, Sonar & Navigation*, vol. 6, pp. 1-14 (2011).

Soares Filho, W.; Manoel de Seixas, J.; Pereira Caloba, L., Principal component analysis for classifying passive sonar signals. *International Symposium on Circuits and Systems*, Sydney, Australia, vol. 2, pp. 592-595 May (2001).

Van Trees H. L., *Detection, Estimation and Modulation Theory*, John Wiley & Sons, Inc. (2001).

Vassali, M. R. ; Seixas, J. M. de and Calôba, L. P.: A Neural Particle Discriminator Based on a Modified Art Architecture. *IEEE International Symposium on Circuits and Systems*, Phoenix, v. II. p. 121-124 (2002).

Yan D. P. and Peach J., Comparison of Blind Souce Separation Algorithms, *Advances in Neural Networks and Applications*, 18-21 (2000).

Yang Z., Qing C., Huang D., An improved empirical AM/FM demodulation method. *International Conference on Wavelet Analysis and Pattern Recognition*, vol. 3, pp. 1049-1053 (2007).

Waite A. D., *Sonar for practical Engineers*, John Wiley, 3rd Ed. (2003).

# 6

# Sonar Model Based Matched Field Signal Processing

Nikolai Kolev and Georgi Georgiev
*NV Naval Academy*
*Bulgaria*

## 1. Introduction

Passive underwater sonar array systems were subject to development in the past decades for accomplishing tasks in naval surveillance and geo-acoustical investigation of the ocean bottom. Modern sonar passive array systems are using adaptive matched field signal processing techniques. Passive sonar array signal processing based on ocean physical propagation model is known as "Matched Field Processing" (MFP). It is a specific application of the more general case of space-time adaptive signal processing in the area of sonar array signals processing (Klemm, 2002). It is a cross-correlation technique developed for matching the values of the computed with a propagation model sound pressure with the measured values at the output of a passive sonar vertical array. The maximum in the cross-correlation or ambiguity surface gives an estimation of the position of the underwater source. This technique is first proposed by Clay in 1966 and later on theoretically and experimentally developed by Hinich and Bucker in 1972 and 1975 (Sullivan & Middleton, 1993). In 1980 and 1981 Klemm introduced in the matched field signal processing algorithms the Capon's minimum variance and Burg's maximum entropy estimators for better resolution in shallow water environments (Sullivan & Middleton, 1993). The acoustic field in an ocean horizontally stratified waveguide may be presented as a sum of "normal waves" or "normal modes"[1], each satisfying the boundary conditions and propagating down the layer with its own speed, wave constant and attenuation. The coherent modal sum is used in matched field signal processing for localization of an underwater source in range and depth. It can be used for localization of the source also with a horizontal array because the angular modal spread is proportional to the direction of arrival relative to the axis of the array (Klemm, 2002). The matched field processing for source position localization is presented in most of the investigations as a parameter estimation problem. It is assumed that the detection has been carried out and the frequency of the source is known. Some authors consider joint detection and localization problem [Booth et al., 2000]. In the beginning MFP of narrowband signals was used which was later modified with the more realistic for practical applications broadband case where MFP results for several frequencies are averaged. The aim of this chapter is to introduce the reader to the principles of matched field signal processing and present narrowband MFP experimental results. An available on the web normal mode propagation code and vertical sonar array signals data from a controlled source localization experiment are used. A comparison is made between MFP

[1] standing waves

results of three known matched field processors applied to vertical array sonar experimental data.

## 2. Model based matched field processing of vertical sonar array signals

In conventional antenna array systems signal processing the usual methods of beamforming include phase control of the signals in receive or transmit mode in order to form and steer antenna beam pattern in given direction (see Van Trees, 2002). In MFP the steering vector is determined by the predicted with a propagation model sound pressure for a range of source coordinates in a waveguide with known depth sound speed profile (SSP) and bottom acoustical parameters. A flow diagram of MFP is given on fig. 1.

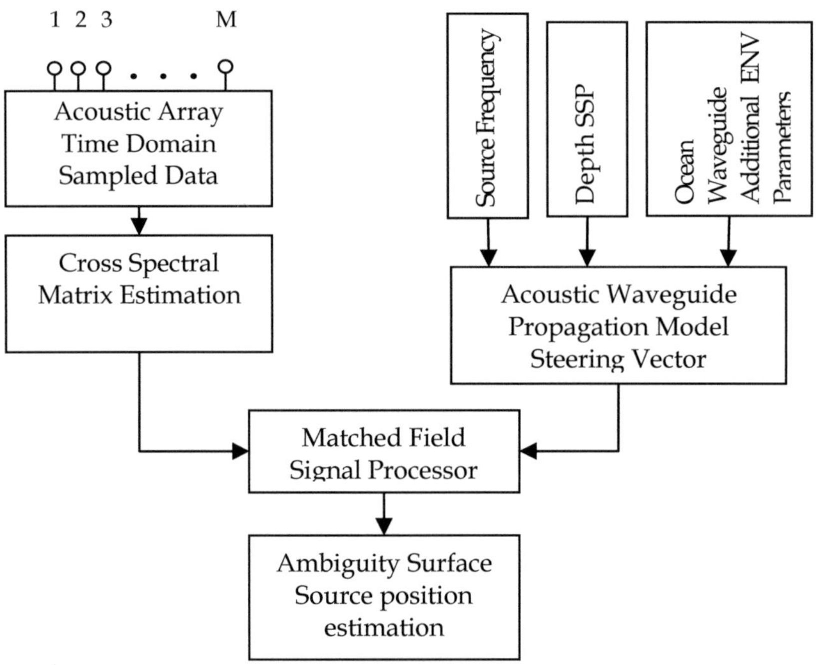


Fig. 1. Matched Field Processing Flow Diagram for Source Position Estimation

From the time domain array sampled signals the cross-spectral density matrix (CSDM) is estimated. As a steering vector for the matched filed processor the calculated sound pressure level with normal mode propagation model is used. An ambiguity surface is formed through matching or cross-correlating of the steering vector with the received sound pressure field. The maximum of the ambiguity surface gives the most likely position of the sound source in range and depth. There are several general signal processing algorithms applicable for MFP.

### 2.1 Cross-spectral density matrix estimation

Cross spectral density matrix[2] (CSDM) is formed through estimation of the cross-spectral functions of the signal at the output of the sonar array. The cross-spectral function may be

[2] The name covariance matrix is also used

defined with a Furier transform of the cross-correlation function or directly via finite Furier transform of the input time domain signal (Bendat, 2010). If sample records are available from a stationary random processes in the time domain[3] $\{x_k(t)\}, \{y_k(t)\}$ for a finite time interval $0 \le t \le T$ the cross-spectal function of the two processes is defined:

$$S_{xy}(\omega,T,k) = \frac{1}{T} X_k^*(\omega,T) Y_k(\omega,T) \tag{1}$$

where

$$X_k(\omega,T) = \int_0^T x_k(t) e^{-j\omega t} dt \tag{2}$$

$$Y_k(\omega,T) = \int_0^T y_k(t) e^{-j\omega t} dt \tag{3}$$

The functions $X_k(\omega,T)$, $Y_k(\omega,T)$ represent finite Furier transforms of the time domain samples $x_k(t)$ $y_k(t)$, (*) is the complex conjugate operator, k is sample index. An estimate of the cross spectral density function is

$$S_{xy}(\omega,k) = \lim_{T\to\infty} E\left[S_{XY}(\omega,T,k)\right] \tag{4}$$

In practise the record length is always finite and the expectation operation $E[\ ]$ is taken from a finite number of ensemple elements. If the signals at the outputs of the M elements antenna array are stationary the data can be presented after the finite Furier transforms in the frequency domain with the following matrix:

$$\begin{matrix} X_1(\omega_1) & X_1(\omega_2) & X_1(\omega_3) & \cdots & X_1(\omega_N) \\ X_2(\omega_1) & X_2(\omega_2) & X_2(\omega_3) & \cdots & X_2(\omega_N) \\ X_3(\omega_1) & X_3(\omega_2) & X_3(\omega_3) & \cdots & X_3(\omega_N) \\ \vdots & \vdots & \vdots & \cdots & \vdots \\ X_M(\omega_1) & X_M(\omega_2) & X_M(\omega_3) & \cdots & X_M(\omega_N) \end{matrix} \tag{5}$$

The row in (5) represents the frequency spectrum of the signal from a given hydrophone (channel).
The array cross-spectral matrix for the n-th frequency bin is:

$$\begin{matrix} X_1(\omega_n)X_1^*(\omega_n) & X_1(\omega_n)X_2^*(\omega_n) & X_1(\omega_n)X_3^*(\omega_n) & \cdots & X_1(\omega_n)X_M^*(\omega_n) \\ X_2(\omega_n)X_1^*(\omega_n) & X_2(\omega_n)X_2^*(\omega_n) & X_2(\omega_n)X_3^*(\omega_n) & \cdots & X_2(\omega_n)X_M^*(\omega_n) \\ X_3(\omega_n)X_1^*(\omega_n) & X_3(\omega_n)X_2^*(\omega_n) & X_3(\omega_n)X_3^*(\omega_n) & \cdots & X_3(\omega_n)X_M^*(\omega_n) \\ \vdots & \vdots & \vdots & \cdots & \vdots \\ X_M(\omega_n)X_1^*(\omega_n) & X_M(\omega_n)X_2^*(\omega_n) & X_M(\omega_n)X_3^*(\omega_n) & \cdots & X_M(\omega_n)X_M^*(\omega_n) \end{matrix} \tag{6}$$

[3] For example signals at the output of two hydrophones from the array

Where (*) is the complex conjugate operator. The cross-spectral matrix for an M element array is square MxM matrix for a given frequency $\omega_n$. The estimation is defined with the following expectation operator for the frequency bin $\omega_n$:

$$\hat{S}_X(\omega_n) = \mathrm{E}\{X(\omega_n)X^*(\omega_n)\} \tag{7}$$

An estimate of the CSDM $\hat{S}_X(\omega_n)$ is formed through averaging ensemble cross-spectrum data. On figure 2 a data cube of CSDM estimates is presented for three frequency bins.

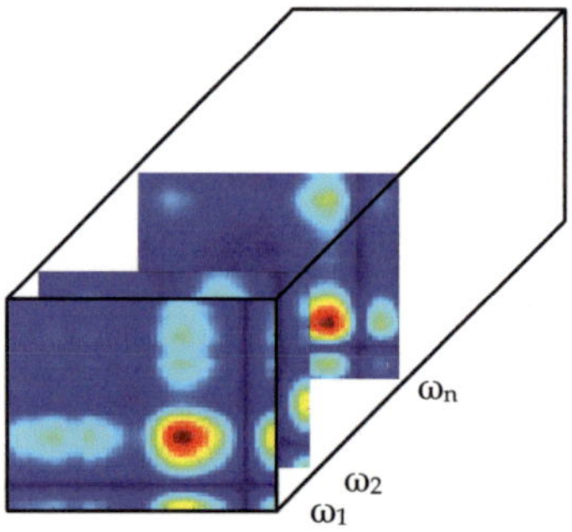


Fig. 2. Data cube of a real CSDM estimates for different frequency bins.

### 2.2 Steering vector calculation

The frequency range and the parameters of the ocean waveguide determine the sound propagation model for the matched field processing steering vector. In the application a low frequency source positioned in a shallow water ocean waveguide is considered which specifies the normal mode propagation model. For sound propagation numerical calculations in a range-independant environment the normal mode representation of the acoustical pressure actuated by a harmonic source in a horizontally stratified medium is given by the following infinite radical sum (Ferla et al., 1993):

$$p(r,z) = \frac{i}{4\rho(z_s)\sqrt{8\pi r}} \sum_{m=1}^{\infty} \Psi_m(z_s)\Psi_m(z) H_0^1(k_m r) \tag{8}$$

Where r is the range, z the depth, ρ the density, $z_s$ the source depth, $\Psi_m$ the mode amplitude, $H_0^1$ the zero-order Hankel function of the first kind, $k_m$ is the eigenvalue. In the solution a large argument asimptotic approximation of the Hankel function is used. When the number of modes is limited to M the final expression for the sound pressure is:

$$p(r,z) = \frac{j}{\rho(z_s)\sqrt{8\pi r}} e^{-j\frac{\pi}{4}} \sum_{m=1}^{M} \Psi_m(z_s)\Psi_m(z) \frac{e^{jk_m r}}{\sqrt{k_m}}, \tag{9}$$

where dependence of the pressure – p from distance – r and depth z is obvious.

Each of the modes has unique wave number, attenuation, phase speed and group speed. The sum of the modes for different source positions and distances from the source forms unique vertical distribution of the pressure which is subject to matching with the experimentally measured pressure (see fig.7). The calculated sound pressure is used as a steering vector in the matched field processor.

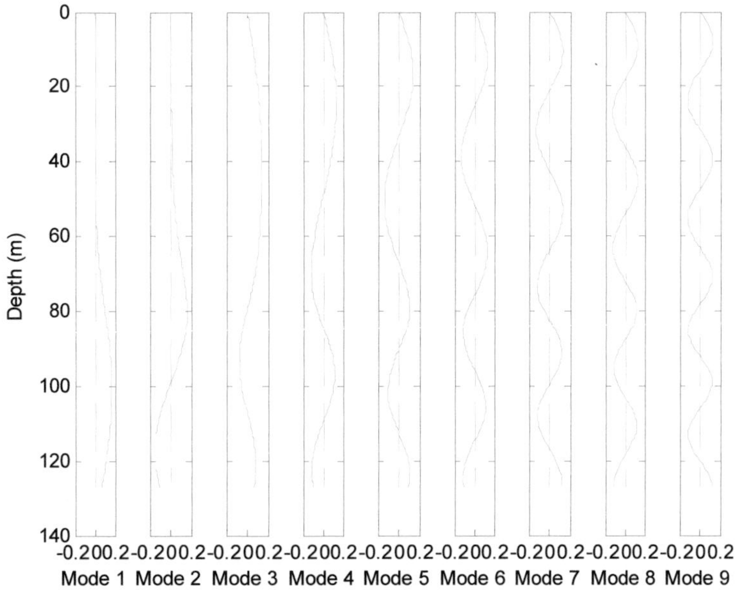


Fig. 3. Graphics of nine normal modes representing the sound field in a shallow water waveguide considered in the examples.

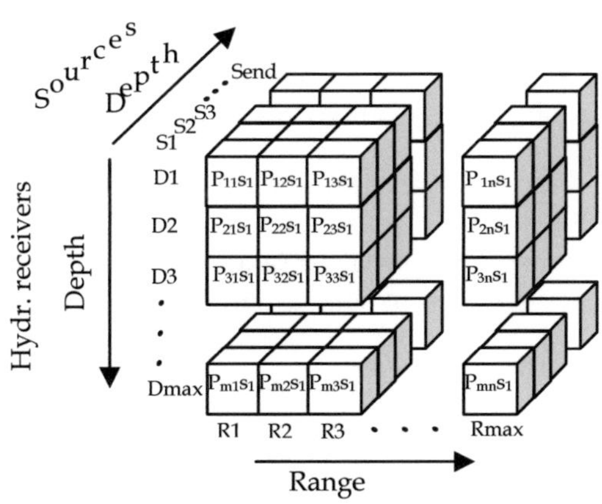


Fig. 4. Data cube of the calculated sound pressure level

The steering vector is the sound pressure level for a given possible position of the known source in depth . This forms a sound pressure level data cube presented on fig. 4. For every possible position of the source in depth $S_1, S_2, S_3, \dots, S_{end}$ the sound pressure level is calculated in a 2D region. For every distance from the source $R_1, R_2, R_3, \ldots, R_{\max}$ a depth sound pressure vector $P_1, P_2, P_3, \ldots, P_m$ is formed for the known depth hydrophone receiver positions $D_1, D_2, D_3, \ldots, D_{\max}$. This vector is later used as a steering vector for correlation with the estimated CSM and forming the ambiguity surface. A maximum in the ambiguity surface gives an estimation of the source position in range and depth. On figure 5 a data cube with real pressure fields respectively for sources: S1, S10 and S68 considered in chapter 3.

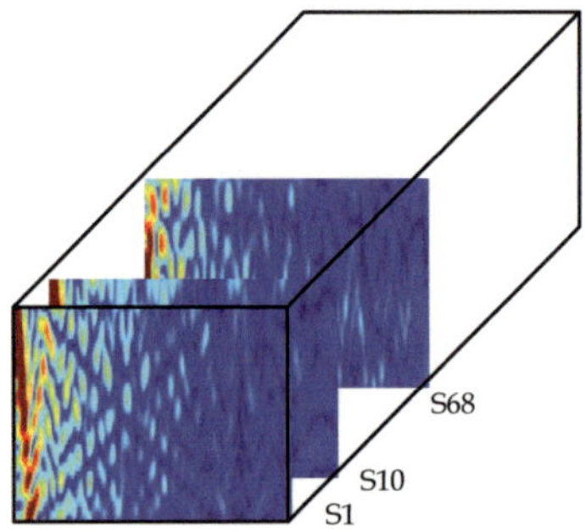


Fig. 5. Data cube of the calculated sound pressure fields

### 2.3 Matched field signal processors

Two matched field processors are subject to investigation in the application. The first one is the classical or Bartlett matched field processor and the second one is the so called minimum variance distortion less processor (MVDR).

#### 2.3.1 Bartlett (conventional) matched field signal processor

In conventional beamforming the sonar array pattern is steered through phase control of the signals at the output of the antenna array (see fig. 1) through steering vector. In matched field processing the steering vector represents the computed pressure in the points of the array elements. The processor output or as in our case a matched field filter output is:

$$Y(\omega) = v^H(\omega, r, z) X(\omega) \tag{10}$$

Where $v^H(\omega, r, z)$ is the pressure vector, $X(\omega)$ is the signal at the input of the procerssor and $Y(\omega)$ is the output signal. We want to estimate the power of the signal at the output of the filter:

$$P(\omega, r, z) = E\left[v^H(\omega, r, z) X X^* v(\omega, r, z)\right] = v^H(\omega, r, z) S_X(\omega) v(\omega, r, z) \tag{11}$$

where $S_X(\omega)$ corresponds to CSDM given with (7). This conventional processor is known as a Bartlett processor. Bartlett matched field processor forms ambiguity surface as a direct dot product of model replica vectors $\vartheta(\omega, r, z)$ and the sample CSDM - $S_X(\omega)$. The maximum of the power estimate in the ambiguity surface gives the probable position of the source.

$$P_{BART} = v^H(\omega, r, z) S_X(\omega) v(\omega, r, z) \tag{12}$$

This processor is a benchmark for the other more sophisticated processors because it has much power and low probability of no detection.

### 2.3.2 MVDR matched field signal processor

The MVDR processor is derived with quadratic constraints (Van Trees, 2002). An optimal filter is given with the block diagram on fig. 6. $X(\omega)$ and $Y(\omega)$ represent the signals at the input and at the output of the filter in the frequency domain. $W^H(\omega)$ is the filter frequency response.

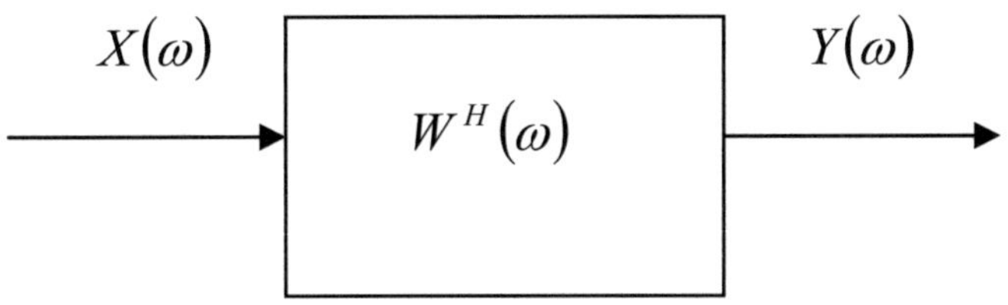


Fig. 6. General optimal filter (processor) structure.

The signal vector at the input can be written as:

$$X_S(\omega) = F(\omega)v(\omega,r,z), \tag{13}$$

where $F(\omega)$ is the frequency-domain snapshot of the source signal and $v(\omega,r,z)$ is the ocean waveguide manifold vector. It is required that in the absence of noise $Y(\omega) = F(\omega)$. The constraint of no distortion[4] implies $W^H(\omega)v(\omega,r,z) = 1$ in order to receive the input signal at the output of the processor without distortion (with unity gain). Minimization of the mean square of the output noise leads to MVDR beamformer first derived by Capon. Optimal weight of the processor is given (Van Trees, 2002):

$$W_{MVDR}^H(\omega) = \frac{v^H(\omega,r,z)S_N^{-1}(\omega)}{v^H(\omega,r,z)S_N^{-1}(\omega)v(\omega,r,z)} \tag{14}$$

where $S_N^{-1}(\omega)$ is the inverse of noise CSDM for a given angular frequency, $v^H(\omega,k_S)$ is the hermitian transpose operator. When instead of $S_N(\omega)$ the signal CSDM $S_X(\omega)$ is used in (14) the estimator is known as minimum power distortionsless response (MPDR) (Van Trees, 2002). Usually it is more difficult to estimate $S_n(\omega)$ to form the weight, because it is colored with multipath ocean waveguide interference. It is not possible however also to estimate a pure signal matrix $S_X(\omega)$ beacause it is mixed with noise. The optimum weight of MPDR[5] has the form:

$$W_{MPDR}^H(\omega) = \frac{v^H(\omega,r,z)S_X^{-1}(\omega)}{v^H(\omega,r,z)S_X^{-1}(\omega)v(\omega,r,z)} \tag{15}$$

---

[4] Or no gain

[5] In the majority of the literature it is also referred as MVDR (Van Trees, 2002)

Later in the text we are using the name MVDR for weights (14), (15) and specify if the CSDM of the signal or noise is used. This estimator (beamformer) in the literature is known also as Minimum Variance Estimator (MVE) and Maximum Likehood Estimator (Klemm, 2002). The power of MFP-MVDR is given with (Van Trees, 2002):

$$P_{MVDR} = W^H(\omega) S_X(\omega) W(\omega) \tag{16}$$

where $S_X(\omega)$ is the signal CSDM and $W(\omega)$ is the optimal weight defined with (14) or (15). When we have acceptable for array processing signal to noise ratio (SNR) it is relatively easy to estimate $S_X(\omega)$ and use (15). Subject to optimization in the described processors is the match between the function of the computed sound pressure over the array elements and the measured output signal power for a given frequency, given with CSDM. On fig. 7 the computed normalized values of the pressure and the measured signal power are superimposed at distance and depth of good match .

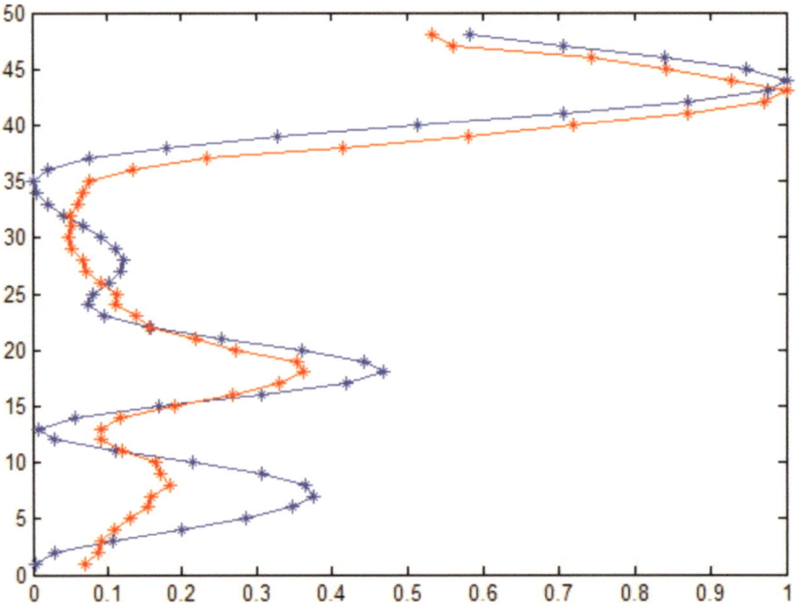


Fig. 7. Predicted and measured signal power (diagonal values of CSDM) for a frequency bin of interrest over the array at distance and depth of good match (red - measured signal power, blue - computed squared pressure).

One way to build robust to some degree of mismatch weight vector and improve the numerical stability in matrix inversion is through diagonal loading of the CSDM. This is the so called method of "diagonal loading" [6] (Van Trees, 2002). It is also referred in the literature as a "white-noise-constraint" (WNC). Adding noise to diagonal elements expands the noise

[6] Van Trees shows that imposition of quadratic constrain leads to diagonal loading

space and eliminates small eigenvalues in the inverse CSDM. Usually the CSDM is not well conditioned matrix[7] and the diagonal loading is a way toward more stable matrix inversion procedure. The MVDR weight with diagonal loading of the CSDM is (Van Trees, 2002):

$$W_{MVDRDL}(\omega) = \frac{v^H(\omega,r,z)\left[S_X(\omega)+\sigma^2 I\right]^{-1}(\omega)}{v^H(\omega,r,z)\left[S_X(\omega)+\sigma^2 I\right]^{-1} v(\omega,r,z)} \quad (17)$$

Another way to overcome problems with CSDM inversion is to apply singular value decomposition (Van Trees, 2002). With singular value decomposition the square CSDM is presented in the following form:

$$S_X = U\Lambda V^* \quad (18)$$

where $\Lambda$ is diagonal matrix of the eigen values:

$$\Lambda = \begin{vmatrix} \lambda_1 & & & & \\ & \lambda_2 & & & \\ & & \cdot & & \\ & & & \cdot & \\ & & & & \lambda_M \end{vmatrix} \quad (19)$$

where $\lambda_1 > \lambda_2 > ... > \lambda_M$ and U and V* are unitary with size MxM. Choosing the rank "r" of the decomposed CSDM directly imposes constraints on the spatial signal subspace and is powerful instrument for multipath interference rejection (r is smaller than the number of sensors M which in our case is 48). The inverse matrix is formed with the following matrix multiplication:

$$S^{-1}{}_X = \begin{vmatrix} U_{11} & \cdot & \cdot & U_{1r} \\ \cdot & \cdot & & \\ \cdot & & \cdot & \\ U_{M1} & & & U_{Mr} \end{vmatrix} \begin{vmatrix} 1/\lambda_1 & & 0 \\ & \cdot & \\ 0 & & 1/\lambda_r \end{vmatrix} \begin{vmatrix} V_{11} & \cdot & \cdot & V_{1M} \\ \cdot & \cdot & & \\ \cdot & & \cdot & \\ V_{r1} & & & V_{rM} \end{vmatrix} \quad (20)$$

The optimal weight for MVDR is formed:

$$W_{MVDRPINV}(\omega) = \frac{v^H(\omega,r,z)\left(U\Lambda^{-1}V^* + \varepsilon^2 I\right)}{v^H(\omega,r,z)\left(U\Lambda^{-1}V^* + \varepsilon^2 I\right)v(\omega,r,z)} \quad (21)$$

where the inverse matrix[8] is formed with rank reduction and diagonal loading is applied (adding noise to the inverse matrix through diagonal loading) , controlled by the parameter ε. A similar form of this processor weight was used in (Kolev & Georgiev, 2007).

---

[7] The analysis of the experimental averaged CSDM showed that the ratio between largest singular value to the smallest is more than 1 000 000

[8] Also known in the literature as pseudo-inverse

The rank in the application is determined if the corresponding singular value is under 1/100 from the maximum value. On fig. 8 the singular values of a CSDM are plotted and the rank is determined to be r=5.

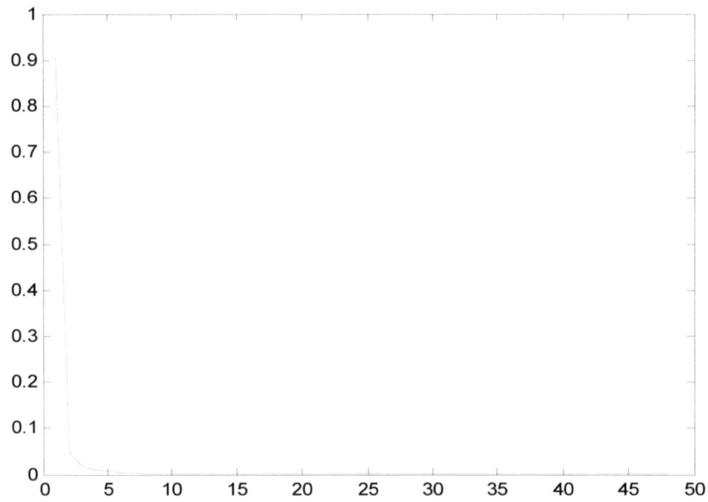


Fig. 8. Typical eigenvalues of CSDM determining the matrix rang – r=5.

## 3. Vertical sonar array MFP experimental results

Passive array sonar data from SACLANTC 1993 North Elba experiment available on Internet (Saclant sonar data, 1996) is used for processing. Additional information about the experiments and signal processing examples is given in (Gingras & Gerstoft,1995, Gerstoft & Gingrass, 1996). The array data was collected in shallow-water off the Italian west coast by the NATO SACLANT Center in La Spezia, Italy. SACLANT Center has made this data available to the public for the purposes of fostering signal processing research. The original SACLANT time series has been converted to a series of MATLAB files each of which contains a matrix data file from the 48 sensors. Each file represents about 1 minute of data The vertical array consists of 48 hydrophones with spacing 2 m between elements at total aperture length 94 m (18.7 m to 112.7 m in depth). On 27-th of October a source was towed in the surveillance area at a speed approximately 3.5 kn. The source emitted PRN signal with a center frequency of 170 Hz. For this source 10 data files are available. The data for the moving source was subject to MFP. The available on the site sound speed profiles were used for replica computations with KRAKEN normal mode program (Porter, 2007). Technical manual of the software was used from (Porter, 1992). On fig. 9 a datagram of sampled time domain array signals and their spectrum is given.
The CSDM was formed from averaged FFT data. The FFT length was 4096 resulting in 0.2441 Hz FFT binwidth for 1000 Hz sampling data rate. A Hamming window with 50 % overlapping was used for weighting. It was observed from fig. 11 the relatively low level of the power of CSDM in the frequency window of interest with center frequency 170 Hz compared to the low frequency noise (50 – 100 Hz). For estimation of signal CSDM the frequency bin for 169.9 Hz is used.

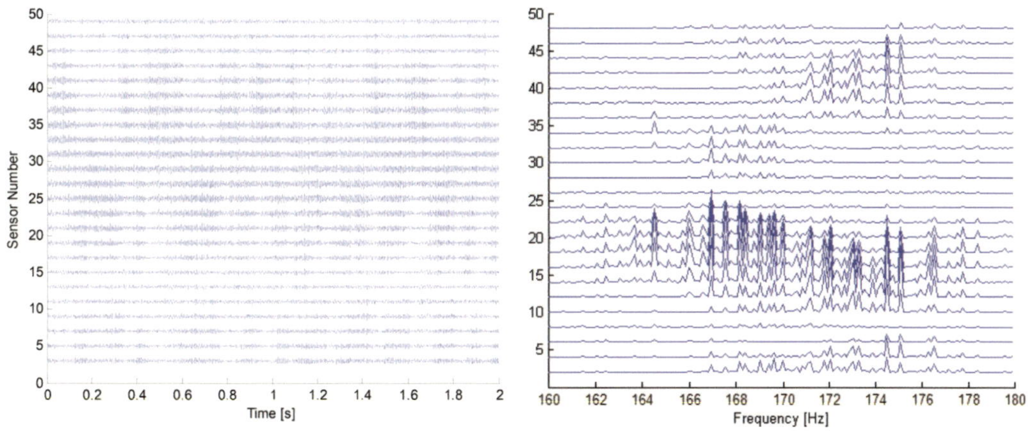


Fig. 9. A datagram of time domain array sampled multichannel signals and their power spectrum.

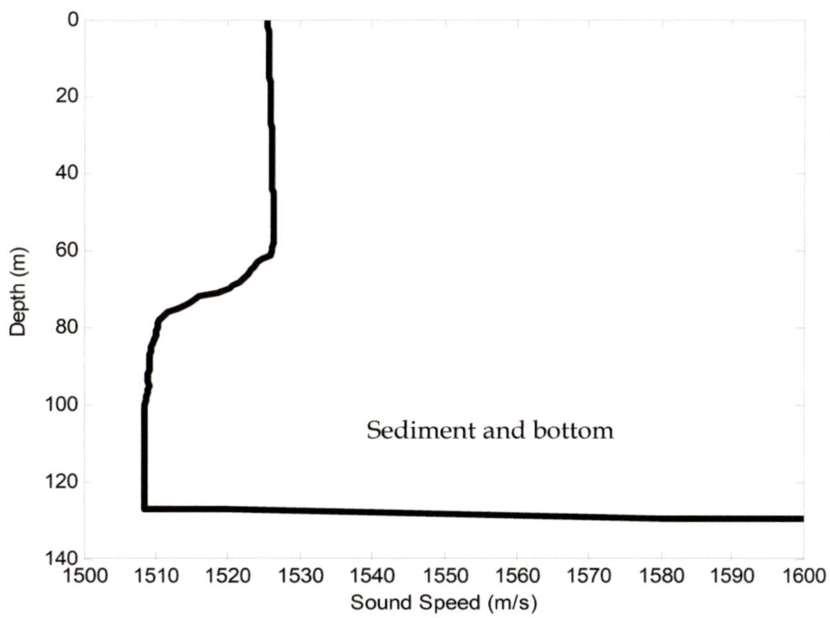


Fig. 10. Measured Sound Speed Profile of SCALANTC 1993 North Elba experiment.

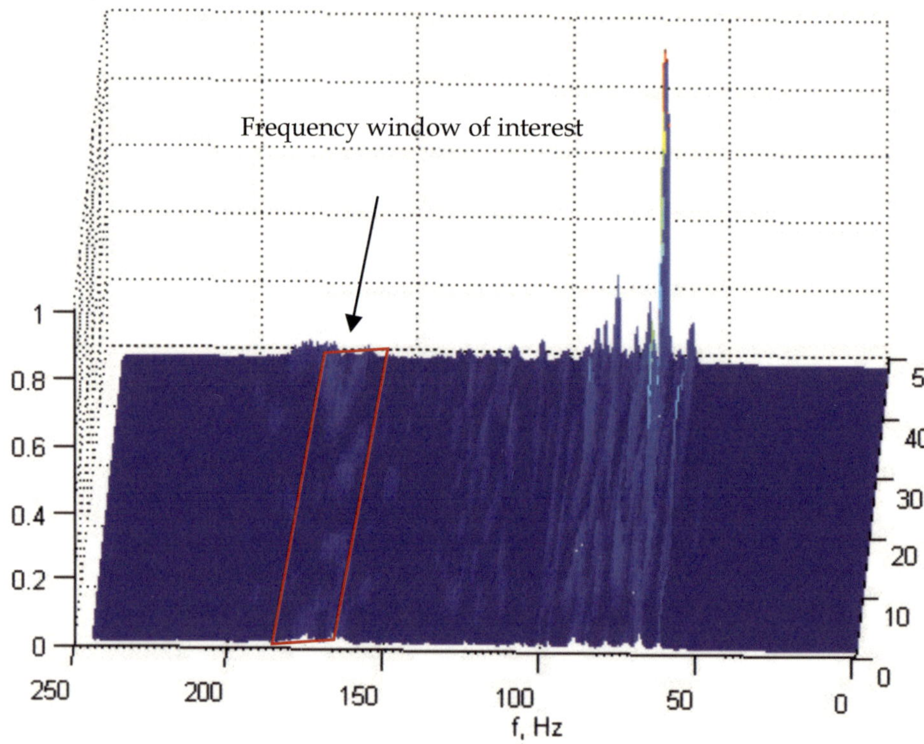


Fig. 11. CSDM for 1 minute moving source array signal.

On fig. 12 the ambiguity surfaces of three matched field processors: Bartlet, MVDR with diagonal loading and MVDR with reduced rank SVD and pseudoiverse matrix are plotted. The position of the source is determined with the maximum in the ambiguity surface. All data was equally treated with rank of the reduced matrix r = 5, and diagonal loading constant $\varepsilon^2 = \sigma^2 = 10$. Comparison of the three processors side lobes for an equal depth cut of the ambiguity surfaces is given on fig. 13. The level of the side lobes is lower when introducing spatial constraints with rank reduction of the inverse matrix. In the source localization results on fig. 12 the Bartlett processor is used as a benchmark. Even it has more side lobes in the ambiguity surface there is global maximum corresponding to the source position. During the processing of the available array data it was established that the accuracy of passive localization with the narrow band MFP is very sensitive to environment parameters change. Little inaccuracy of SSP or bottom geo-acoustical parameters leads to substantial errors in localization. The images on fig. 12 are for the moving source data recorded on the 27-th of October. The available environment file on the web site is used in the computations (Saclant Sonar Data, 1996). If in the environment file the SSP is changed with the data from another file - "sspprofiles.txt" for the 27-th of October more than one and halve thousand meters offset in the source range occurs.

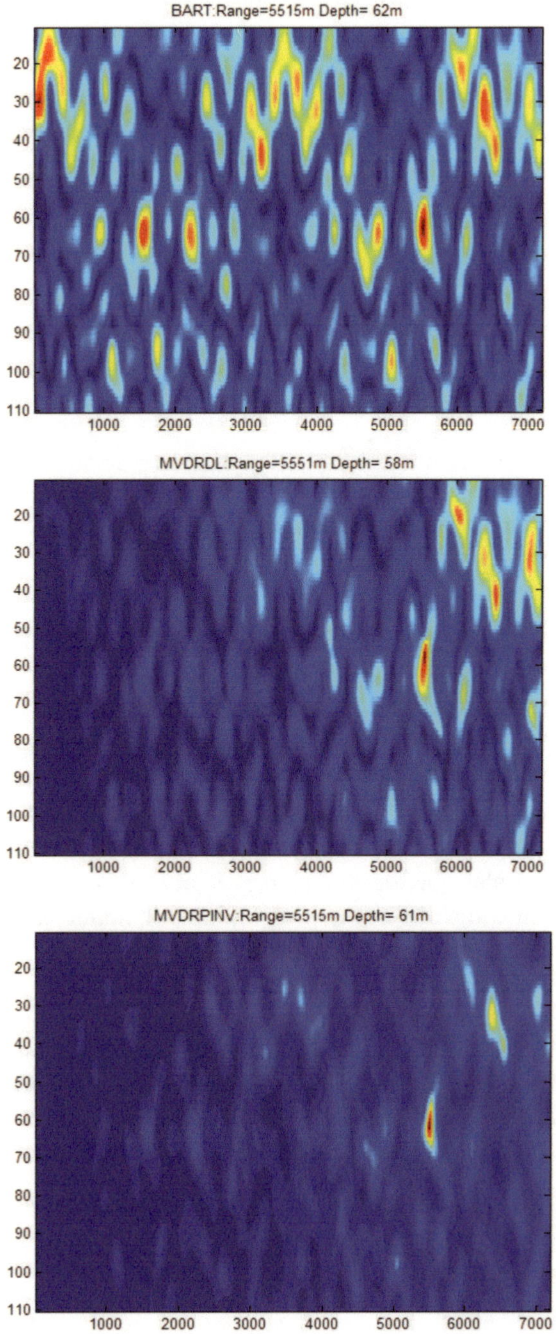


Fig. 12. Comparison of ambiguity surfaces localization results of the three processors

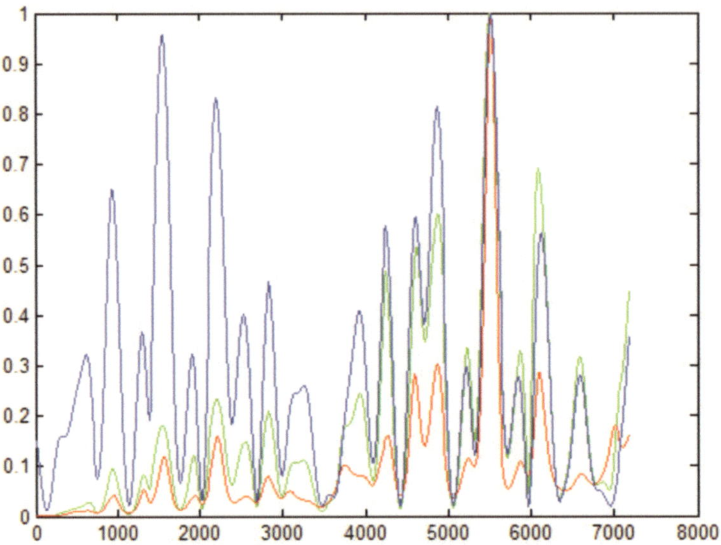


Fig. 13. Comparison of the normalized power of the three processors MFP-Bartlet - blue, MFP-MVDRDL - green and MFPPINV - red.

## 4. Conclusion

The basics of matched field signal processing were presented. Three matched field processors were applied on real experimental vertical array sonar data for source localization purposes. The MFP-MVDR processor has a lower side lobe levels especially applied with rank reduction and pseudo-inverse matrix diagonal loading (see fig. 13). For the investigated ranges and shallow water waveguides however the sound structure in range is reverberant and repetitive resulting in multiple ambiguity surface local maximums or false alarms. The current state of the technology leads to development of underwater sensor networks with application of bistatic and multistatic sonar concepts. Results from measurements with a bistatic sonar experimental setup are presented in (Kolev et al. 2009), (Kolev et al. 2010). The interest for developing adaptive beamforming methods is as with uniform but also with non-uniform and random sparse (ad-hoc) arrays with distance between nodes bigger than a half-wavelength (Hodgkiss, 1981, Gerstoft, P & Hodgkiss, 2011). MFP methods are subject to development for application in these underwater networks.

## 5. Acknowledgment

The authors are grateful to SACLANT centre scientists who have made experimental sonar array data and normal mode propagation codes available on the web.

## 6. References

Bendat J., Persol A.(2010), *Random Data: Analysis and Measurement Procedures*, John Wiley & Sons, 2010. ISBN : 978-0-470-24877-5

Booth O., Abawi A., Schey P., Hodgkiss W., (2000) Detectability of Low-Level Broad-Band Signals Using Adaptive Matched-Field Processing with Vertical Array Aperture Arrays. IEEE Journal of Ocean Engineering, Vol 25, No 3, JULY 2000. ISSN 0364-9059.

Ferla, M.; Porter, M. & Jensen F., (1993) C-SNAP: Coupled SACLANTCEN normal mode propagation loss model. SACLANTCEN Memorandum SM-274, SACLANT Undersea Research Centre, La Spezia, Italy, Dec. 1993.

Gerstoft, P. & Hodgkiss, W., (2011) Improving beampatterns of 2D random arrays using convex optimization, JASA Express Letters, March 2011.

Gerstoft P., Hodgkiss W., W., Kuperman., W., Song., H. (2003), Adaptive Beamforming of a Towed Array During a Turn. IEEE Journal of Ocean Engineering, Vol. 28, No. 1. January 2003, pp. 44 - 54. ISSN 0364-9059.

Gerstoft P., Gingras D. (1996), Parameter estimation using multifrequency range-dependent acoustic data in shallow water. J. Accoust. Soc. Am. 99 (5) May 1996.pp. 3589-3596. ISSN: 0001-4966

Gingras D., Gerstoft P.(1995), Inversion for geometric and geoacoustic parameters in shallow water: Experimental results. J. Accoust. Soc. Am. 97 (6) June 1995.pp. 3589-3596. ISSN: 0001-4966

Hodgkiss, W. Random Acoustic Arrays. (1981) In: *Underwater Acoustics and Signal Processing, Proceedings of the NATO Advanced Study Institute,* L. Bjørnø, (Ed.), 327-347, Reidel Publishing Company, Dordrecht, Holland, 1981, ISBN 9789027712554.

Klemm R. (2002), *Principles of space-time adaptive processing.* The Institution of Electrical Engineers, 2002. ISBN 0 85296 172 3.

Klemm R. (1993), Interrelations Between Matched-Field Processing and Airborn MTI Radar. IEEE Journal of Ocean Engeneering, Vol. 18, No. 3. July 1993., pp. 168- 180. ISSN 0364-9059.

Kolev N., Georgiev G.(2007), Reduced Rank Shallow Water Matched Field Processing for Vertical Sonar Array Source Localization. 15-th International Conference on Digital Signal Processing, Cardiff, UK, pp. 87- 90, 2007. Digital Object Identifier: 10.1109/ICDSP. 2007. 4288525.

Kolev N., Iliev I., Ivanov P.(2009), Experimental investigation of an underwater sensor network with radio control on the basis of 802.11. Proceedings of the National scientific conference "ACOUSTICS 2009", pp. 81-87. ISSN 1312-4897.

Kolev N., Kaloyanchev P., Iliev I., Ivanov P. (2010), A Multistatic system for underwater surveillance, Proceedings 5-th international scientific conference HEMUS 2010, Plovdiv, Bulgaria, 2010, pp.245-254, ISSN 1312-2916.

Porter M. (2007), Kraken Normal Mode Program, retrieved from Ocean Acoustic Library http://oalib.hlsresearch.com/Modes/index.html

Porter M. (1992) The *KRAKEN* Normal Mode Program, NRLIMRI5120-92-6920, Retrieved from http://stinet.dtic.mil.

Saclant Sonar Data (1996), retrieved from Signal Processing Information Base (SPIB) http://spib.rice.edu/spib/saclant.html.

Sullivan J., Middleton D. (1993), Estimation and Detection Issues in Matched Field Processing, IEEE Journal of Ocean Engineering, Vol. 18, July 1993, pp. 156-167. ISSN 0364-9059.

Van Trees H. (2002), *Optimum Array Processing,* John Wiley & Sons, 2002. ISBN 0-471-09390-4

# 7

# Image Processing Techniques for the Detection and Classification of Man Made Objects in Side-Scan Sonar Images

Esther Dura
*Institute of Robotics and Technologies of Information and Communication,*
*University of de València*
*Spain*

## 1. Introduction

This is a review chapter that surveys past work in, and the recent status of image processing and other related techniques involved in the detection and classification of man made objects in side scan sonar images. Side scan sonar is a readily, available and cheap device which has found increasing applications, specially for military purposes such as Computer Aided Detection (CAD) and Classification (CAC) of mines. Therefore the main focus of the chapter is on this topic. The list of references is sufficiently complete to include most past and recent publications in the open refereed literature.

Although side scan sonar displays many features similar to an optical sensor from a purely image processing point of view, the basics of the physics and formation of the images are crucial for understanding the difficulties found when detecting and classifying mine like objects (MLO's) in side-scan sonar images. Therefore in the first part of the chapter a brief review of the principles of the side-scan sonar, image formation process and characteristics of the images are explained. Different types of sonar images as well as diagrams showing the the process of generating an image from a single diagram ping will be provided.

The classification and detection of MLO's is traditionally carried out by a skilled human operator. This analysis is difficult due to the large variability in the appearance of the side-scan images as well as the high levels of noise usually present in the images. With the recent advances of Autonomous Underwater Vehicle (AUV) automatic techniques, CAD/CAC of mines, are now required to replace a human operator.

In the literature the computer aided detection/classification (CAD/CAC) problem is not well defined as detection involves an element of classification (mine/not mine), therefore these terms must be defined. For the purpose of this work, we will consider detection as the process of identifying a mine and classification will be a further step where the aim is to determine the shape of such a mine. Therefore the second part of the chapter is divided into two main sections: 1) detection 2) classification of MLO's.

In the last part of this chapter a review will be done on the current state of fusion of multiple algorithms aiming to overcome the limitations and weaknesses of every single CAC/CAD algorithm reviewed in the previous section.

## 2. Principles of the side-scan sonar: image formation process and characteristics

The section starts by introducing the basic principles of sonar. The following sections present the fundamental side-scan sonar characteristics along with the image construction, characteristics of the images, frequency and resolution of the images.

### 2.1 Basic principles

Devices which use underwater sound for communication or observation are generally referred to as SONAR systems. This term was coined after the Second World War to provide analogy to the equivalent electromagnetic-echo location system of radar and is an acronym for "SOund Navigation And Ranging".

In general the basic principles of a sonar involves the transmission of a pulse energy into the water medium and the subsequent reception of any returned energy reflected from objects or seabed.

Basically the sonar generates a short electrical pulse, in the form of an acoustic wave centred at particular frequency, length and energy, by the transmitter. This electrical signal is transformed by the transducer, which is normally a piezo-electric ceramic, into mechanical vibration energy. This vibration is transferred into the water as an oscillating pressure, the pulse. The pulse travels trough the water until it is reflected back or scattered by the seafloor or any object. The energy reflected back, which is mechanical energy, is converted by the transducer into electrical energy. This energy is then detected and amplified by the receiver of the sonar.

There is a master unit, with a control function, in charge of synchronizing the operations and control timing for the transmission and reception of the electrical signals. The control unit normally has a unit to display the received data.

It should be noted that what the sonar is measuring is the time that it takes for the transmitter sonar pulse to travel from the transducer to the target and return. It is not measuring the depth or distance.

### 2.2 Side-scan sonar characteristics

The fundamental purpose of a side-scan survey is to provide images which map a visible representation (intensity of marking) of the strength of acoustic back scattering, from the sea floor onto a two-dimensional image medium, by the process illustrated in 1. These sensors are usually mounted onto a separate body which is towed through the water behind the survey vessel. Alternatively the transducers may be mounted onto Remote Operated Vehicles (ROV) or Autonomous Underwater Vehicles (AUV) allowing more accurate positioning and motion of the vehicle. The characteristic of the side-scan sonar comes from implementing the basic principles mentioned. The main feature of this sensor, as can be seen in figure 2 is that is a side-ways looking device. Each pulse of acoustic energy emitted causing echoes from an area of the sea bottom perpendicular to the direction of travel of the tow fish. The transducers are normally shaped and controlled to produce a beam for each emitted pulse which is narrow in the horizontal direction and wide in the vertical direction as illustrated in figure 2. Due to the narrow horizontal beam, returned energy is received from one strip of the seafloor. The wide vertical across trace beam permits the ensonification of a large area of the seafloor. Another of the characteristics of the side-scan sonar is that sometimes two channels are used to gather information at the same time from the seabed on either side of the tow fish.

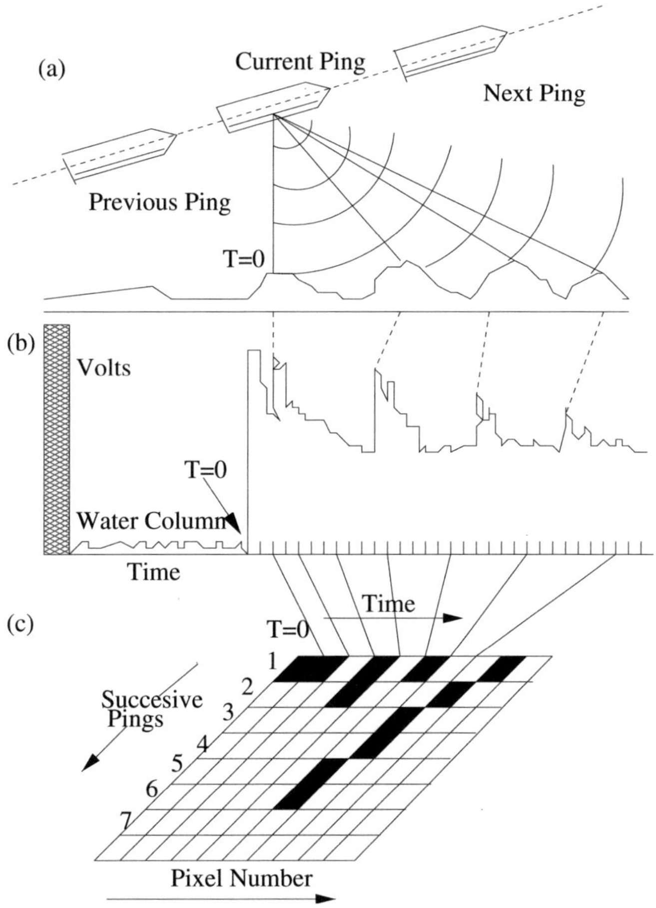


Fig. 1. Three-phase diagram showing the process of generating images from a single side-scan sonar ping. In the top diagram (a) the outgoing pulse from an individual ping is reflected back from the seafloor directly under the fish, and the internal side-scan clock (T=0) is started. (b) The hatched region represents the outgoing pulse, and the low amplitude returns are the time when the pulse is the two-way travel time in the water-column. After the return of the first bottom bounce, subsequent returns appear as peaks and valleys in the transducer voltage. (c) Peaks and valleys are then integrated and translated into pixels values.

### 2.3 Beamwidth

It has been pointed out that the side-scan sonar has a beam that is narrow in the horizontal plane and broad in the vertical plane. For a typical system this must be 1 degree horizontal and 40 degree vertical beam.

The beamwitdh can give some idea of the resolution which a sonar will achieve. It is also very important to consider the overall beam pattern of a particular sonar. This will be a truer representation of the the expected behaviour.

The shape of the beam is the result of the transducer design. A side-scan transducer it normally consists of a line array of crystal elements. Each point on the faces of the crystals acts as a sound radiator. It can be though of each infinitesimal point sending out a sound

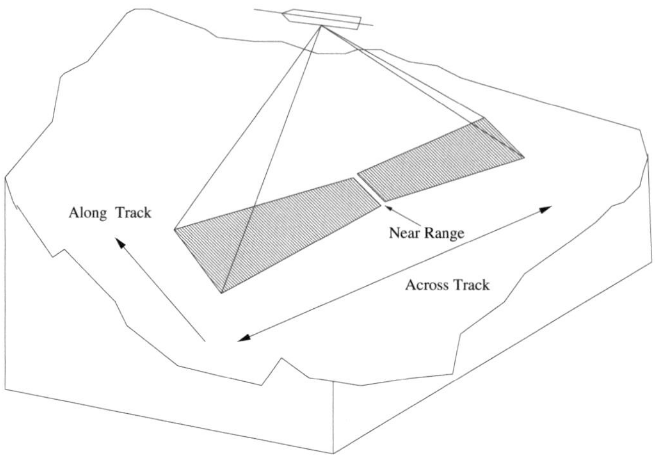


Fig. 2. Schematic diagram of the acoustic 'footprint' of a side-scan sonar system. The gray area represents the swath

pulse that spreads out in all the directions. Looking at just two of these points, as depicted in 3, it can be seen that some distance from the transducer the pressure disturbances from each of the points will meet and will either add or substract from each other depending on the phase. This process is going on for all the points of the transducer face. The net effect of all additions and substractions is to produce the beam pattern. Along the axis of the beam the pressure contributions are reinforced, while on the sides they tend to cancel. For a line of array the beamwidth can be expressed as $50.6\lambda/L$, where $\lambda$ is the wavelength of the acoustic pulse($\lambda =$ [sound velocity/$frequency$])) and L is the array length.

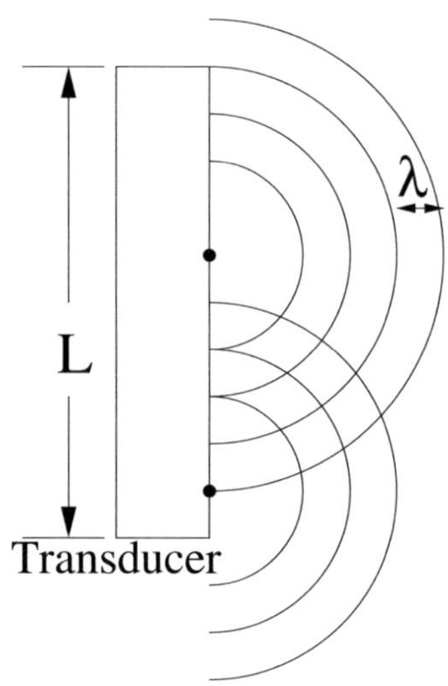


Fig. 3. Beam formation process

### 2.4 Image construction

The interpretation of the side-scan sonar image, requires an understanding of the image formation process. As the transducers are towed along, they gather sequential lines of data returned from each pulse, and these lines ('A' scan) are displayed sequentially down a vertical trace to generate an image. This systematic sideways scanning is the basic principle of side-scan sonar. The scanning occurs along track and across track. The data gathered along track is a function of the beam width, the pulse repetition rate and the tow speed. Across track the intensity is received with successively increasing two way travel times or time of flight. The intensity received is dependent on attenuation of sound in water, the direction and angle from which the target was ensonified and the reflectance properties of the seafloor. With rock and gravel acting as stronger reflectors than soft sediments such as mud and sand.The intensity of the return is displayed against the two way travel time, or time of flight.

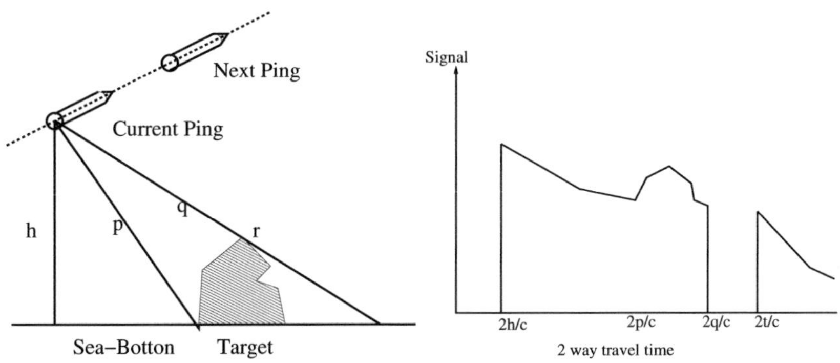


(a) Geometry of side-scan sonar system (b) Corresponding 'A' scan

Fig. 4. Geometry of side-scan sonar system and corresponding 'A' scan

Figure 4(a) illustrates the geometry of the side-scan sonar system and figure 4(b) displays the returned intensity against the two way travel of time in the form of a 'A' scan. An 'A' scan is simply one line of a sonar images corresponding to the returned energy from a narrow strip of the seabed due to the reflections from one emitted pulse.

At the beginning of the trace there is a blank area, as the pulse propagates through the water column without returning any echo. The first bottom return is the first echo to return from the sea bottom closest to the transducer. For a relatively flat seabed the first return is from the seabed directly below the transducer and it occurs at approximately time 2h/c seconds where h is the height of the transducer in metres and c is the velocity of sound. The first return is then followed by successive echos at successively increasing slant range, where the slant range is the actual distance from the sonar to the point of the seabed from which the sound was reflected. These points are followed by successive echoes at increasing slant ranges as the sound wave propagates, as illustrated in figure 4 (b).

Once the "spike" of the high-amplitude bottom bounce is received, the side-scan processor begins to divide the transducer voltage time series, which is produced by the subsequent bottom return signals, into equally spaced "time" slices. Because of the geometric effect illustrated in figure 1 and 2, these time slices represent extremely narrow regions of the seabed for the early returns and much wider regions for the later returns. Within each time slice, the varying voltage of the transducer represents the acoustic energy from a fairly large area of the seafloor, and are much larger than that represented by the pixel size of the final image. The Voltage within each individual time slice is averaged (see figure 1 (b)) and then converted to a single digital number that is assigned to a specific pixel location as illustrated in figure 1(c). In practice, the conversion from uncorrected transducer voltage to spatially correct pixel value is more complicated than this description. The signal received from the seabed return

is normally passed through an amplifier with a time varying gain (TVG). This compensates for the effects of absorption of sound by water and the geometric effects of spreading and scattering.
The process described, varies significantly from system to system, and requires a variety of corrections to become an intelligible image. For most side-scan systems, there are approximately 1024 pixels per side, or 2048 total pixels in the full swath. Depending on the system, each pixel value is usually an 8-bit integer (ranging from 0 to 255, or 256 possible shades of gray), which represent the value of the received acoustic echo after detection and the electronic low-pass filtering associated with time slice averaging.

## 3. Characteristics of side-scan sonar images

As mentioned above the side-scan sonar images are typically displayed as grayscale images, with dark and bright areas representing features of the seabed and water column.
The orientation of the target relative to the direction of the incoming pulse will influence the intensity of the reflected signal and consequently the intensity in the image. The closer the inclination of the surface normal of the target to the direction of the incoming pulse, the greater the energy. Objects protruding above the seabed will create high intensity returns, highlight (see figure 5) but will prevent the sound from reaching the seafloor for some distance behind them. This will produce an acoustic shadow in the images (see figure 5 ) and will appear on the trace as blank area. Shadows can also be generated by depressions on the seafloor or by the self shadowing of the seafloor.

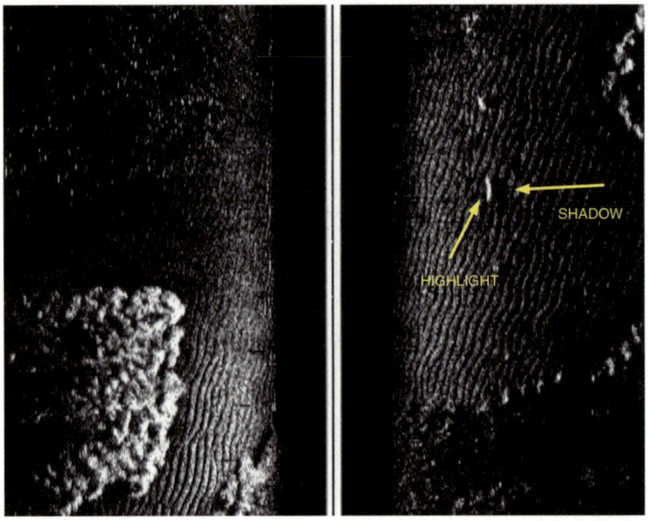


Fig. 5. Example of side-scan sonar images containing man-made objects: a mine like object is identified by a highlight followed by a shadow region

Shadows are one of the primary features which provide three dimensional information and their position and shape contain valuable information for the accurate interpretation of the images.
The side-scan sonar images, as it was mentioned, are essentially a 'picture of the seafloor' but they are usually distorted. In order to become recognizable, image pixels need to be corrected for a variety of effects. These include slant range correction (compensating for the equal time slice interval, which result in unequal distance slice interval), absorption of the sound in sea

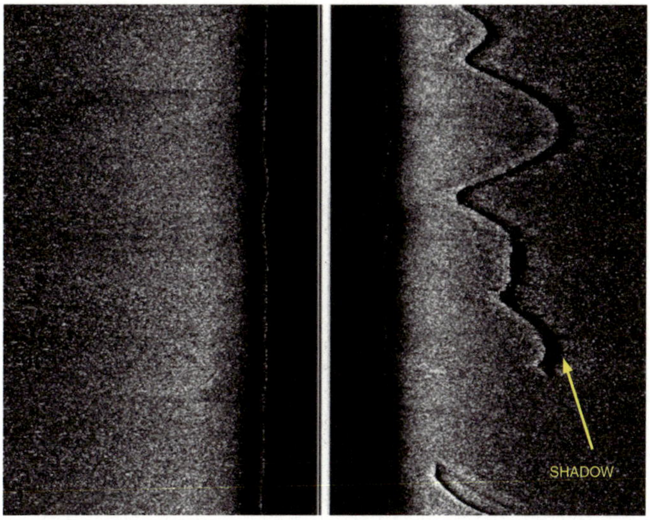


Fig. 6. Example of side-scan sonar images containing a man-made object: a pipe

water, the geometric effect of spreading (compensated by TVG), noise (potentially associated with telemetry errors, multipath effect which occurs when signals arriving at the same time as normal backscatter signals from beyond the target are superimposed resulting in the elimination of the shadow in the record, and side-lobes returns for both horizontal and vertical plane. Horizontal side lobes can direct significant energy in other directions which could then reflect of the seabed/target not perpendicular to the direction of travel and produce reflected energy. On the other hand in the vertical plane we want the main beam to give uniform energy only over the are we are interested in. If we have side lobes they may be for instance be pointing towards seasurface. This can produce backscattered data from sea surface which can interfere with backscattered signal from seabed. Because sea surface is a good reflector this produce noise. This is probably a much larger cause of noise than horizontal effects which may cause more blurring of the image than noise. External interference is caused by other acoustic devices operated at the same time which are added to the process Cervenka & de Moustier (1993), and variable ship speed. The final image after the corrections has a 1:1 aspect ratio and one that has the sonar targets roughly positioned at the same location on the chart as they are in the seafloor. Further details of these corrections can be obtained from Johnson & Helferty (1990)Somers & Stubbs (1984)Mazel (1985)Bell (1995)Cervenka & de Moustier (1993)

## 4. Detection of mines

The general approach for detecting targets is a two-tier process: 1) Detection of possible MLO's (regions of interest (ROI)). 2) Classification into mine or not-mine like objects with a low detection rate of false alarms. Both stages are crucial in order to get a good detection rate. For the first stage, detection, several approaches based on segmentation techniques and matched filters are reviewed.

The various techniques used in the literature for the second stage, classification, can be divided into three categories: unsupervised, semi-supervised and supervised algorithms. A comprehensive review of these techniques will be done. These techniques require the extraction of mine features, therefore a review of the most discriminant features used in the literature will also be explained.

### 4.1 Detection of possible MLO's

As stated before this first step involves identifying regions of interest that may contain a mine. Two main approaches are used in the literature for this purpose: 1) segmentation and 2) matched filtering.

#### 4.1.1 Segmentation

*Segmentation* is the process of classifying pixels as belonging to a certain class. In side-scan sonar mages the classes of interest normally are: highlight and shadow. Because of the shadow cast by a side-scan sonar appears more consistent than the highlight, some of the most successful algorithms rely only on the shadow information. The main techniques used for segmentation are: 1) thresholding 2) clustering and 3) Markov Random Fields.
*Thresholding* is the simplest method on image segmentation. During the thresholding process, individual pixels in a side-scan image are marked as shadow pixels if their value is greater than some gray pixel value (assuming shadow to be darker than the background) and as no-shadow pixels otherwise Quidu et al. (2000). Some approaches use two thresholds values to segment images into shadow, highlight and background regions.
*Clustering* a procedure to determine the intrinsic grouping in a set of unlabeled data, has also been used to segment the images into three categories (shadow, highlight and background). In this technique, a feature vector for each pixel of the image is extracted and then a similarity metric is used to cluster vectors having similar features Guillaudeux et al. (1996)
Unlike previous methods, *Markov Random Fields* provides a reliable framework for incorporating pixel dependencies into the segmentation (i.e a pixel surrounded by a shadow is most likely to belong to shadow itself). This ability to model inter-spacial dependencies between pixels has ensured the use of MRF models for a range of applications. In the context of side-scan sonar images where there is a large variation in the appearance of the images, more complicated models have been used Mignotte & Collet (1999)Reed et al. (2003). These models include a priori knowledge: object highlight generally lies close to shadow regions. One of these studies, Reed et al. (2003), introduced the size and appearance information as a priori information into the model. In this study, after the MRF segmentation, a further post-segmentation step that provided an accurate and robust method for extracting the shadow and highlight was carried out by using a cooperative statistical snake. The model segments the object-highlight and the shadow region by considering the image as being composed of three different statistical regions. The main advantage that this method presented compared to other models, was that using a priori information on the relationship between the object-highlight and shadow, accurate segmentation was achieved on seabed types where other models failed. Details of this implementation can be found in Reed et al. (2003)
An example of MRF segmentation on side-scan images containing mines can be visualized in figure 7(b).

#### 4.1.2 Matched filtering

It is a technique for finding small parts of an image which match a template. This is done by convolving a known template with an image to detect the presence of the template in the image. The identification of mine-size regions in the sonar image has been carried out, as explained inDobeck (1997), by convolving a template that contains four distinct regions:(pre-shadow, highligth, dead zone, shadow and post-target) with the image. After that a threshold is applied to the post-processed images and neighbour pixels over a threshold are grouped together to identify possible MLO's. The threshold varies between the different

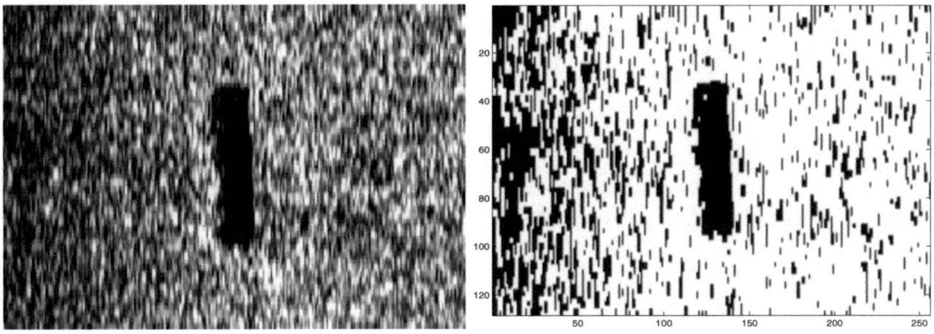


(a) Original side-scan image

(b) Segmentation result using the MRF model proposed by Mignotte & Collet (1999)

Fig. 7. MRF segmentation

detectors, and is fixed according to the desired sensitivity of each detector. Details of this implementation and similar approaches can be found in several works Dobeck (1997)Dobeck (2000)Hasanbelliu et al. (2009)

### 4.2 Classification into mine or not-mine

By using any of the techniques explained above the regions of interest of the image, that may contain MLO's, are identified. Afterwards these regions of interest have to be classified into mine or not mines. This classification procedure normally requires the extraction of mine features, therefore a review of the most discriminant features used in the literature will first be explained in the following section.

#### 4.2.1 Feature extraction

In pattern recognition feature extraction is a special form for reducing dimensionality of an image. For side-sonar images the aim of feature extraction is to extract some characteristics that describe a region of interest that may contain a MLO. The main feature used in the literature Aridgides et al. (2001a)Dura et al. (2005)Dobeck (1995)Zerr & Stage (1996)Quidu et al. (2000) for extracting features fall into two categories: 1)shape features 2) gray-level features

Shape features characterize the appearance and geometry of an object. MLO as opposed to non man-made object cast regular shadows and highlight of anticipated dimensions. The following features are mainly used in the literature for extracting some features from the shadow and highlight information:

1)Area: is the surface area of an object , $O$,(shadow or highlight), defined as:

$$Area = \sum_{i,j} O(i,j)$$

where O(i,j ) has a value of one for a pixel in the object and zero if not.

2)Elongation represents the ratio of major axis to that of the minor axis. It is computed from second order central moments as:

$$Elongation = \frac{\sqrt{4\mu_{11}^2 + (\mu_{20} - \mu_{02})^2}}{\mu_{20} + \mu_{02}}$$

where $\mu_{pq}$ stands for the central moment of order $p+q$ which is computed as:

$$\mu_{pq} = \sum_{i=0}^{M}\sum_{j=0}^{N}(i-i_g)^p(j-j_g)^q I(i,j)$$

where $(i_g, j_g)$ is the position of the center of mass of the shadow or highlight and $I(i,j)$ is the digital image. This position is calculated as:

$$i_g = \frac{m_{10}}{m_{00}}$$

and

$$i_g = \frac{m_{01}}{m_{00}}$$

and the two-dimensional moment $m_{pq}$ of order $p+q$ is defined as

$$m_{pq} = \sum_{i=0}^{M}\sum_{j=0}^{N} i^p j^q I(i,j)$$

3) Circularity or shape factor: is a measure of circularity or the compactness of a shape and can be calculated as:

$$Compactness = \frac{4.\pi}{p^2}$$

4)Orientation: the orientation of an object can be defined as:

$$Orientation = \frac{1}{2}tan^{-1}\frac{2\mu_{11}}{\mu_{20}-\mu_{02}}$$

5)Eccentricity: Is the ratio of the length of the longest chord of the shape to the longest chord perpendicular to it.
6)Rectangularity: This shows how well a region is approximated by a rectangle. The rectangularity measure $f_{rect}$ is the ratio of the area of a region, $A$, to area of the smallest rectangle, $A_{rectangle}$, that encloses it:

$$f_{rect} = \frac{A}{A_{rectangle}}$$

7)Number of zero crossing of the curvature at different scales: for a fixed length, a small number of curvature zero crossing suggests as simple regular contour; a high number, suggests a tortous, irregular, frequently turning contour. Therefore scanning the curvature scale space of a given set of contours from fine to coarse scales, regular shapes can be identified.
The number of zero crossing of an image is obtained by applying the convolution operator $\nabla^2 G$ (which is the Laplacian of a two dimensional Gaussian G(i,j)) over the image $I(i,j)$ as:

$$I(i,j)' = \nabla^2 G(i,j) * I(i,j)$$

where

$$\nabla^2 G(i,j) = (\frac{r^2-\sigma^2}{2\pi\sigma^2})exp\frac{-r^2}{2\sigma^2}$$

where

$$r = (i^2 + j^2)^{\frac{1}{2}}$$

Some individual features and others that relate shadow and highlight information has also been used such as:
Shadow high profile, ratio of the highlight to shadow area, ratio of highlight to shadow height, minimum distance between highlight and shadow, horizontal aligment between shadow and highlight
When the quality and resolution of the images is low, these are not well characterize by the profile (shape features) of the shadow and highlight. Therefore gray level features extracted from the shadow and highlight are also used for discriminating targets from clutter. Among them the most discriminative features used are:
1) *Average shadow* strength, which is a measure of the object's shadow darkness
2) *Average highlight strength*, which is a measure of the object's highlight brightness
3)The *variance* of the *shadow*
4)The *variance* of the *highlight*
5)*Contrast* between *shadow* and *highlight*, which is the absolute difference of the average shadow strength and average highlight strength
6)*Contrast* between *shadow* and *background*, which is the absolute difference of the average shadow strength and average background strength
7)*Contrast* between *highlight* and *background*, which is the absolute difference of the average highlight strength and average background strength

#### 4.2.2 Classification

At this step classification is the detection of MLO's in side-scan images. There various techniques examined can be broadly divided into three groups: supervised, semi-supervised and unsupervised.

*Supervised*

With these techniques, one typically requires an a priori-set of training data consisting of a set of features and associated binary labels( mine/clutter) and a testing set to validate the results. A supervised learning algorithm analyzes the training data and produces an inferred function, which is called a classifier. To constitute a training set, known targets (e.g mines) must be emplaced in a given environment and side-scan data collected with all nonemplace scatters are assumed to be clutter. In the context of sonar images, the difficulty of this supervised classification procedure resides 1) in the very large number of mine types, mine deployments and orientations, and mines history; 2)the significance dependence on the nature of the training data, specially the dependence of the imagery on the sea bottom environment. The variability of 1) and 2) makes it impossible to constitute a training set that is robust to all type of mines and environments to be encountered. An algorithm trained for one type of sonar setting may perform poorly when used in another environment. Besides, the point at which a 'training data set' becomes sufficiently large is difficult to define. To overcome this problem some researchers Reed et al. (2004) and Coiras et al. (2007) have generated their data set of synthetic side scan images with inserted random mines at random locations and orientations. The mines inserted had realistic shadows and highlights that took into account the angle of incidence and topography of seabed.
Pioneering research on supervised detection/classification of MLO's was carried out by Dobeck (1997).They used a K-nearest neural network (KNN) and an optimal discriminatory filter classifier (ODFC). The KNN technique involved a two-layer neural network, which

classifies features according to the proximity of the features to a 'feature vector center'. These classifiers were then combined to yield the final result. Results and details can be found in Dobeck (1997)
An adaptive filter has also been used for the detection/classification of mines based on a bayesian classifier (simple probabilistic classifier based on applying Baye's theorem, with strong independence assumptions) known as the log-likehood ratio test (LLRT)Fernandez et al. (1993). A given feature vector was assigned as belonging to either of two classes on the basis of the LLRT. This algorithm addressed the main shortcoming of the bayesian classifiers (the determination of the multidimensional distributions essential for the computation of the LLRT) by mapping the sets of learning vectors to a space of orthogonal features in order to yield histograms. These histograms were then used to get the log-likehood ratio and summed to obtain the final results.
Linear and quadratics classifiers has also been employed for supervised classification by Fawcett (2001).A linear classifier separates objects or events by a linear function whereas the quadratic classifier separate objects or events by a quadratic surface. Unlike previous methods, in the work proposed by Fawcett (2001) instead of extracting features from the regions of interest the whole image was used as a feature vector. Principal components technique was used to identify the most discriminant features of the image. Details of the classifiers, implementation and results can be found in Fawcett (2001)
Recent machine learning techniques based on kernel-based algorithms such as Support Vector Machines (SVM)Vapnik (1995) and Relevance Vector Machines (RVM)Tipping & Smola (2001) have been investigated. These kernel-based learning algorithms are based on mapping data from an original input space to a kernel feature space of higher dimensions to solve a linear problem in that space. The advantages for relevance vector machines over support vector machines is the availability of probabilistic predictions, using arbitrary kernel functions and not requiring to set many parameters. Details of the implementations and results can be found in Dura et al. (2005)Couillard et al. (2008)
It is important to highlight, that is not always necessary to use all the features; sometimes using a smaller is better than using a large set of features which are correlated. Therefore some all the supervised techniques reviewed used some optimisation procedures before the training process to determine the best combination of features.

*Semi-supervised*

Semi-supervised is a class of machine learning technique that make use of both labeled and unlabeled data for training. The amount of labeled data required for training is tipically very small compare to the amount of unlabeled data. The cost of adquiring data with the associate label data is expensive and may make a set of data infeasible, whereas the adquisition of adquiring unlabeled data is inexpensive. Therefore semi-supervised techniques introduce and important advantage: the cost for mine hunting operations is reduced. In the context of side-scan sonar this is very important as labeling the data is very expensive, a diver or unmanned underwater vehicle with a camera has to label it.
An active-learning algorithm based on semi-supervised techniques was first proposed by Dura et al. (2005). The algorithm, kernel-based, was developed with the goal of enhancing mine detection/classification of mines without requiring a priori data set. It was assumed that divers or unmanned underwater vehicles with a camera were used to determine the binary labels of a small set for a given side-scan collection. This set of data and associated label were used to train the algorithm. Information-theoric concepts were used to adaptively construct the kernel classifier and guide which data and associate label were most informative

in the context of of algorithm training (this information content is computed without a priori knowledge of the labels itself). In this work authors demonstrated that the number of data for which the associate label was required was very small relative to the number of potential targets in a given image.

*Unsupervised*

Most of the current automated systems, as stated before, require training data and thus produce poor results when the training data differ from the test set. The success of this systems depend on the similarity of the training and testing set of data. This has led research into unsupervised techniques that requires no training data. The main advantage of these systems is that they are able to cope with the large variability in conditions and seabeds seen in side-scan sonar images. Also, very important, the cost of mine-hunting operations is reduced. One of the most complete unsupervised and sucesfull systems implemented so far is the one implemented by Reed et al. (2003)Reed et al. (2004). The system was composed of two consecutive and complementary phases: 1) a MRF algorithm was employed to segment the raw side-scan image into regions of object highlight, shadow and background. A post-processing procedure was then applied to remove false alarms. Objects that were too large or small were removed. The height (calculated by using the shadow length and navigational data) was also taken into account to remove false alarms. 2)In a second phase a cooperating statistical snake model was use to consider each of the detected MLO's. The model assumed the highlight and shadow regions to be statistically separated, therefore it was enforced a dependency between the two snakes. Also their movement was constrained. If snakes expanded beyond MLO dimensions the MLO was identified as false alarm and removed. Good detection rates was obtained with this two-step unsupervised algorithm.
Another approach with tackled this problem was the one proposed by Mignotte et al. (2000). In this work a set of deformable template model which allow linear transformation were used to separate natural objects from man made objects in an image. The detection was based on a objective function measuring how well a given instance of a template fits the contents of the segmented image (previously the image was segmented using a MRF). If the result of the objective function was less than a certain threshold then the desired object was assumed to be present and the final configuration revealed the shape of the object.

## 5. Classification of MLO's

Once the mine has been the detected the following step is the classification. As stated before, classification is the process of recognizing the shape of a mine (type of mine). In sonar imagery, MLO's produce a shadow which represents a regular geometry shape. In particular the shadow cast by spherical mine almost always is an ellipse with different vertical and horizontal axis lengths. For cylindrical mines the associate shadow may be a rhomboid, rectangle or ellipse.
The classification problem has not been widely addressed in the literature. The few approaches that deal with this problem fall into two different groups: mono-view and multi-view classification depending on whether they make use of a single view or several views for determining the shape.

### 5.1 Mono-view classification

In general classical models Castellano & Gray (1990)Quidu et al. (2000)Delvigne (1992) consisting of feature extraction and classification have widely been used for mono-view classification purposes. First using a presegmented shadow a mine a set of features are

extracted from the shadow. Afterwards the set of features are normally presented as input to the classifier. Althougth feature based are appealing the performance of the classifiers depend to a significant extent upon the feature extraction.
A totally different approach based on available properties of the shape (as a prior model) and an observation model (likehood model) was proposed by Mignotte et al. (2000). In such terms they proposed two prototype templates, square and ellipse, along with a set of transformations, to take into account the shape variability of different for every type of mines. The classification of an object was based on an objective function measuring how well an instance template fitted the content of the segmented image.
Along similar lines, Balasubramania & Stevenson (2001) did some interesting work on model fitting. In this work it was assumed that the shadows from targets such as cones, cylinders and rocks were close to an ellipse. Hence the shadow shapes were modelled as ellipses. To this end the edges of the shadow regions were extracted and the elliptical parameter fitting was performed using Karhonen-Loeve method. Then the parameters were used as features to describe the ellipse. Althought this approach is relevant for spherical mine-like shapes, it is not the best to provide good separation class.
In the work proposed by Dura et al. (2008)they also advocated for a model-fitting approach by modelling the mine-like shadow with a superllipse. Superellipse provide a compact and interesting approach for representing a variety of shapes. By simply varying the squarness of the function shapes such as ellipses, rectangles, parallelograms, ovals and pinched diamonds can be easily generated. Thus, based on these observations, a classification procedure was proposed based on the squareness parameter. The procedure extracted the contour of the shadow given by an Unsupervised Markovian segmentation algorithm. Afterwards a superellipse was automatically fitted by minimising an appropiate metric with the Nelder Mead Simplex optimization technique. Some results can visualized in figure 8
Another approach has recently been investigated by Reed et al. (2004) and Coiras et al. (2007). A synthetic database of side-scan sonar images was generated with a sonar simulation under different conditions: seabed types, mine orientations and sizes. Then a classifier Coiras et al. (2007) was trained on the features extracted from the synthetic images generated. Afterwards real side-scan sonar images were classified. In Reed et al. (2004) instead of using a supervised classifier, the Hausdorff technique was implemented to measure the resemblance between the features of the synthetics images generated and the real images.

### 5.2 Multi-view classification

Sometimes is possible to obtain an accurate classification relying on a single view of an object. However some uncertainty of the object true class remains. In particular for sonar images if more than view of an object is provided at different angles this uncertantity can be reduced.
The fusion of multiples images for classifying an object has been investigated by various authors. One of the most extensive work on multi-view classification was the one undertaken by Fawcett et al. (2010). In this work they extracted two features sets corresponding to two different view side-scan images and they investigated two approaches for fussing this information: 1)fuse-feature and 2)fuse- classification. In the first approach the two feature sets were combined to form a large feature vector (CF). Then a kernel based classifier was trained a tested with the resulting extended feature vector. The second approach consisted of fussing the two individual-aspect classification of the two feature vectors using Dempster-Shafer (DS) Theory. DS, frequently used as alternative to Bayesian theory and fuzzy logic for data fusion, allows to combine evidence from different sources and arrive to a degree of belief (represented

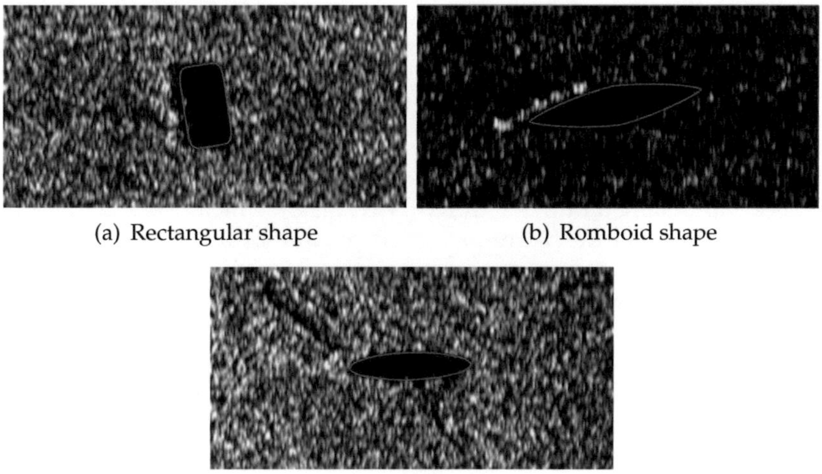

(a) Rectangular shape (b) Romboid shape

(c) Elliptical shape

Fig. 8. Results of the Superellipse fitted algorithm proposed by Dura for different shadows shapes

by a belief function) that takes into account all the available evidence. The belief function is derived from a mass function, which is analogous to well known probability density function. In this work, three different approaches were examined to calculated DS masses for each of the two looks of the classifier. In the first approach the histogram of the output labels for a single-aspect kernel regression classifier were used to empirically determine a simple analytical function which converted the values of the multiclas outputs into a set of masses. In the second approach they used a confusion matrix obtained from the single aspect classifier to specify the DS masses for various objects, for each of the looks, given the single aspect classifications. For the last approach, a nonempirical mass assigment based upon the relative values of the classifier outputs were considered.
Along similar lines Zerr et al. (2001) Stage & Zerr (1996)andReed et al. (2004) have also investigated the classification of a target by fusing several views using DS theory. However in the work investigated by Reed et al. (2004) the mass functions were generated from a fuzzy functions membership algorithm based on fuzzy logic.

## 6. Fusion of detection algorithms

The detection algorithms described in section 4.2.2 have their own weakness and strengths. This is due to the fact that each algorithm is based on different statistical properties and therefore emphasizes different characteristics of the data. Thus a combination of them may increase the probability of detection of MLO's and consequently reduce the number of false alarms.
Various methods of fusion of algorithms have been studied. In the work presented by Aridgides et al. (2001b), the results of different detection algorithms developed by three research teams (Naval Surface Warafe (NSWC), Coastal Systems Station (CSS), Raytheon and Lockeed Martin ) were fused. Three different strategies were examined: 1)Logic-based fusion , 2)m-out-of-n fusion. 3) Log-Likehood Radio Test (LLRT)-based fusion algorithm. The logic-base fusion strategy was based on a set variety of rules including Boolean operators, AND, OR and their combinations. The m-out-of-n fusion was based of a particular instance

of this, '2-out-of-3'. This means that a target was included in the overall result if at least 2 out of the 3 algorithms tested also detected it. The LLRT-based strategy, originally developed by Fernandez et al. (1993)to perform perfusion, utilized the three confidence output vectors to form a three dimensional vector which were then processed through an orthogonalization and matrix extraction procedure to yield histograms for each orthogonal feature. These histograms were then used to obtain the log-likehood ratio and sum to obtain the overall result (detection). Best perfomance was obtained utilizing LLRT-based fusion which resulted in a 3:1 false alarm reduction improvement over the '2-out-of-3' strategy and 4:1 improvement over logic-based function.

Another different approach which also combined the output of the three detection algorithms previously mentioned, was the one investigated by Ciany & Huang (2000). The fusion algorithm received the two dimensional coordinates and confidence value for the detection mines and a geometrical clustering algorithm was applied. The resulting clusters were then processed via cluster confidence processing to produce the final fused results , which were the position of the mines. This procedure was applied as it was assumed that valid mines would be close by whereas false alarms would appear in random position of the image

Fusion detection algorithms based on score results of each individual algorithm have also been proposed. This can be performed by a number of ways such as was suggested by Dobeck (2005) 1) majority voting where the detections can be conditioned on thresholds applied to scores, 2) computing the sum of the algorithms scores and comparing the sum to a threshold, 3) computing a linear combination of the scores and comparing the weighted sum to a threshold. In this work was demonstrated that one can afford to run individual algorithms with higher probability of detection and higher probability of false alarm that would normally tolerate, in the knowledge that the fusion process will bring the false alarm rate down.

## 7. Conclusions

In this chapter the techniques involved in the detection and classification of MLO's on side-scan sonar images have been reviewed. The main components of CAD/CAC systems have been examined. These components are: 1)Image formation and characteristics of the image 2)Detection 3) Classification and 4) Fusion of different algorithms for detection of mines. For each component successful image processing techniques as well as related areas were examined .

However some questions remain: are the current automatic systems reliable enough to detect and classify mines without the assistance of a human operator?Do they perform well under different environment conditions?

## 8. Acknowledgments

This work has been supported by grant DPI2008-06691 of Spanish Ministry of Science and Innovation.

## 9. References

Aridgides, T., Fernandez, M. & Dobeck, G. (2001a). Fusion of adaptive algorithms for the classification of sea mines using high resolution side scan sonar in very shallow water, *OCEANS, 2001. MTS/IEEE Conference and Exhibition*, Vol. 1, pp. 135 –142 vol.1.

Aridgides, T., Fernandez, M. & Dobeck, G. (2001b). Fusion of adaptive algorithms for the classification of sea mines using high resolution side scan sonar in very shallow water, *OCEANS, 2001. MTS/IEEE Conference and Exhibition*, Vol. 1, pp. 135 –142 vol.1.

Balasubramania, R. & Stevenson, M. (2001). Pattern Recognition for Underwater Mine Detection, *CAC/CAD 2001 Conference, Halifax, Canada.*

Bell, J. M. (1995). *A model for the simulation of sidescan sonar,* PhD thesis, Department Computing and Electrical Engineering, Heriot Watt University, Scotland, UK.

Castellano, A. R. & Gray, B. C. (1990). Autonomous Interpretation of side-scan sonar returns, *Proceedings of the IEEE Symposium on Autonomous Underwater Vehicle Technology Conference,* pp. 248–253.

Cervenka, P. & de Moustier, C. (1993). Sidescan sonar image processing techniques, *Oceanic Engineering, IEEE Journal of* 18(2): 108 –122.

Ciany, C. & Huang, J. (2000). Computer aided detection/computer aided classification and data fusion algorithms for automated detection and classification of underwater mines, *OCEANS 2000 MTS/IEEE Conference and Exhibition,* Vol. 1, pp. 277 –284 vol.1.

Coiras, E., Mignotte, P.-Y., Petillot, Y., Bell, J. & Lebart, K. (2007). Supervised target detection and classification by training on augmented reality data, *Radar, Sonar Navigation, IET* 1(1): 83 –90.

Couillard, M., J.A.Fawcett, Myers, V. & Davidson, M. (2008). Support vector machines for classification of underwater targets in sidescan sonar imagery, technical memorandum drdc atlantic tm 2008-190, *Technical report,* Defence Reserach and Development Canada.

Delvigne, J. C. (1992). Shadow classification using neural networks, *Proceedings of 4th Undersea Defence Conference,* pp. 214–221.

Dobeck, G. (1995). Sea mine detection and classification using side-looking sonar, *Proc. SPIE,* Vol. 2496, pp. 442 – 453.

Dobeck, G. (1997). Automatic detection/classification of sea bed mines in sonar imagery, *Proceedings of SPIE. In detection and Remediation Technologies for Mines and Minelike Targets V* 3079: 90–110.

Dobeck, G. (2000). Algorithm fusion for the detection and classification of sea mines in the very shallow water region using side scan sonar, *Proceedings of SPIE. In detection and Remediation Technologies for Mines and Minelike Targets V* 4038: 348–354.

Dobeck, G. (2005). A probabilistic model for score-based algorithm fusion, *OCEANS, 2005. Proceedings of MTS/IEEE,* pp. 2429 – 2434 Vol. 3.

Dobeck, G. (2006). The k-nearest neighbor attractor-based neural network and the optimal linear discriminatory filter classifier, *OCEANS 2006,* pp. 1 –6.

Dura, E., Bell, J. & Lane, D. (2008). Superellipse fitting for the recovery and classification of mine-like shapes in sidescan sonar images, *Oceanic Engineering, IEEE Journal of* 33(4): 434 –444.

Dura, E., Zhang, Y., Liao, X., Dobeck, G. & Carin, L. (2005). Active learning for detection of mine-like objects in side-scan sonar imagery, *Oceanic Engineering, IEEE Journal of* 30(2): 360 – 371.

Fawcett, J. A. (2001). Image-based classification of sidescan sonar detections, *CAD/CAC 2001 Conference, Halifax, Canada.*

Fawcett, J., Myers, V., Hopkin, D., Crawford, A., Couillard, M. & Zerr, B. (2010). Multiaspect classification of sidescan sonar images: Four different approaches to fusing single-aspect information, *Oceanic Engineering, IEEE Journal of* 35(4): 863 –876.

Fernandez, M., Aridgides, A. & Bourdelais, J. (1993). Algorithm for sonar-based signal identification, *OCEANS '93. Engineering in Harmony with Ocean. Proceedings,* Vol. 3, pp. 438–443.

Guillaudeux, S., Daniel, S. & Maillard, E. (1996). Optimization of a sonar image processing chain: a fuzzy rules based expert system approach, *OCEANS '96. MTS/IEEE. 'Prospects for the 21st Century'. Conference Proceedings*, Vol. 3, pp. 1319 –1323 vol.3.

Hasanbelliu, E., Principe, J. & Slatton, C. (2009). Correntropy based matched filtering for classification in sidescan sonar imagery, *Systems, Man and Cybernetics, 2009. SMC 2009. IEEE International Conference on*, pp. 2757 –2762.

Johnson, H. P. & Helferty, M. (1990). The Geological interpretation of Side-Scan Sonar, *Review of Geophysics* 28(4): 357–380.

Mazel, C. (1985). *Sidescan Sonar Training Manual*, Klein Associates, Salem.

Mignotte, M. & Collet, C. (1999). Three-Class Markovian Segmentation of High-Resolution Sonar Images, *Computer Vision and Image Understanding* 76(3): 191–204.

Mignotte, M., Collet, C., Perez, P. & Bouthemy, P. (2000). Hybrid Genetic Optimization and Statistical Model-Based Approach for the Classification of Shadows Shapes in Sonar Imagery, *IEEE Transactions on Pattern Analysis and Machine Intelligence* 22(2): 689–700.

Quidu, I., Malkasse, J., Burel, G. & Vilbe, P. (2000). Mine Classification using a Hybrid Set of Descriptors, *Proceedings of OCEANS 2000 MTS/IEEE Conference and Exhibition*, pp. 291–297.

Reed, S., Petillot, Y. & Bell, J. (2003). An automatic approach to the detection and extraction of mine features in sidescan sonar, *Oceanic Engineering, IEEE Journal of* 28(1): 90 – 105.

Reed, S., Petillot, Y. & Bell, J. (2004). Model-based approach to the detection and classification of mines in sidescan sonar, *Applied Optics* 43(2): 237– 246.

Somers, M. L. & Stubbs, A. (1984). Sidescan sonar, *IEE Proceedings* 131(3): 243–257.

Stage, B. & Zerr, B. (1996). Application of the dempster-shafer theory to combination of multiple view sonar images, *Proceedings. Undersea Defence Technology*, pp. 148–152.

Tipping, M. & Smola, A. (2001). Sparse bayesian learning and the relevance vector machine, *Journal of Machine Learning Research* 1(1): 211–24.

Vapnik, V. (1995). *The Nature of Statistical Learning Theory.*, Springer-Verlag.

Zerr, B., Bovio, E. & Stage, B. (2001). Automatic mine classification approach based on AUV manoeuverability and cots side scan sonar, *Proceedings of Goats 2001 Conference, La Spezia, Italy*.

Zerr, B. & Stage, B. (1996). Three-dimensional reconstruction of underwater objects from a sequence of sonar images, *Proceedings of the IEEE International Conference on Image Processing*, pp. 927–930.

# 8

# SONAR Images Denoising

Alexandru Isar[1], Ioana Firoiu[2], Corina Nafornita[1] and Sorin Moga[3]
[1]*Politehnica University Timisoara,*
[2]*ALCATEL-LUCENT Timisoara*
[3]*Institut TELECOM; TELECOM Bretagne, UMR CNRS 3192*
*Lab-STICC; Université européenne de Bretagne,*
[1,2]*Romania*
[3]*France*

## 1. Introduction

The proliferation of SONAR images produced by different equipments: multibeam echo sounders, side scan sonar, forward looking imaging SONAR…, [Lurton, 2002], and the large number of processing methods, created the necessity of expert systems for assisting the decision making process. An example is the SonarScope, an IFREMER expert system [Ifremer, 2010]. The basic functionality of such an expert system is the representation and the analysis of sonar data, organized as a "multilayer" structure defined by its various attributes (bathymetry, image, angles, and raw data from auxiliary sensors …). These data can be represented and processed using various techniques either classical (signal and/or image processing) or specific to SONAR. The goal of an expert system for SONAR images is to achieve three main tasks: Quality Control, Data Processing and Data Interpretation. There are two classes of signal processing techniques used in an expert system for SONAR images. The first one represents the so-called "image conditioning" methods. The first category of conditioning methods refers to the echoes acquisition, meaning the acquisition of depth, of the across and along-track distance, of the received beam angle, of the numbers of the transmitted and received beam, or of the two-way travel time of the acoustic pulse. The echo acquisition has some features like for example the bottom detection mode. The second category of conditioning method refers to the correction of acquired data. There are techniques to complete the missing data (caused by some acquisition imperfections due to unexpected displacements of the platform where the acoustical transducers are mounted or by the non-uniform movement or deviations from the course of the vessel which tracts the platform) and techniques for the correction of the differences between the directivity characteristics of the sensors. The third type of conditioning techniques consists in data organization and management. SONAR data are arranged as a set of arrows called pings (swath), which correspond to all the soundings acquired in a ping cycle. At the beginning of the ping cycle, each sensor's value is logged: gyro, pitch, roll, positions, etc.
The fourth category of conditioning methods refers to the formation of SONAR images. A SONAR image is thus linked with the sensor's time series, pertaining to that given dataset and in synchronization with each ping. The time series can be plotted as curves, displayed in conjunction with the image. There are few geometric formats used for the SONAR

images: PingBeam, LatLong, PingSamples and PingSwat. The fifth category of conditioning methods refers to the assembling of few neighboring individual images. This could be a technique to produce mosaics or digital terrain models (DTM). Finally the sixth conditioning category of methods refers to the angular correction and despecklisation of SONAR images. The last category of image processing methods represents the aim of this chapter. The second class of signal processing methods applied in SONAR expert systems is represented by the so called intelligent image processing techniques: segmentation, textures analysis, classification,...The aim of this chapter is the despecklisation of SONAR images. The SONAR images are perturbed by speckle. It is of multiplicative nature. The aim of a denoising algorithm is to reduce the noise level, while preserving the image features. The noisy image is:

$$f = s \cdot v \tag{1}$$

where s is the noise-free input image and $v$ the speckle noise. We present the most frequently used despecklisation techniques starting with the classical ones, Lee, Kuan and Frost filters and continuing with the pure statistical despecklisation method proposed in [Walessa&Datcu, 2000] to arrive at the most modern ones which act in the wavelet domain.

## 2. Denoising methods

The field of natural images denoising methods is very large. A lot of articles dedicated to denoising methods were already written. Most of these articles are focused on the case of additive noise.

There are two ways of reducing the speckle to an additive noise. The first one involves the separation into a sum between the signal and a signal-dependent noise:

$$f = s + s \cdot (v - 1) \tag{2}$$

while the second one simply takes advantage of the properties of the logarithm:

$$\ln f = \ln s + \ln v \tag{3}$$

Assuming that usually $f$ is not stationary, in the first approach the additive noise is not stationary as well, [Argenti et al., 2009]. There are two categories of denoising methods for the additive noise, the methods acting in the spatial domain and the methods which act in a transformed domain. We proceed with the presentation of the first category.

### 2.1 Denoising based on differential equations with partial derivatives

A first category of images denoising methods acting in the spatial domain are based on differential equations with partial derivatives. These denoising methods do not take into account any a priori information about the image to be processed, being non parametric. The aim of those methods is to consider the denoising like a decomposition of the acquired image into two components, the noiseless part and the noise. This decomposition can be realized by the projection of the acquired image on two very different vector spaces. The projection can be done by the minimization of a cost function. The result corresponds to the solution of the system of equations which is obtained imposing the zero value to the partial derivatives of the cost function. This is a system of partial differential equations. The simplest projection on the noiseless space is realized averaging the acquired image. The corresponding denoising method works well for the homogeneous regions of the acquired image if the noise is zero mean. Tacking into account the fact that the averager is a low-pass filter, this denoising method distorts the edges and some of the textured regions by oversmoothing.

### 2.2 Denoising by non-local averaging

Another very modern non parametric denoising method is based on non-local (NL) averaging [Buades, 2007]. The NL-means algorithm tries to take advantage of the high degree of redundancy of any natural image. Every small area (or window) in a natural image has many similar areas in the same image. In a very general sense, one can define as „neighborhood of a pixel *i*" any set of pixels *j* in the image so that a window around *j* looks like a window around *i*. All pixels in that neighborhood can be used for predicting the value at *i*. Given a discrete noisy image $x = \{x(i) | i \in I\}$, the estimated value $NL_x(i)$ is computed as a weighted average of all the pixels in the image, $NL_x(i) = \sum_{j \in I} \beta(i,j) x(j)$, where the weights $\{\beta(i,j)\}_j$ depend on the similarity between the pixels *i* and *j* and satisfy the usual conditions $0 \le \beta(i,j) \le 1$ and $\sum_j \beta(i,j) = 1$. The non-locality of the average prevents from oversmoothing.

### 2.3 Denoising using statistical filters

Another category of denoising methods are the parametric ones. These methods take into account statistical models for the noiseless component of the acquired image and for the noise. One of the best parametric denoising methods uses maximum a posteriori (MAP) filters. The MAP estimation of $w$, based on the observation $y=w+n$, (where $n$ represents the additive noise) is given by the MAP filter equation:

$$\hat{w}(y) = \underset{w}{\operatorname{argmax}} \{\ln(p_{\mathbf{n}}(y - w) p_{\mathbf{w}}(w))\} \tag{4}$$

where $p_a$ denotes the probability density function (pdf) of $a$.

#### 2.3.1 Classical despecklisation filters

Kuan [Kuan et al., 1987] considered a multiplicative speckle model and designed a linear filter based on the minimum mean square error (MMSE) criterion, optimal when both the scene and the detected intensities are Gaussian distributed.

The Lee filter [Lee, 1981] is a particular case of the Kuan filter due to a linear approximation made for the multiplicative noise model.

The Frost filter [Frost et al., 1982] is a Wiener filter adapted to multiplicative noise. The parameters of the Kuan, Lee and Frost filters are: the size of the rectangular windows used for the estimation of the local standard deviation of the useful component of the acquired image and its number of looks.

### 2.4 Denoising in a transform's domain

Another excellent denoising method for the case of additive noise makes the object of [Katkovnik et al.,2006]. There are a lot of original contributions highlighted in this reference:

the local polynomial analysis of images, the role of anisotropic analysis windows, etc... Some very clever associations of those notions conduct to very efficient local denoising methods applied in the spatial domain or in the field of the shape adaptive Digital Cosinus Transform (SADCT).

### 2.4.1 Denoising into the wavelets domain

A set of denoising methods for additive noise act in the wavelets domain. These methods have three steps: the computation of a wavelet transform (WT); the filtering of the detail wavelet coefficients; the computation of the corresponding inverse WT (IWT). The usefulness of the filtering in the wavelet domain comes from the sparseness of the WTs. There are only few wavelet coefficients with high magnitude, which concentrate the most of the energy of the noiseless component of the input image. The other wavelet coefficients have small magnitude and can be considered noise. They can be discarded without producing high distortions.

A first category of denoising methods applied in the wavelets domain is based on non-parametric techniques [Donoho&Johnstone, 1994] and uses the hard or the soft thresholding filters. The parent of these methods is Professor David Donoho from Stanford University. His goal was to estimate the noiseless component of the acquired image by the minimization of the min-max approximation error. He studied the case of the Discrete Wavelet Transform (DWT). In the following we will make a general presentation of the wavelet transforms and we'll continue with the denoising techniques based in these transforms.

#### 2.4.1.1 Wavelet transforms

There are few wavelet transforms (WT) which were used for images denoising. The first one was the DWT.

*2.4.1.1.1 The DWT*

All the WTs are characterized by two features: the mother wavelets, MW and the primary resolution, PR (number of iterations). The importance of their selection is highlighted in [Nason, 2002]. An appealing particularity of the 2D DWT is the interscale dependency of the wavelet coefficients. The main advantage of the implementation of the 2D DWT is its flexibility, as it inherits some of the classes of mother wavelets developed in the framework of the 1D DWT, like the Daubechies, Symmlet or Coiflet families. The implementation of 1D DWT is presented in figure 1. It implements an orthonormal multiresolution analysis [Mallat, 99]. Each of the iterations of the algorithm used for the computation of the 2D DWT implies several operations. First, the lines of the input image (obtained at the end of the previous iteration) are passed trough two different filters (a lowpass filter having the impulse response $m_0$ and a high-pass filter $m_1$) resulting two different sub-images. Then the lines of the two sub-images obtained at the output of the two filters are decimated with a factor of 2. Next, the columns of the two images obtained are low-pass filtered with $m_0$ and high-pass filtered with $m_1$. The columns of those four sub-images are also decimated with a factor of 2. Four new sub-images, representing the result of the current iteration (which corresponds to the current decomposition level (or scale)), are obtained. These sub-images are called subbands. The first sub-image, obtained after two lowpass filterings, is called approximation sub-image (or LL subband). The other three are named detail sub-images: LH, HL and HH. The LL sub-image represents the input for the next iteration. In the following, the coefficients of the DWT will be denoted with ${}_xD_m^k$, where $x$ represents the image who's DWT is computed (considered as a bivariate random signal), $m$ represents the current scale and $k$ = 1 - for the subband LH, $k$ = 2 - for HL, $k$ = 3 - for HH and $k$ = 4 - for LL. These coefficients are computed using the following relation:

$$ {}_xD_m^k[n_1,p_1] = \left\langle x(\tau_1,\tau_2), \psi_{m,n_1,p_1}^k(\tau_1,\tau_2) \right\rangle, $$

where the wavelets are real functions and can be factorized as:

$$ \psi_{m,n,p}^k(\tau_1,\tau_2) = \alpha_{m,n}^k(\tau_1) \cdot \beta_{m,p}^k(\tau_2), $$

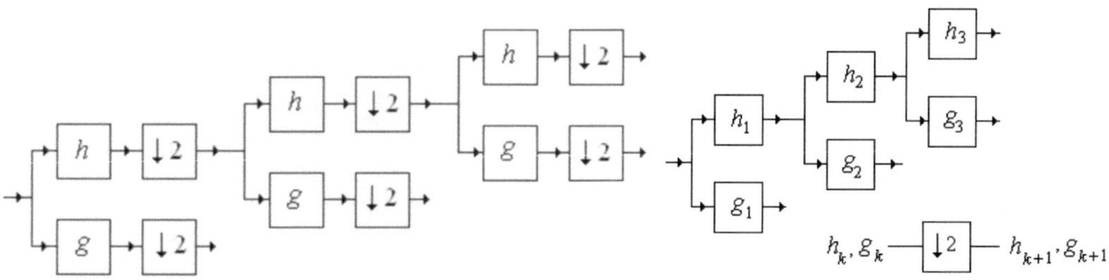


Fig. 1. (left) 1D DWT implementation. $h$ – lowpass filter, $g$ - highpass filter. They form a pair of quadrature mirror filters [Mallat, 99]. The filters are followed by decimators with decimation factor equal with 2; (right) UDWT implementation. $h_k$ – lowpass filters, $g_k$ - highpass filters.

and the two factors can be computed using the scale function $\varphi(\tau)$ and the mother wavelets $\psi(\tau)$ with the aid of the following relations:

$$ \alpha_{m,n}^k(\tau) = \begin{cases} \varphi_{m,n}(\tau), & k = 1, 4 \\ \psi_{m,n}(\tau), & k = 2, 3 \end{cases}, \quad \beta_{m,p}^k(\tau) = \begin{cases} \varphi_{m,p}(\tau), & k = 2, 4 \\ \psi_{m,p}(\tau), & k = 1, 3 \end{cases}, $$

where:

$$ \varphi_{m,n}(\tau) = 2^{-\frac{m}{2}} \varphi\left(2^{-m}\tau - n\right) \text{ and } \psi_{m,n}(\tau) = 2^{-\frac{m}{2}} \psi\left(2^{-m}\tau - n\right). $$

Taking into account the last equations it can be written:

$$ \psi_{m,n,p}^k(\tau_1,\tau_2) = 2^{-m} \psi^k\left(2^{-m}\tau_1 - n, 2^{-m}\tau_2 - p\right) \text{ where } \psi^k(\tau_1,\tau_2) = \psi^k{}_{0,0,0}(\tau_1,\tau_2). $$

An iteration of the 2D DWT is presented in figure 2. The main disadvantages of the 2D DWT are the poor directional selectivity and the shift sensitivity. Separable filtering along the rows and columns of an image produces four images at each level. The LH and HL bandpass sub-images can select mainly horizontal or vertical edges respectively, but the HH sub-image contains components from diagonal features of either orientation. This means that the separable real 2D DWT has 'poor directional selectivity'. This is illustrated in fig. 3. From left to right we can observe the vertical details (HL), the horizontal details (LH) and the diagonal details (HH). Prof. Kingsbury, in [Kingsbury, 1999] explains this limitation by the fact that real highpass row filters select both positive and negative horizontal high frequencies and, consequently, the combined HH filter must have pass-bands in all four quadrants of the 2-D frequency plane.

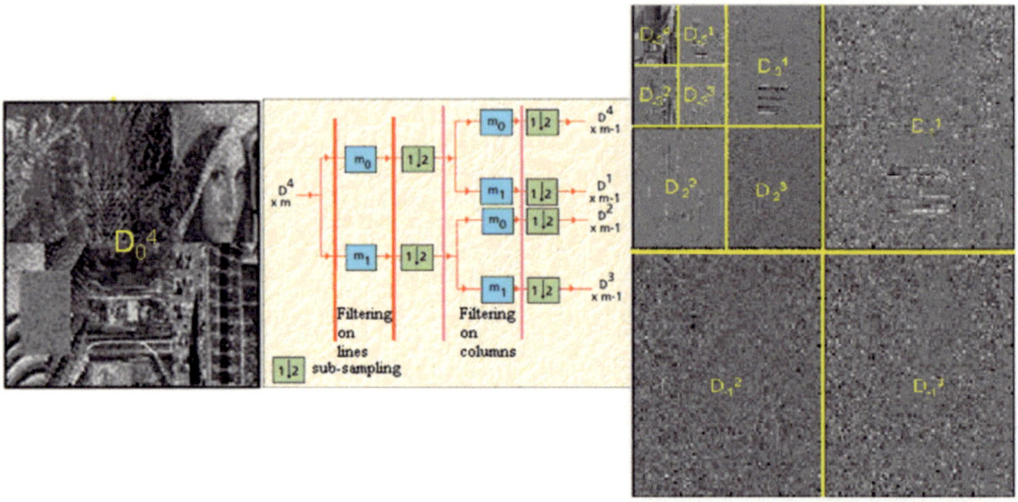


Fig. 2. An iteration of the 2D DWT (middle). The structure of the result obtained applying three iterations of the 2D DWT to the image in the left part of the figure can be seen in the image from the right part of the figure.

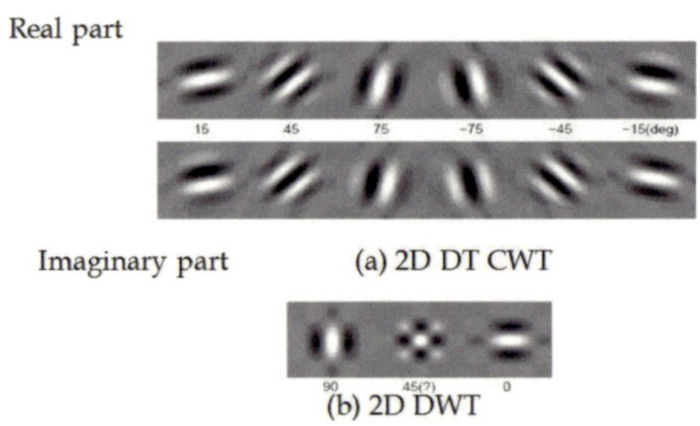


Fig. 3. Basis functions of 2D Q-shift complex wavelets (a) and 2D real wavelet filters (b), all illustrated at level 4 of the transforms. The complex wavelets provide 6 directionally selective filters, while real wavelets provide 3 filters, only two of which have a dominant orientation.

On the other hand, a directionally selective filter for diagonal features with positive gradient must have pass-bands only in quadrants 2 and 4 of the frequency plane, while a filter for diagonals with negative gradient must have pass-bands only in quadrants 1 and 3. The poor directional properties of real separable filters make it difficult to generate steerable or directionally selective algorithms, based on the separable real DWT. Other WTs which overcomes one or both these disadvantages are: the Undecimated Discrete Wavelet Transform (UDWT), the Double Tree Complex Wavelet Transform (DTCWT) and the Hyperanalytic Wavelet Transform (HWT).

*2.4.1.1.2 The 2D UDWT*

A particular wavelet transform is the undecimated wavelet transform (UWT), also known as the 'Stationary Wavelet Transform' and can be implemented using the 'Algorithme à Trous'. The UDWT refers to the discrete implementation of this transform. The 'Algorithme à Trous', first introduced in [Holschneider et al., 1989] for music synthesis applications, is similar to a non orthonormal multiresolution algorithm for which the discrete wavelet transform is exact. An implementation of the UDWT is presented in figure 1. Due to the absence of decimators in the UDWT's implementation, each coefficient sequence from any level of decomposition has the same length as the original, in other words, if the original signal has $N$ samples, the UDWT $L$-level representation is of size $N\times(L+1)$, making from the UDWT a highly redundant transform. But eliminating the decimators is obtained a translation invariant WT. The 2D UDWT is shift-invariant (due to the absence of decimators) but it has a poor directional selectivity and is very redundant.

*2.4.1.1.3 The 2D DTCWT*

In 1998, Kingsbury first introduced the DT CWT [Kingsbury, 1998] that relies on the observation that approximate shift invariance can be achieved with a real DWT by doubling the sampling rate at each level of the tree. For this to work, the samples must be evenly spaced. The sampling rates can be doubled by eliminating the down-sampling by 2 after the level 1 filters. This is equivalent to having two parallel fully-decimated trees $a$ and $b$, like in fig. 4, provided that the delays of $H_{0b}$ and $H_{1b}$ are one sample offset from $H_{0a}$ and $H_{1a}$. Kingsbury found that, to get uniform intervals between samples from the two trees below level 1, the filters in one tree must provide delays that are half a sample different (at each filter input rate) from those in the other tree. This statement is also supported by Selesnick who, in [Selesnick, 2001], gives an alternative derivation and explanation of the same result. The implementation of such a transform is done using two mother wavelets, one for each tree, one of them being (approximately) the Hilbert transform of the other. On one hand, the dual-tree DWT can be viewed as an overcomplete wavelet transform with a redundancy factor of two. On the other hand, the dual-tree DWT is also a complex DWT, where the first and second DWTs represent the real and imaginary parts of a single complex DWT. In order to have a visual aspect of the DT CWT, we present in figure 4, the Q-shift version of the DT CWT as it is given in [Kingsbury, 2001]. In order to examine the shift invariance properties of a transform, Kingsbury [Kingsbury, 2001] proposes a method based on the retention of just one type (details or approximations), from just one level of the decomposition tree. For example one might choose to retain only the level-3 detail coefficients and set all the others to zero. If the signal reconstructed from just these coefficients is free of aliasing then it can be said that the transform is shift invariant at that level. The degree of shift invariance of two implementation schemes (one for the DT CWT and the other for the classical DWT) is compared in fig. 5. In each case the input is given by 16 shifted versions of a unitary step signal. Each unitary step is passed through the forward and inverse version of the chosen wavelet transform. The figure shows the input steps and the components of the inverse transform output signal, reconstructed from the wavelet coefficients at each of levels 1 to 4 in turn and from the scaling function coefficients at level 4. Summing these components reconstructs the input steps perfectly. Good shift invariance is shown when all the 16 output components from a given level have the same shape, independent of shift. It is easily observed that the DT CWT has outstanding performances in this direction compared to the severe shift dependence of the normal DWT.

Extension of the DT CWT to two dimensions is achieved by separable filtering along columns and then rows. However, if column and row filters both suppress negative frequencies, then only the first quadrant of the 2-D signal spectrum is retained. It is well known, from 2-D Fourier transform theory, that two adjacent quadrants of the spectrum are required to represent fully a real 2-D signal. Therefore in the DT CWT the image is also filtered with complex conjugates of the row (or column) filters in order to retain a second (or fourth) quadrant of the spectrum. This then gives 4:1 redundancy in the transformed 2-D signal. A schematic representation of the 2D DT CWT based on the even-odd implementation was given by [Jalobeanu et al., 2003]. At level $m = 1$, the 2D DT CWT is simply a non-decimated wavelet transform (using a pair of odd-length filters $h^o$ and $g^o$) whose coefficients are re-ordered into 4 interleaved images by using their parity. This defines the 4 trees $T = A, B, C$ and $D$. If $a$ and $d$ denotes approximation and detail coefficients ($a^0 = x$, the input image), we have:

$$\begin{array}{lllll}
\text{Tree } T & A & B & C & D \\
\left(a_T^1\right)_{x,y} & \left(a^0 * h^o h^o\right)_{2x,2y} & \left(a^0 * h^o h^o\right)_{2x,2y+1} & \left(a^0 * h^o h^o\right)_{2x+1,2y} & \left(a^0 * h^o h^o\right)_{2x+1,2y+1} \\
\left(d_T^{1,1}\right)_{x,y} & \left(a^0 * g^o h^o\right)_{2x,2y} & \left(a^0 * g^o h^o\right)_{2x,2y+1} & \left(a^0 * g^o h^o\right)_{2x+1,2y} & \left(a^0 * g^o h^o\right)_{2x+1,2y+1} \\
\left(d_T^{1,2}\right)_{x,y} & \left(a^0 * h^o g^o\right)_{2x,2y} & \left(a^0 * h^o g^o\right)_{2x,2y+1} & \left(a^0 * h^o g^o\right)_{2x+1,2y} & \left(a^o * h^o g^o\right)_{2x+1,2y+1} \\
\left(d_T^{1,3}\right)_{x,y} & \left(a^0 * g^o g^o\right)_{2x,2y} & \left(a^0 * g^o g^o\right)_{2x,2y+1} & \left(a^0 * g^o g^o\right)_{2x+1,2y} & \left(a^0 * g^o g^o\right)_{2x+1,2y+1}
\end{array}$$

For all other scales ($j > 1$), the transform involves an additional pair of filters, even-length, denoted $h^e$ and $g^e$. There must be a half-sample shift between the trees to achieve the approximate shift invariance. Therefore, different length filters are used for each tree, i.e. it is necessary to combine $h^e$, $g^e$ with $h^o$, $g^o$, the 4 possible combinations corresponding to the 4 trees:

$$\begin{array}{lllll}
\text{Tree } T & A & B & C & D \\
\left(a_T^{j+1}\right)_{x,y} & \left(a_A^j * h^e h^e\right)_{2x,2y} & \left(a_B^j * h^e h^o\right)_{2x,2y+1} & \left(a_C^j * h^o h^e\right)_{2x+1,2y} & \left(a_D^j * h^o h^o\right)_{2x+1,2y+1} \\
\left(d_T^{j+1,1}\right)_{x,y} & \left(a_A^j * g^e h^e\right)_{2x,2y} & \left(a_B^j * g^e h^o\right)_{2x,2y+1} & \left(a_C^j * g^o h^e\right)_{2x+1,2y} & \left(a_D^j * g^o h^o\right)_{2x+1,2y+1} \\
\left(d_T^{j+1,2}\right)_{x,y} & \left(a_A^j * h^e g^e\right)_{2x,2y} & \left(a_B^j * h^e g^o\right)_{2x,2y+1} & \left(a_C^j * h^o g^e\right)_{2x+1,2y} & \left(a_D^j * h^o g^o\right)_{2x+1,2y+1} \\
\left(d_T^{j+1,3}\right)_{x,y} & \left(a_A^j * g^e g^e\right)_{2x,2y} & \left(a_B^j * g^e g^o\right)_{2x,2y+1} & \left(a_C^j * g^o g^e\right)_{2x+1,2y} & \left(a_D^j * g^o g^o\right)_{2x+1,2y+1}
\end{array}$$

The trees are processed separately, as in a real transform. The combination of odd and even filters depends on each tree. The transform is achieved by a fast Filter Bank (FB) technique, of complexity $O(N)$. The reconstruction is done in each tree independently, by using the dual filters. To obtain $a^0$, the results of the 4 trees are averaged. This ensures the symmetry between them, thus enabling the desired shift invariance. The complex coefficients are obtained by combining the different trees together. If the subbands are indexed by $k$, the detail subbands $d^{j,k}$ of the parallel trees $A$, $B$, $C$ and $D$ are combined to form complex subbands $z_+^{j,k}$ and $z_-^{j,k}$, by the linear transform:

$$z_{+}^{j,k} = \left(d_A^{j,k} - d_D^{j,k}\right) + i\left(d_B^{j,k} + d_C^{j,k}\right) \quad z_{-}^{j,k} = \left(d_A^{j,k} + d_D^{j,k}\right) + i\left(d_B^{j,k} - d_C^{j,k}\right) \tag{5}$$

Complex filters in multiple dimensions can provide true directional selectivity, despite being implemented separately, because they are still able to separate all parts of the m-D frequency space. For example the 2D DT CWT produces six bandpass sub-images of complex coefficients at each level, which are strongly oriented at angles ±15°, ±45°, ±75°, as illustrated by the level 4 impulse responses in fig. 3.

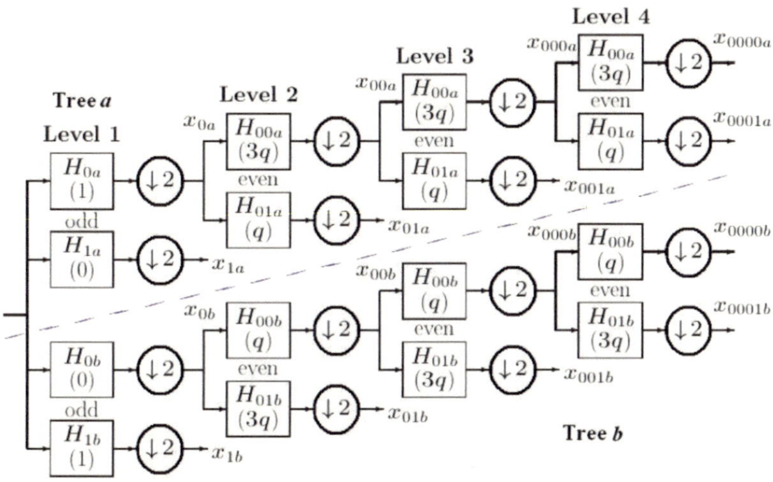


Fig. 4. The Q-shift version of the DT CWT as it is given in [Kingsbury, 2001].

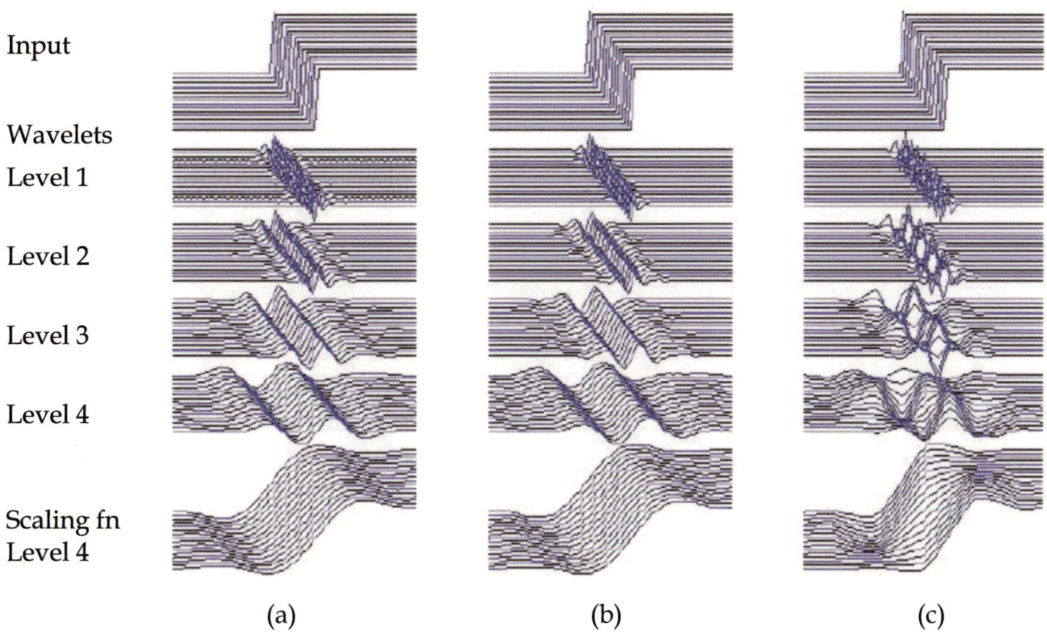


Fig. 5. A comparison of the shift invariance of three WTs: ADWT (a), DTCWT (b) and DWT (c).

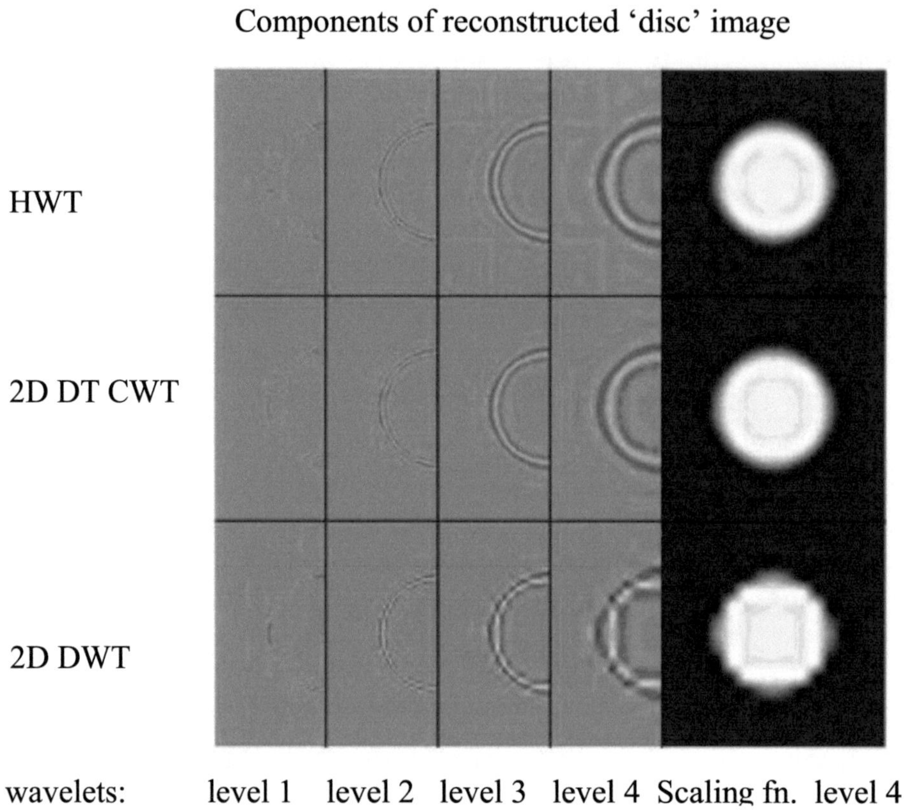


Fig. 6. Wavelet and scaling function components at levels 1 to 4 of an image of a light circular disc on a dark background, using the HWT (upper row), the 2D DT CWT (middle row) and 2D DWT (lower row). Only half of each wavelet image is shown.

In order to obtain these directional responses, it is necessary to interpret the scaling function (lowpass) coefficients from the two trees as complex pairs (rather than as purely real coefficients at double rate) so that they can be correctly combined with wavelet (highpass) coefficients, which are also complex, to obtain the filters oriented at ±15° and ±75° (see 5). The main property of the 2D DT CWT is the quasi shift invariance, as shown by Kingsbury in [Kingsbury, 2001] i.e. the magnitudes $|z_{\pm}|$ are nearly invariant to shifts of the input image. The shift invariance is perfect at level 1, and approximately achieved beyond this level: the transform algorithm is designed to optimize this property. In fig. 6, the shift-dependence properties of the 2D DT CWT were compared with the 2D DWT. The input is now an image of a light circular disc on a dark background. This circular form is suited for the analysis of the shift dependence in 2D as neighbor pixels from the contour of the disc can be interpreted as obtained through 2D shifts. The rows of images, from left to right in fig. 6, show the components of the output image, reconstructed from the 2D DT CWT wavelet coefficients at levels 1, 2, 3 and 4 and from the scaling function coefficients at level 4. The lower row of images show the equivalent components when the fully decimated 2D DWT is used instead. In the lower row, we see substantial aliasing artifacts, manifested as irregular edges and stripes that are almost normal to the edge of the disc in places. Contrast this with the row of 2D DT CWT images, in which artifacts are virtually absent. The smooth and continuous images here demonstrate good shift invariance because all parts of the disc edge are treated equivalently; there is no shift dependence.

*2.4.1.1.4 The HWT*

In [Abry, 1994] is demonstrated that approximate shiftability is possible for the DWT with a small, fixed amount of transform redundancy. In this reference is designed a pair of real mother wavelets such that one is approximately the Hilbert transform of the other. This wavelet pair defines an analytic discrete wavelet transform (ADWT) presented in figure 7 a). A complex wavelet coefficient is obtained by interpreting the wavelet coefficient from one DWT tree as being its real part, whereas the corresponding coefficient from the other tree is interpreted as its imaginary part. The implementation of the ADWT is presented in figure 7 b). We first apply a Hilbert transform to the data. The real wavelet transform is then applied to the analytical signal associated to the input data, obtaining complex coefficients. The two implementations of the ADWT presented in figure 7 are equivalent because:

$$\begin{aligned} d_{ADWT}[m,n] &= \langle x(t),\psi_{m,n}(t)+i\mathrm{H}\{\psi_{m,n}(t)\}\rangle = \langle x(t),\psi_{m,n}(t)\rangle - i\langle x(t),\mathrm{H}\{\psi_{m,n}(t)\}\rangle = \\ &= \langle x(t),\psi_{m,n}(t)\rangle + i\langle \mathrm{H}\{x(t)\},\psi_{m,n}(t)\rangle = \langle x(t)+i\mathrm{H}\{x(t)\},\psi_{m,n}(t)\rangle \end{aligned} \tag{6}$$

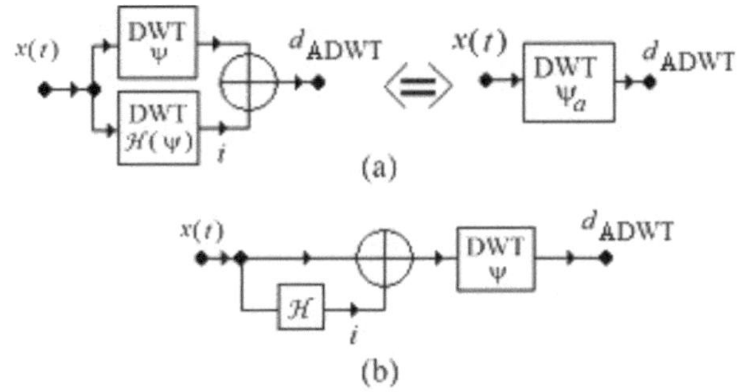


Fig. 7. Three equivalent implementations of the ADWT. (a) The double tree of real DWTs is equivalent with a complex DWT, which can be reduced at the input (b).

In fact neither the DTCWT nor the proposed implementation of ADWT correspond to perfect analytic mother wavelets, because the exact digital implementation of a Hilbert transform pair of mother wavelets with good performance is not possible in the case of the first transform (due to the fractional delay between the two trees required) and because the digital Hilbert transformer is not a realizable system in the case of the second transform. The DTCWT requires special mother wavelets (the implementation of the ADWT proposed in figure 7 b) can be realized using classical mother wavelets like those conceived by Daubechies) but can assure a higher degree of shift invariance. These two transforms have in the 1D case a redundancy of 2. In order to evaluate the shift-invariance performance of ADWT we make a visual evaluation of the degree of shift invariance by taking the same test professor Kingsbury has proposed in [Kingsbury, 2001]. We have used this test to make a comparison between ADWT, DT CWT, and the classical DWT. As can be seen, we obtain, using Daubechies 10-taps filter, results comparable with those obtained by Kingsbury. From fig. 5 it can be observed that the DWT is not shift-invariant; the lines of coefficients corresponding to different shifts are not parallel, while the ADWT and DT CWT are quasi shift-invariant.

All the 1D WTs already mentioned have simpler or more complicated 2D generalizations. The generalization of the analyticity concept in 2D is not obvious, because there are multiple definitions of the Hilbert transform in this case. In the following we will use the definition of the analytic signal associated to a 2D real signal named hypercomplex signal. So, the hypercomplex mother wavelets associated to the real mother wavelets $\psi(x,y)$ is defined as:

$$\psi_a(x,y)=\psi(x,y)+i\mathrm{H}_x\{\psi(x,y)\}+j\mathrm{H}_y\{\psi(x,y)\}+k\mathrm{H}_x\left\{\mathrm{H}_y\{\psi(x,y)\}\right\} \quad (7)$$

where $i^2=j^2=-k^2=-1$, and $ij=ji=k$, [Davenport, 2010]. The HWT of the image $f(x,y)$ is: $HWT\{f(x,y)\}=\langle f(x,y),\psi_a(x,y)\rangle$. Tacking into account relation (7) it can be written:

$$\begin{aligned}&HWT\{f(x,y)\}=DWT\{f(x,y)\}+iDWT\{\mathrm{H}_x\{f(x,y)\}\}+\\&+jDWT\left\{\mathrm{H}_y\{f(x,y)\}\right\}+kDWT\left\{\mathrm{H}_y\{\mathrm{H}_x\{f(x,y)\}\}\right\}=\\&\langle f_a(x,y),\psi(x,y)\rangle=DWT\{f_a(x,y)\}.\end{aligned} \quad (8)$$

So, the HWT of the image $f(x,y)$ can be computed with the aid of the 2D DWT of its associated hyper complex image. Consequently, the HWT of the image $f(x,y)$ can be computed with the aid of the 2D DWT of its associated hyper complex image. The HWT implementation uses four trees, each one implementing a 2D DWT. The first tree is applied to the input image. The second and the third trees are applied to 1D Hilbert transforms computed across the lines ($\mathrm{H}_x$) or columns ($\mathrm{H}_y$) of the input image. The fourth tree is applied to the result obtained after the computation of the two 1D Hilbert transforms of the input image.

The enhancement of the directional selectivity of the HWT is realized like in the case of the 2D DTCWT, by the separation of the real and imaginary parts of the complex coefficients belonging to each subband of the WT (eq. 5). The HWT implementation is presented in figure 8.

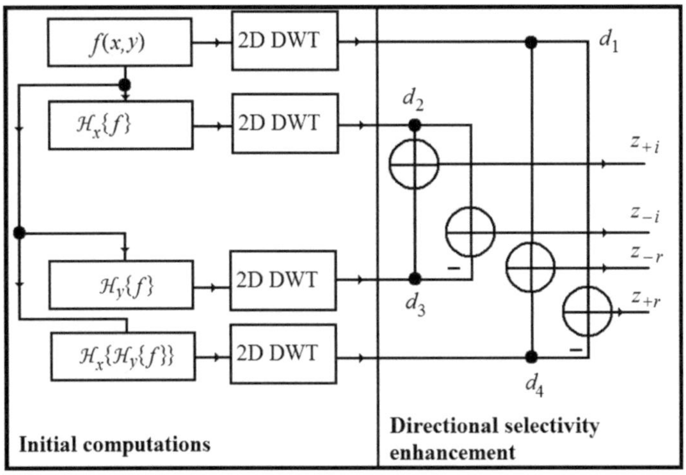


Fig. 8. HWT implementation architecture.

Let us consider, for example, the case of the diagonal detail subbands, (HH), presented in figure 9. We selected a particular input image, $f(x,y)=\delta(x,y)$, to appreciate the frequency responses associated to different transfer functions represented in figure 8. More precisely, the example in figure 9 refers to the transfer functions that relate the input $f$ with the outputs $z_{-r}$ and $z_{+r}$. The spectrum of the input image $\mathrm{F}\{\delta(x,y)\}(f_x,f_y)$ is constant. For the subband HH of each 2D DWT we have two preferential orientations, corresponding to the two diagonals ($\pm\pi/4$). So, the 2D DWT cannot separate these two orientations. But the spectra of the coefficients obtained after linear combinations, for example $z_{-r}$ and $z_{+r}$, $\mathrm{F}\{{}^{HH}z_{-r}\}(f_x,f_y)$ and $\mathrm{F}\{{}^{HH}z_{+r}\}(f_x,f_y)$, have only one preferential direction, the second diagonal respectively the first one.

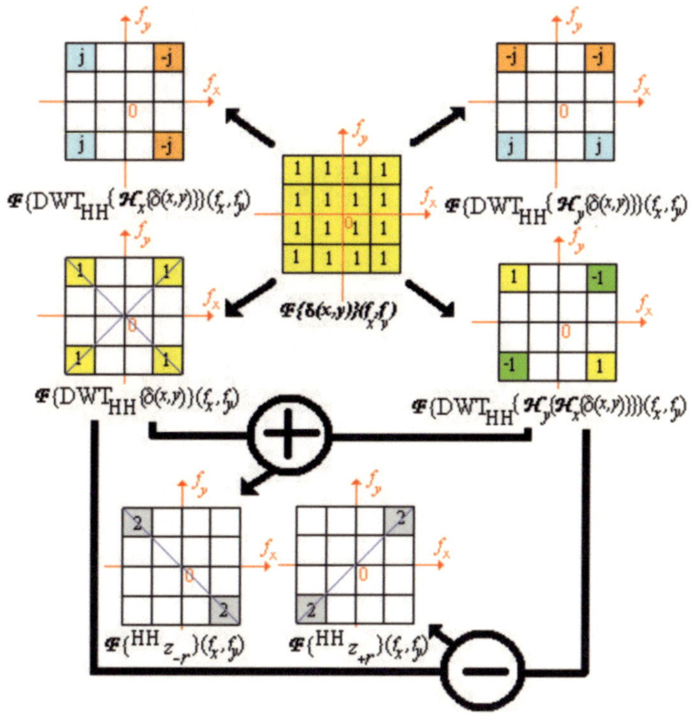


Fig. 9. The strategy of directional selectivity enhancement in the HH subband. The frequency responses of the systems that transform the input image $f$ into the output diagonal detail coefficient sub-images $z_{-r}$ and $z_{+r}$ represented in figure 12.

In conclusion, by using the HWT these directions can be separated. The same strategy can be used to enhance the directional selectivity in the other two subbands: LH and HL, obtaining the preferential orientations: $\pm\operatorname{atan}(2)$ and $\pm\operatorname{atan}(1/2)$. A comparison of the directional selectivity of the 2D DWT and HWT, implemented as proposed in figure 8, is presented in figure 10. We have conceived a special input image, in the frequency domain, to conduct this simulation. Its spectrum, represented in figure 10, is oriented following the directions: 0, $\pm\operatorname{atan}(1/2)$, $\pm\pi/4$, $\pm\operatorname{atan}(2)$ and $\pi$. Like the 2D DTCWT, the HWT implemented as proposed in figure 8, has six preferential orientations: $\pm\operatorname{atan}(1/2)$, $\pm\pi/4$ and $\pm\operatorname{atan}(2)$. The 2D-DWT has only three preferential orientations: 0, $\pi/4$ and $\pi/2$, it does not make the difference between the two principal diagonals. The better directional selectivity of the proposed implementation of HWT versus the 2D DWT can be easily observed, comparing the corresponding detail sub-images in figure 10. For the diagonal detail sub-images, for example, the imaginary part of the HWT rejects the directions: $-\mathrm{atan}(1/2)$, $-\pi/4$ and $-\mathrm{atan}(2)$, whereas the 2D DWT conserves these directions. Concerning the 2D shift invariance, in figure 6 is presented a comparison between the HWT and the 2D DTCWT.

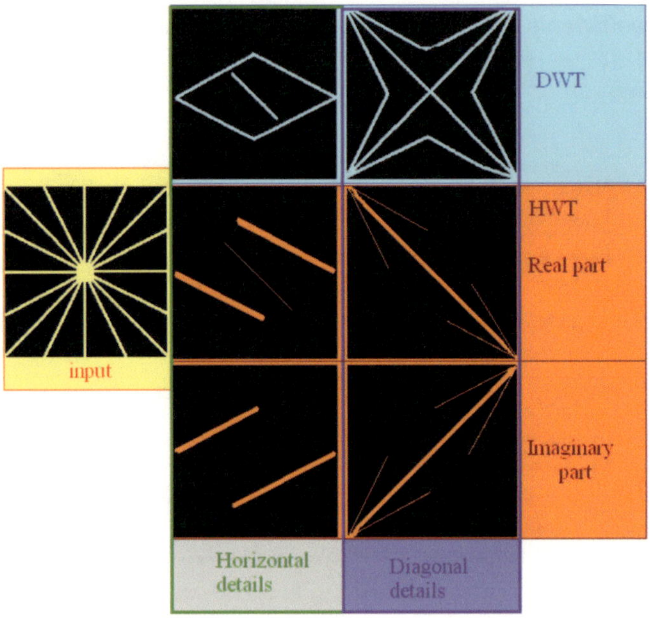


Fig. 10. The absolute values of the spectra of horizontal and diagonal detail sub-images obtained after the first iterations of 2D DWT and HWT (proposed implementation). In the HWT case, the real and imaginary parts of complex coefficients are separated.

The two complex WTs outperform the 2D DWT. The behavior of the HWT is quite similar with the comportment of the 2D DTCWT. One of the goals of the present chapter is to compare the efficiency of those WTs in despecklisation. A second goal of this chapter is to compare the efficiency of the filters applied in the wavelet domain for SONAR despecklisation.

**2.4.1.2 MAP Filters in the wavelets domain**

As was already said, after the transformation of the multiplicative noise into an additive one based on equations (2) or (3), the WT of the signal obtained is computed. Tacking into account the linearity of the WT, it can be written the equation $y=w+n$, where the first term in the right hand member represents the WT of the noiseless component and the second term represents the WT of the noise. Then the MAP filter equation (4) must be solved. Its solution depends on the selection of the models of pdfs for the noiseless component and for the noise. There are two categories of MAP filters: marginal (both pdfs are univariate functions) and joint (both pdfs are multi-variate functions).

The simpler MAP filter is the marginal zero order Wiener filter. It is constructed considering that the noiseless component of the acquired image and the noise are Gaussian distributed.

*2.4.1.2.1 The zero order Wiener filter*

We will derive first the joint zero order Wiener filter considering that both pdfs are $d$-dimensional zero mean Gaussians:

$$p_{\mathbf{w}}(\mathbf{w})=\frac{1}{(2\pi)^{\frac{d}{2}}\left|\mathbf{C}_{\mathbf{w}}\right|^{\frac{1}{2}}}\cdot e^{-\frac{\mathbf{w}^{T}\cdot\mathbf{C}_{\mathbf{w}}^{-1}\cdot\mathbf{w}}{2}} \text{ and } p_{\mathbf{n}}(\mathbf{n})=\frac{1}{(2\pi)^{\frac{d}{2}}\left|\mathbf{C}_{\mathbf{n}}\right|^{\frac{1}{2}}}\cdot e^{-\frac{\mathbf{n}^{T}\cdot\mathbf{C}_{\mathbf{n}}^{-1}\cdot\mathbf{n}}{2}} \tag{9}$$

where the covariance matrix of vector **a** was denoted by $\mathbf{C}_{\mathbf{a}}$ and its determinant by $\left|\mathbf{C}_{\mathbf{a}}\right|$, for $d$=1. Then, the MAP filter equation (4) becomes:

$$\hat{\mathbf{w}}(\mathbf{y})=\underset{\mathbf{w}}{\operatorname{argmax}}\left\{\ln\left(\frac{1}{(2\pi)^{d}\cdot\left|\mathbf{C}_{\mathbf{w}}\right|^{\frac{1}{2}}\cdot\left|\mathbf{C}_{\mathbf{n}}\right|^{\frac{1}{2}}}\cdot e^{-\frac{(\mathbf{y}-\mathbf{w})^{T}\cdot\mathbf{C}_{\mathbf{n}}^{-1}\cdot(\mathbf{y}-\mathbf{w})}{2}}\cdot e^{-\frac{(\mathbf{w})^{T}\cdot\mathbf{C}_{\mathbf{w}}^{-1}\cdot(\mathbf{w})}{2}}\right)\right\} \tag{10}$$

or:

$$\hat{\mathbf{w}}(\mathbf{y})=\underset{\mathbf{w}}{\operatorname{argmax}}\left\{\ln\frac{1}{(2\pi)^{d}\cdot\left|\mathbf{C}_{\mathbf{w}}\right|^{\frac{1}{2}}\cdot\left|\mathbf{C}_{\mathbf{n}}\right|^{\frac{1}{2}}}-\frac{(\mathbf{y}-\mathbf{w})^{T}\cdot\mathbf{C}_{\mathbf{n}}^{-1}\cdot(\mathbf{y}-\mathbf{w})}{2}-\frac{\mathbf{w}^{T}\cdot\mathbf{C}_{\mathbf{w}}\cdot\mathbf{w}}{2}\right\}$$

problem which is equivalent with the equation obtained by putting the derivative of the argument of the right hand side member equal with zero:

$$\frac{\partial}{\partial\mathbf{w}}\left\{(\mathbf{y}-\mathbf{w})^{T}\cdot\mathbf{C}_{\mathbf{n}}^{-1}\cdot(\mathbf{y}-\mathbf{w})+(\mathbf{w})^{T}\cdot\mathbf{C}_{\mathbf{w}}^{-1}\cdot(\mathbf{w})\right\}=0 \tag{11}$$

Considering the following forms of the $d$-dimensional vectors:

$$\mathbf{y}=[y_1,\ y_2,\ldots,y_d]\ ;\ \mathbf{w}=[w_1,\ w_2,\ldots,w_d]$$

where the elements are zero mean independent random variables $a_k$ , $k$=1,2,...,$d$, with variances ${}_a\sigma_k^2$ , the expression of the covariance matrix becomes:

$$\mathbf{C}_{\mathbf{a}}=\prod_{k=1}^{d}{}_a\sigma_k^2\cdot\mathbf{I}$$

where **I** represents the $d\times d$ unitary matrix. Its determinant is expressed by:

$$\left|\mathbf{C}_{\mathbf{a}}\right|=\prod_{k=1}^{d}{}_a\sigma_k^2={}_aP$$

So, the multi-variate pdf of the Gaussian random process from (9), for independent univariate components can be put in the equivalent form:

$$p_{\mathbf{a}}(\mathbf{a})=\frac{1}{(2\pi)^{\frac{d}{2}}\left|\mathbf{C}_{\mathbf{a}}\right|^{\frac{1}{2}}}\cdot e^{-\frac{\mathbf{a}^{T}\cdot\mathbf{C}_{\mathbf{a}}^{-1}\cdot\mathbf{a}}{2}}=\frac{1}{(2\pi)^{\frac{d}{2}}\,{}_{a}P^{\frac{1}{2}}}\cdot e^{-\frac{\sum_{k=1}^{d}\frac{a_k^2}{{}_{a}\sigma_k^2}}{2}} \tag{12}$$

If all the independent random variables $a_k$, $k=1,2,\ldots,d$ , have the same variances ${}_{a}\sigma^2$ , then the last equations became:

$$\mathbf{C}_{\mathbf{a}}=({}_{a}\sigma)^{2d}\cdot\mathbf{I},\ \left|\mathbf{C}_{\mathbf{a}}\right|={}_{a}^{i}P=({}_{a}\sigma)^{2d},\ \mathbf{C}_{\mathbf{a}}^{-1}=({}_{a}\sigma)^{-2}\cdot\mathbf{I},$$
$$p_{\mathbf{a}}(\mathbf{a})=\frac{1}{(2\pi)^{\frac{d}{2}}({}_{a}\sigma)^{d}}\cdot e^{-\frac{\|\mathbf{a}\|^2}{2({}_{a}\sigma)^2}}. \tag{13}$$

Substituting (13) into equation (11), the following system of equations is obtained.

$$\frac{\partial}{\partial w_k}\left\{\sum_{k=1}^{d}\frac{(y_k-w_k)^2}{{}_{n}\sigma_k^2}+\sum_{k=1}^{d}\frac{w_k^2}{{}_{w}\sigma_k^2}\right\}=0,\ k=1,2,\ldots,d\ \Leftrightarrow\ \frac{2}{{}_{n}\sigma_k^2}(w_k-y_k)+\frac{2}{{}_{w}\sigma_k^2}w_k=0\ \Leftrightarrow$$
$$\Leftrightarrow\ \hat{w}_k=\frac{\frac{1}{{}_{n}\sigma_k^2}}{\frac{1}{{}_{n}\sigma_k^2}+\frac{1}{{}_{w}\sigma_k^2}}y_k\ \Leftrightarrow\ \hat{w}_k=\frac{{}_{w}\sigma_k^2}{{}_{w}\sigma_k^2+{}_{n}\sigma_k^2}y_k.$$

The solution can be also written in matrix form:

$$\hat{\mathbf{w}}=\mathbf{C}_{\mathbf{w}}\cdot(\mathbf{C}_{\mathbf{w}}+\mathbf{C}_{\mathbf{n}})^{-1}\cdot\mathbf{y} \tag{14}$$

This is the expression of the **joint zero order Wiener estimator.** For $d$=1, the last equation can be written in the particular form:

$$_{w}P={}_{w}\sigma_1^2=\sigma^2,\ {}_{n}P={}_{n}\sigma_1^2=\sigma_n^2,\ \hat{w}_1=\frac{\sigma^2}{\sigma^2+\sigma_n^2}y_1\ \text{or}\ \hat{w}=\frac{\sigma^2}{\sigma^2+\sigma_n^2}y \tag{15}$$

This is the input-output relation of the marginal zero order Wiener filter. Unfortunately, the two variances in the last equation are not known a priori. The noise variance can be estimated globally, using the diagonal detail sub image obtained at the first decomposition level, $D_1^3$ (denoted by HH) of a 2D DWT applied to the acquired image, using the following equation:

$$\hat{\sigma}_n^2=\left[\frac{\text{median}(|y_i|)}{0.6745}\right]^2,\ y_i\in\text{subband HH} \tag{16}$$

Generally, the variance $\sigma^2$ changes in space. This is the reason why is preferable to estimate locally this quantity. Practically the Wiener filtering is realized pixel wise (for each pixel of coordinates $(\alpha,\beta)$ ). For this purpose is used a window, for example of rectangular form, centered on the pixel of coordinates $(\alpha,\beta)$ and of size (2$P$+1)× (2$P$+1) denoted by $F_P(\alpha,\beta)$, and the variance is locally estimated inside the window. First is estimated the local expectation of the acquired image:

$$ {}_{l}\hat{\mu}[\alpha,\beta] = \frac{1}{(2P+1)^2} \sum_{(k,l)\in F_P(\alpha,\beta)} y[k,l] \tag{17}$$

and next its local variance :

$$ {}_{l}\hat{\sigma}_y^2[\alpha,\beta] = \frac{1}{(2P+1)^2} \sum_{(k,l)\in F_P(\alpha,\beta)} \left( y[k,l] - {}_{l}\hat{\mu}[\alpha,\beta] \right)^2 \tag{18}$$

Using these values, the local variance of the noiseless component is computed by:

$$ {}_{l}\hat{\sigma}^2[\alpha,\beta] = \max\left(0, {}_{l}\hat{\sigma}_y^2[\alpha,\beta] - \hat{\sigma}_n^2\right) \tag{19}$$

The zero order Wiener filter can be applied in association with the 2D DWT or 2D DT CWT or HWT. Its main disadvantage is the fact that the model considered for the noiseless image is not adequate because the wavelet coefficients have a heavy tail distribution (due to the sparseness of the WTs). Due to the sparseness of the WTs of the noiseless component, its distribution is heavy tailed. Indeed, there are only few wavelet coefficients with high magnitude; the majority of the wavelet coefficients have small magnitudes producing the heavy tails of the distribution. This affirmation is verified with the aid of the following experiment.

The histograms of different subbands obtained applying the HWT to the image Lena are computed. The results are represented in figure 11. The linear dependencies of the two branches of the logarithms of the histograms prove that the pdfs of the real and imaginary parts of the HWT coefficients correspond to exponential laws (which are heavy tailed):

$$ \log\left(K_1 \cdot e^{-|x|}\right) = K_2 - |x| \tag{20}$$

where $K_1$ and $K_2$ represent two constants. The Gaussian model supposed for the construction of the marginal zero order Wiener filter does not correspond to the linear dependencies of the two branches of the histograms:

$$ \log\left(K_1 \cdot e^{-K_2 x^2}\right) = K_3 - K_2 x^2 \neq K_3 - |x| \tag{21}$$

So, the hypothesis that the real and imaginary parts of the useful HWT coefficients are distributed following Laplace (exponential) laws can be made.

Taking into consideration the histograms of the noiseless wavelet coefficients it can be observed that their repartition is heavy-tailed. The majority of the wavelet coefficients have small magnitude (close to zero). So, the repartition of the wavelet coefficients can be considered of Laplace type.

On the basis of this hypothesis, another MAP filter named adaptive soft thresholding can be constructed.

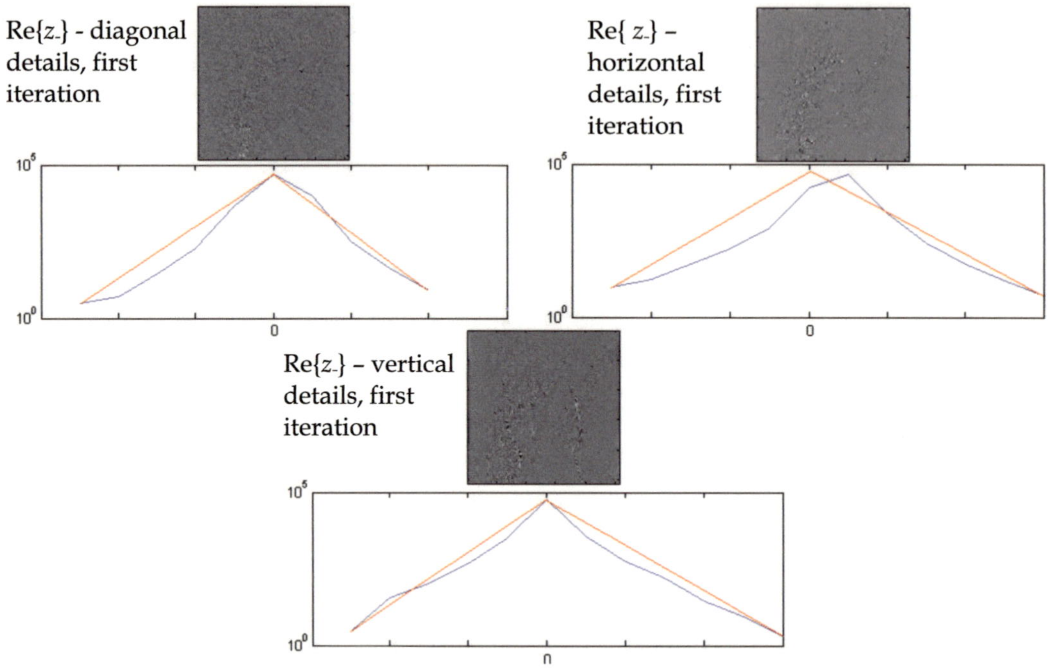


Fig. 11. The histograms of some subbands of the HWT of the image Lena computed using the mother wavelets Dau_20 are represented semi logarithmically (on the vertical axis are represented the logarithms of the values of the histograms) in blue. The corresponding linear dependencies are represented in red.

*2.4.1.2.2 The adaptive soft-thresholding filter*

If $n$ is Gaussian distributed and $w$ has a Laplacian distribution then the MAP filter becomes an adaptive soft thresholding filter (STF). The hypotheses for this type of marginal MAP filter are:

$$p_w(w) = \left(1/\sqrt{2}\sigma\right)\cdot e^{-\frac{\sqrt{2}|w|}{\sigma}} \text{ and } p_n(n) = \left(1/\sqrt{2\pi}\sigma\right)\cdot e^{-\frac{n^2}{2\sigma_n^2}} \tag{22}$$

The MAP filter equation (4) becomes:

$$\hat{w}(y) = \underset{w}{\operatorname{argmax}}\left\{\ln\left(1/2\sqrt{\pi}\sigma\sigma_n\right) - (y-w)^2/2\sigma_n^2 - \sqrt{2}|w|/\sigma\right\}. \tag{23}$$

To maximize the argument of the right hand side, the following equation must be solved:

$$\begin{aligned} &d/dw\left(\ln\left(1/2\sqrt{\pi}\sigma\sigma_n\right) - (y-w)^2/2\sigma_n^2 - \sqrt{2}|w|/\sigma\right) = 0 \Leftrightarrow \\ &\Leftrightarrow (y\text{-}w)/\sigma_n^2 - \left(\sqrt{2}/\sigma\right)\operatorname{sgn}(w) = 0 \end{aligned} \tag{24}$$

For $w$>0 it becomes:

$$(y\text{-}w)/\sigma_n^2 - \sqrt{2}/\sigma = 0 \Leftrightarrow w = y - \sqrt{2}\sigma_n^2/\sigma \tag{25}$$

For $w<0$ the equation (24) becomes:

$$(y - w) / \sigma_n^2 + \sqrt{2} / \sigma = 0 \Leftrightarrow w = y + \sqrt{2}\sigma_n^2 / \sigma \quad (26)$$

The condition $w>0$ implies the condition $y > \sqrt{2}\sigma_n^2 / \sigma$ and the condition $w<0$ implies the condition $y < -\sqrt{2}\sigma_n^2 / \sigma$. Taking into account the fact that any real number can be expressed as $y = \text{sgn}(y) \cdot |y|$, the solution of the equation (24) can be put in the following form:

$$\hat{w} = \begin{cases} \text{sgn}\, y\left(|y| - \sqrt{2}\sigma_n^2 / \sigma\right), & |y| < \sqrt{2}\sigma_n^2 / \sigma \\ 0, & |y| > \sqrt{2}\sigma_n^2 / \sigma \end{cases} \quad (27)$$

equivalent with the final form:

$$\hat{w} = \text{sgn}\, y\left(|y| - \sqrt{2}\sigma_n^2 / \sigma\right)_+ \quad (28)$$

where:

$$(X)_+ = \begin{cases} X, & \text{for} \quad X > 0 \\ 0, & \text{otherwise} \end{cases}. \quad (29)$$

This MAP filter is a STF with the threshold's value equal to $\sqrt{2}\sigma_n^2 / \sigma$. Because the quantity σ varies from subband to subband, the threshold's value is variant. In this respect the last MAP filter can be considered adaptive. Of course the two variances in equation (28) could be estimated using equations (16) and (19).

*2.4.1.2.3 Inter-scale dependence of wavelet coefficients*

The wavelet coefficients are characterized by interscale dependence.

The cross-correlation of two wavelet coefficients belonging to the 2D DWT of an image $f$, located in the same subband $k$ at the scales $m_1$ and $m_2 = m_1 + q$ and having the geometrical coordinates $(n_1 = 2^q n_1', p_1 = 2^q p_1')$ and $(n_2, p_2)$ respectively is given by the following relation:

$$\begin{gathered} R_{{}_f D_{m_1}^k \, {}_f D_{m_2}^k}\left(2^q n'_1 - n_2, 2^q p'_1 - p_2\right) = 2^{2m_1+q} R_f\left(2^{m_1+q}(n_2 - n_1'), 2^{m_1+q}(p_2 - p_1')\right) \\ * R_{\psi^k}(n_2 - n_1', p_2 - p_1'). \end{gathered} \quad (30)$$

which represents an inter-scale and intra-band dependency. If the mother wavelet $\psi^k$ generates by translations and dilations an **orthogonal** basis of $L^2(R^2)$ then its autocorrelation has the following property:

$$R_{\psi^k}(n,p) = \delta[n,p] \quad (31)$$

and the expression of the inter-scale and intra-band dependency of the coefficients cross-correlation becomes:

$$R_{{}_f D_{m_1}^k \, {}_f D_{m_2}^k}\left(2^q n'_1 - n_2, 2^q p'_1 - p_2\right) = 2^{2m_1+q} R_f\left(2^{m_1+q}(n_2 - n_1'), 2^{m_1+q}(p_2 - p_1')\right). \quad (32)$$

This cross-correlation is function of the autocorrelation of the input signal. If we refer to pairs of wavelet coefficients located at two consecutive decomposition levels ($q$=1, $m_2$= $m_1$+1) at the same coordinates ($n_1 = 2n_2, p_1 = 2p_2$) which are named parent coefficient (${}_fD^k_{m_1+1}(n_2,p_2)$) and child coefficient (${}_fD^k_{m_1}(n_1,p_1)$) the expression of their inter-scale dependence becomes:

$$R_{{}_fD^k_{m_1}\,{}_fD^k_{m_1+1}}(n_2,p_2) = 2^{2m_1+1}R_f(0,0) \tag{33}$$

which means that the child and the parent coefficient are correlated. So, if the magnitude of the parent coefficient is large (small) then the magnitude of the child coefficient is also large (small). Taking into consideration that the computation of the HWT reduces at the computation of the 2D DWTs of four related images $f$, $\mathrm{H}_x\{f\}$, $\mathrm{H}_y\{f\}$ and $\mathrm{H}_y\{\mathrm{H}_x\{f\}\}$, as can be seen in figure 8, a similar inter-scale dependence can be observed in the case of the pair of parent and child HWT coefficients:

$$R_{z_{+i}z_{+i}}[m_1,m_1+1,k,n_2,p_2]=2^{2m_1+1}\{R_{\mathrm{H}x}(0,0)+R_{\mathrm{H}x\mathrm{H}y}(0,0)+R_{\mathrm{H}y\mathrm{H}x}(0,0)+R_{\mathrm{H}y}(0,0)\}. \tag{34}$$

The MAP filters which take into account this dependence are multivariate and have better performance. Some bivariate probability density functions are constructed starting from the Gaussian scale mixture (GSM) statistical model [Selesnick, 2008].

*2.4.1.2.4 The GSM model*

The construction of the bivariate pdf can be done with the aid of Gaussian Scale Mixtures (GSM). This simple statistical model has been used to model natural signals such as speech and more recently the wavelet coefficients of natural images. The model is given in equation (35). It assumes that each vector of coefficients **w** is specified by a stationary bivariate zero mean Gaussian process **x** and a spatially fluctuating variance $z$.

$$\mathbf{w} = \sqrt{z}\cdot\mathbf{x}, \quad z \in \mathrm{R}, \quad \mathbf{w},\mathbf{x} \in \mathrm{R}^d \tag{35}$$

The multiplier $z$ is usually a function of the surrounding coefficient values (like the local variance of the coefficients within the same scale or a more complex function of the neighboring coefficients within the same and adjacent scales). The result is always leptokurotic (kurtosis ≥ 3), its distribution having long tails. The MMSE estimate with such priors takes the form of a locally adaptive Wiener-like estimator. Usually the number of elements of the vectors **x** and **w** is $d$=2. To model the self-reinforcing property of the coefficients, $z$ must be slowly varying but it does not need to be symmetric in all directions. It has been shown that for slowly varying $z$ this model can successfully simulate the high kurtosis and longer tails of the marginal distributions. The stationary portion of the model **x** is Gaussian distributed over a small neighborhood of wavelet coefficients. It is generally assumed that $z$ varies slowly enough to be considered constant over that neighborhood of coefficients. Under this assumption the model is now a particular form of a spherically invariant random process called a GSM. For a small neighborhood of coefficients at nearby spatial locations and scale, a GSM vector **w** is the product of two independent random variables: a positive scalar $z$ referred to as the hidden multiplier or mixing variable and a Gaussian random vector **x**. The pdf of the Gaussian vector **x** is given by:

$$p_{\mathbf{x}}(\mathbf{x}) = \left[1/\left(2\pi\sigma^2\right)^{d/2}\right]\cdot e^{-\frac{\|x\|^2}{2\sigma^2}} \tag{36}$$

(see (12)). Setting $a=\sqrt{z}$, we obtain $\mathbf{w}=a\cdot\mathbf{x}$ and the pdf of the random vector **w** is given by:

$$p_{\mathbf{w}}(\mathbf{w}) = \int_0^\infty p_a(a)\cdot\left(1/a^d\right)\cdot p_{\mathbf{x}}(\mathbf{w}/a)\,da \tag{37}$$

Tacking into account the relation of $a$ and $z$, the pdf of $a$ can be expressed on the basis of the pdf of $z$:

$$p_a(a) = p_z(z)/(da/dz) = p_z\left(a^2\right)/(1/2a) = 2a\cdot p_z\left(a^2\right) \tag{38}$$

Substituting (38) into (37), the expression of the pdf of **w** becomes:

$$p_{\mathbf{w}}(\mathbf{w}) = \int_0^\infty 2ap_z\left(a^2\right)\cdot\left(1/a^d\right)\cdot p_x(\mathbf{w}/a)\,da \tag{39}$$

It remains to specify the prior probability function $p_z(z)$ for the multiplier $z$. Some propositions of prior probability functions are made in [Selesnick, 2008]. One of these propositions is the Gamma law:

$$p_z(z) = \beta^\alpha/\Gamma(\alpha)\cdot z^{\alpha-1}\cdot e^{-\beta z}\text{ with } \alpha=\beta=(d+1)/2 \tag{40}$$

Substituting (40) into (39), the expression of the pdf of **w** becomes:

$${}^1p_{\mathbf{w}}(\mathbf{w}) = \left[2(\beta)^\alpha/\Gamma(\alpha)\right]\int_0^\infty a\cdot a^{2(\alpha-1)}\cdot e^{-\beta a^2}\cdot a^{-d}\cdot\left(2\pi\sigma^2\right)^{-\frac{d}{2}}\cdot e^{-\frac{\|\mathbf{w}\|^2}{2a^2\sigma^2}}da \tag{41}$$

Finally:

$${}^1p_{\mathbf{w}}(\mathbf{w}) = \left(2/\Gamma\left(\frac{d+1}{2}\right)\right)\cdot\left[(d+1)/2\right]^{\frac{d+1}{2}}\cdot\left(2\pi\sigma^2\right)^{-\frac{d}{2}}\cdot\sqrt{\left\{\pi/\left[(d+1)/2\right]\right\}}\cdot e^{-\frac{\sqrt{d+1}}{\sigma}\cdot\|\mathbf{w}\|} \tag{42}$$

This is a $d$-dimensional spherically-contoured multivariate pdf [Selesnick, 2008]. For $d$=2, this pdf is bivariate:

$${}^1p_{\mathbf{w}}(\mathbf{w}) = \left(3/2\pi\sigma^2\right)\cdot e^{-\frac{\sqrt{3}}{\sigma}\cdot\sqrt{w_1^2+w_2^2}}. \tag{43}$$

With the aid of this bivariate distribution the bishrink MAP filter [Sendur&Selesnick, 2002] can be constructed. This construction will be explained in the following section. For $d$=1, the pdf in (43) is univariate:

$$ {}_{1}p_w(w) = \left(\sqrt{2}/\sigma\right) \cdot e^{-\frac{\sqrt{2}}{\sigma}|w|} \tag{44} $$

This univariate pdf is of the form of the Laplace law; see the hypotheses of the adaptive STF, (22). In consequence the GSM hypothesis, even for the case of a unique scale, is useful for modeling the repartition of the wavelet coefficients. Other proposition of prior probability function made in [Selesnick, 2008] is the exponential $p_z(z) = e^{-z}$ for $z \geq 0$. The GSM model was also used for the conception of one of the best denoising methods [Portilla et al., 2003]. The prior probability function proposed in this reference is $p_z(z) = 1/z$, for $z \geq 0$. It must be remarked that strictly speaking this is not a pdf. Substituting this function in (39) and supposing that the pdf of the Gaussian vector is that given in (36) it can be written:

$$ {}^{4}p_{\mathbf{w}}(w_1,w_2) = -\frac{1}{\pi\|\mathbf{w}\|^2} \cdot e^{-\frac{\|\mathbf{w}\|^2 t^2}{2\sigma^2}} \Big|_0^\infty = \frac{1}{\pi\|\mathbf{w}\|^2} = \frac{1}{\pi\left(w_1^2 + w_2^2\right)}. \tag{45} $$

It results a third spherically-contoured bivariate pdf. A last bivariate pdf for the wavelet coefficients was proposed in [Achim&Kuruoglu, 2005]:

$$ {}^{5}p_{\mathbf{w}}(w_1,w_2) = \frac{\gamma}{2\pi\left(w_1^2 + w_2^2 + \gamma^2\right)^{\frac{3}{2}}}. \tag{46} $$

Using one of these models, in [Selesnick, 2008] is conceived a bivariate MAP filter called bishrink.

**2.4.1.2.5 The bishrink filter**

The noise is assumed i.i.d. Gaussian,

$$ p_{\mathbf{n}}(\mathbf{n}) = \frac{1}{2\pi\sigma_n^2} \cdot e^{-\frac{n_1^2 + n_2^2}{2\sigma_n^2}}. \tag{47} $$

The model of the noise-free image is given in (43):

$$ {}^{1}p_{\mathbf{w}}(\mathbf{w}) = \frac{3}{2\pi\sigma^2} \cdot e^{-\frac{\sqrt{3}}{\sigma}\sqrt{w_1^2 + w_2^2}} $$

a heavy tailed distribution. Substituting these two pdfs in the equation of the MAP filter (4) this equation becomes:

$$ {}^{1}\hat{\mathbf{w}}(\mathbf{y}) = \underset{\mathbf{w}}{\operatorname{argmax}} \left\{ \ln\frac{3}{4\pi^2\sigma^2\sigma_n^2} - \frac{(y_1 - w_1)^2}{2\sigma_n^2} - \frac{(y_2 - w_2)^2}{2\sigma_n^2} - \frac{\sqrt{3}}{\sigma}\sqrt{w_1^2 + w_2^2} \right\}. \tag{48} $$

The system of equations that gives the maximum of the right hand side of the last equation is:

$$\begin{cases} \dfrac{d}{dw_1}\left(\ln\dfrac{3}{4\pi^2\sigma^2\sigma_n^2}-\dfrac{(y_1-w_1)^2}{2\sigma_n^2}-\dfrac{(y_2-w_2)^2}{2\sigma_n^2}-\dfrac{\sqrt{3}}{\sigma}\sqrt{w_1^2+w_2^2}\right)=0 \\ \dfrac{d}{dw_2}\left(\ln\dfrac{3}{4\pi^2\sigma^2\sigma_n^2}-\dfrac{(y_1-w_1)^2}{2\sigma_n^2}-\dfrac{(y_2-w_2)^2}{2\sigma_n^2}-\dfrac{\sqrt{3}}{\sigma}\sqrt{w_1^2+w_2^2}\right)=0 \end{cases} \tag{49}$$

or:

$$\begin{cases} \dfrac{y_1-w_1}{\sigma_n^2}-\dfrac{\sqrt{3}}{\sigma}\cdot\dfrac{w_1}{\sqrt{w_1^2+w_2^2}}=0 \\ \dfrac{y_2-w_2}{\sigma_n^2}-\dfrac{\sqrt{3}}{\sigma}\cdot\dfrac{w_2}{\sqrt{w_1^2+w_2^2}}=0 \end{cases} \tag{50}$$

The system can be put in the following form:

$$\begin{cases} \dfrac{y_1}{\sigma_n^2}=\dfrac{\sqrt{3}}{\sigma}\cdot\dfrac{w_1}{\sqrt{w_1^2+w_2^2}}+\dfrac{w_1}{\sigma_n^2} \\ \dfrac{y_2}{\sigma_n^2}=\dfrac{\sqrt{3}}{\sigma}\cdot\dfrac{w_2}{\sqrt{w_1^2+w_2^2}}+\dfrac{w_2}{\sigma_n^2} \end{cases}$$

or equivalently:

$$\begin{cases} w_1=\dfrac{\sigma\sqrt{w_1^2+w_2^2}}{\sigma\sqrt{w_1^2+w_2^2}+\sqrt{3}\sigma_n^2}y_1 \\ w_2=\dfrac{\sigma\sqrt{w_1^2+w_2^2}}{\sigma\sqrt{w_1^2+w_2^2}+\sqrt{3}\sigma_n^2}y_2 \end{cases} \tag{51}$$

Computing the square of each equation it results:

$$\begin{cases} w_1^2=\dfrac{\sigma^2\left(w_1^2+w_2^2\right)}{\left[\sigma\sqrt{w_1^2+w_2^2}+\sqrt{3}\sigma_n^2\right]^2}y_1^2 \\ w_2^2=\dfrac{\sigma^2\left(w_1^2+w_2^2\right)}{\left[\sigma\sqrt{w_1^2+w_2^2}+\sqrt{3}\sigma_n^2\right]^2}y_2^2 \end{cases} \tag{52}$$

By adding the two equations it can be obtained:

$$w_1^2+w_2^2=\frac{\sigma^2\left(w_1^2+w_2^2\right)}{\left[\sigma\sqrt{w_1^2+w_2^2}+\sqrt{3}\sigma_n^2\right]^2}\left(y_1^2+y_2^2\right) \Leftrightarrow \sqrt{w_1^2+w_2^2}=\left(\sqrt{y_1^2+y_2^2}-\sqrt{3}\frac{\sigma_n^2}{\sigma}\right)_+ \tag{53}$$

Substituting this result in the two equations of the system in (52) it can be written:

$$\begin{cases} {}^{1}\hat{w}_1 = \dfrac{\left(\sqrt{y_1^2+y_2^2}-\sqrt{3}\dfrac{\sigma_n^2}{\sigma}\right)_+}{\sqrt{y_1^2+y_2^2}} y_1 \\ {}^{1}\hat{w}_2 = \dfrac{\left(\sqrt{y_1^2+y_2^2}-\sqrt{3}\dfrac{\sigma_n^2}{\sigma}\right)_+}{\sqrt{y_1^2+y_2^2}} y_2 \end{cases} \tag{54}$$

So the input-output relation of the bishrink filter is:

$$_1\hat{w}_1 = \frac{\left(\sqrt{y_1^2+y_2^2}-\dfrac{\sqrt{3}\sigma_n^2}{\sigma}\right)_+}{\sqrt{y_1^2+y_2^2}} y_1 \tag{55}$$

It can be observed that the bishrink filter is an estimator of the adaptive STF type. In this case the threshold's value is $t = \dfrac{\sqrt{3}\sigma_n^2}{\sigma}$. This estimator requires prior knowledge of the noise variance and of the marginal variance of the clean image for each wavelet coefficient. These quantities can be estimated using the equations (16) and (19). The sensitivity of the bishrink filter with the estimation of the noise standard deviation $\hat{\sigma}_n$ can be computed with the relation:

$$S_{\hat{w}_1}^{\hat{\sigma}_n} = \frac{d\hat{w}_1}{d\hat{\sigma}_n} \cdot \frac{\hat{\sigma}_n}{\hat{w}_1}$$

The input-output relation of the bishrink filter (55) can be put in the following form:

$$\hat{w}_1 = \begin{cases} \dfrac{\sqrt{y_1^2+y_2^2}-\dfrac{\sqrt{3}\hat{\sigma}_n^2}{\hat{\sigma}}}{\sqrt{y_1^2+y_2^2}} y_1, & \text{if } \sqrt{y_1^2+y_2^2} > \dfrac{\sqrt{3}\hat{\sigma}_n^2}{\hat{\sigma}} \\ 0, & \text{otherwise} \end{cases} \tag{56}$$

So, it can be written:

$$S_{\hat{w}_1}^{\hat{\sigma}_n} = \begin{cases} \dfrac{-2\sqrt{3}\hat{\sigma}_n^2}{\hat{\sigma}\sqrt{y_1^2+y_2^2}-\sqrt{3}\hat{\sigma}_n^2}, & \text{if } \sqrt{y_1^2+y_2^2} > \dfrac{\sqrt{3}\hat{\sigma}_n^2}{\hat{\sigma}} \\ 0, & \text{otherwise} \end{cases} \tag{57}$$

The absolute value of this sensitivity is inversely proportional to $\hat{\sigma}_n$. When the value of the estimation of the noise standard deviation is higher then the performance of the bishrink filter is poorer.

Another very important parameter of the bishrink filter is the local estimation of the marginal variance of the noise-free image $\hat{\sigma}$. The sensitivity of the estimation $\hat{w}_1$ with $\hat{\sigma}$ is given by:

$$S_{\hat{w}_1}^{\hat{\sigma}} = \begin{cases} \dfrac{\sqrt{3}\hat{\sigma}_n^2}{\hat{\sigma}\sqrt{y_1^2+y_2^2}-\sqrt{3}\hat{\sigma}_n^2}, & \text{if } \sqrt{y_1^2+y_2^2} > \dfrac{\sqrt{3}\hat{\sigma}_n^2}{\hat{\sigma}} \\ 0, & \text{otherwise} \end{cases} \tag{58}$$

This is a decreasing function of $\hat{\sigma}$. The precision of the estimation based on the use of the bishrink filter decreases with the decreasing of $\hat{\sigma}$. Similar sensitivity analyses can be accomplished for the zero order Wiener filter or for the adaptive soft-thresholding filter, concluding that their worst behavior corresponds to the homogeneous regions of their noise-free input image component.

Secondly, the local variance of a pixel gives some information about the frequency content of the region to which the considered pixel belongs. If the pixels of a given region have low local variances then the considered region contains low frequencies. If these pixels have high local variances then the considered region contains high frequencies.

The bishrink is a local bivariate MAP filter. Its performance depends on the quality of the estimation of a parameter, the local variance of the noiseless component of the acquired image. The quality of this estimate depends on the shape and the size of the estimation window. These estimation windows have different shapes in subbands with different preferential orientations highlighting the better directional selectivity of DTCWT and HWT versus the DWT.

Other MAP filters can also be considered. A MAP filter constructed on the basis of a bivariate Cauchy model is described in [Achim&Kuruoglu, 2005]. The Pearson statistical model is exploited for the construction of the marginal MAP filter presented in [Foucher et al., 2001]. The model of generalized Gaussian (GG) is used in [Argenti et al., 2006].

### 2.5 Despecklisation systems

We will compare in the following the classical despecklisation systems with some modern denoising systems conceived for the reduction of speckle noise in SONAR images based on complex WTs, namely the 2D DT CWT and the HWT.

The modern systems have the architecture presented in figure 12. The block named Sensitivity reduction corrects the drawbacks of the additive noise denoising kernel. One of the disadvantages of homomorphic denoising methods is the introduction of an undesirable bias. The expectations of the result and of the noiseless component of the acquired image are different. Their difference is given by the bias already mentioned. The proposed despecklisation system corrects this bias with the aid of two mean computation systems. The correction is based on the property of the speckle noise to have unitary expectation. The first mean computation system in figure 1 estimates the mean of the acquired image which is equal with the mean of its noise-free component (because the speckle has unitary mean). The second mean computation system in figure 12 computes the mean of the image at the output of the Sensitivity reduction block. This value is extracted and the mean of the noiseless component of the acquired image is added.

Both additive noise denoising kernels which exploit the 2D DT CWT and the HWT use the bishrink filter associated with the corresponding complex WT. The architecture of the corresponding Sensitivity reduction blocks is different for the two complex WTs.

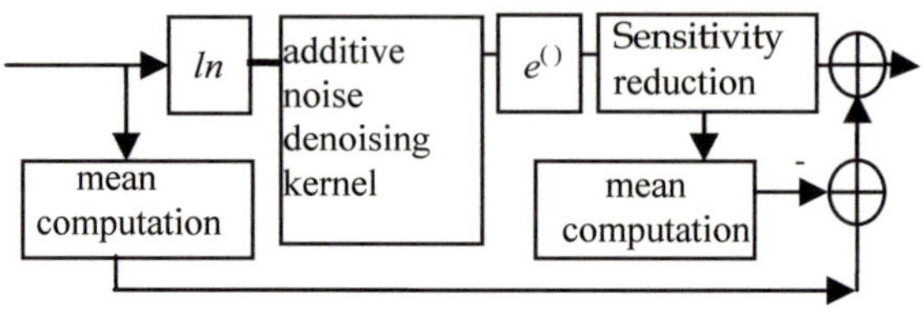


Fig. 12. The architecture of the proposed denoising system. The mean correction mechanism and the additive noise denoising kernel are highlighted.

The bishrink filter can be used for despecklisation in the wavelets domain because its statistical model for the noise is appropriate as can be seen from the following experiment. The speckle can be modeled using a Rayleigh distribution with unitary mean. It is obtained computing the square root of a sum of squares of two white Gaussian noises having the same variance. The first image of figure 13 contains a normalized representation of the pdf in equation (43) particularized for a given variance of the noise. The second image of figure 13 represents the normalized bivariate histogram of the HWT coefficients **of the logarithm of the noise.** It was obtained considering the HWT coefficients $z_{+r}$ corresponding to horizontal details from the first two decomposition levels. The similarity of the surfaces from the two images in figure 13 proves the validity of the bivariate noise statistical model used for the construction of the bishrink filter and the possibility to use this filter in despecklisation applications.

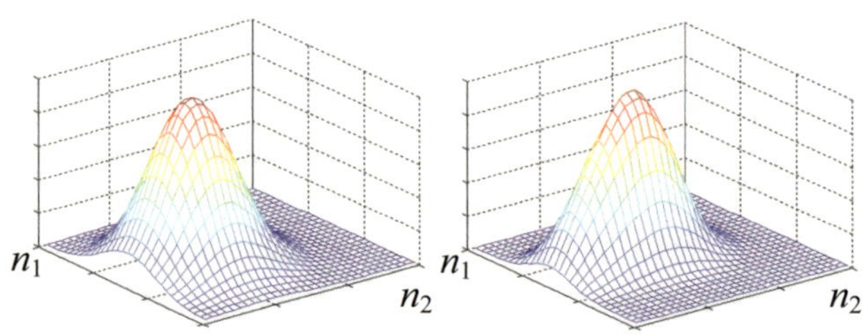


Fig. 13. The representation of the pdf in eq. (49) (left) and the corresponding bivariate histogram (right).

### 2.6 Simulations results

We present two types of simulation results: for synthesized speckle noise and for real SONAR images. The performance of the simulation results reported in this chapter is appreciated on the basis of some quality measures. The first one is the peak signal to noise ratio (PSNR) which is defined in the following. Let $s$ and $\hat{s}$ denote the clean and the denoised images. The root mean square (rms) of the approximation error is computed by:

$$\varepsilon = \sqrt{\frac{1}{N_p}\sum_q \left(s_q - \hat{s}_q\right)^2}$$

where $N_p$ is the number of pixels. The PSNR in dB is given by:

$$\text{PSNR} = 20\log_{10}\left(\frac{255}{\varepsilon}\right)$$

Because its computation requires the knowledge of $s$ the PSNR is computed in the experiments associated with synthesized speckle noise. Another quality measure is represented by the method's noise. For multiplicative noise (as the speckle is) the method's noise is defined by the ratio between the acquired image and the denoising result. This noise must be as similar as possible with the noise which perturbed the noiseless component of the acquired image. As the PSNR, the method noise can be appreciated only in the experiments associated with synthesized speckle. A third measure of quality of the denoising of a homogeneous region is given by the enhancement of the equivalent number of looks (ENL):

$$ENL = \left(\frac{\text{mean}}{\text{standard deviation}}\right)^2.$$

The enhancement of ENL is computed as the ratio of the output and input ENLs of the same region. This is a quality measure which can be appreciated in the case of real SONAR images as well. We have used two types of synthesized speckle. The first type is synthesized using a Rayleigh distribution with unitary mean and it is associated with the image Lena. The second type is synthesized using the method proposed in [Walessa&Datcu, 2000] and is associated with the test image presented in figure 14 [Walessa&Datcu, 2000].

Fig. 14. A test image obtained by the multiplication of a mosaic of real and synthesized images with the speckle noise generated in [Walessa&Datcu, 2000].

To run these simulations on real SONAR images we have used a database obtained from IFREMER Brest France. We are thankful for this opportunity. Besides these three objective quality measures, we considered the visual aspect of the results as well.

### 2.6.1 Synthesized images

We have applied first the classical despecklisation methods to the test image in figure 14 obtaining the results presented in figure 15. The corresponding filters are indexed with two indices, the first one specifying the size of the analysis window (for example 7 for a rectangular window with size 7×7) and the second one the number of looks of the image treated (1 in all the three cases).

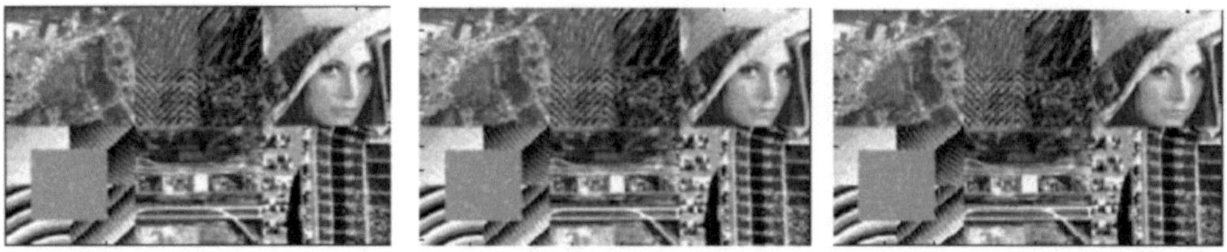

Fig. 15. A comparison of the classical despecklisation filters; from left to right: Kuan 9-1 , Lee 7-1, Frost 5-1.

The PSNRs obtained separating the sub-image Lena from the test image and from the results in figure 15 are indicated in Table I and compared with the results obtained by applying the denoising methods which associates the bishrink filter with the 2D DT CWT (fifth column) and with the HWT (sixth column) respectively.

| Noisy | Lee | Frost | Kuan | 2D DT CWT | HWT |
|---|---|---|---|---|---|
| 21.4 | 27.2 | 27.0 | 28.1 | *31.4* | 31.9 |

Table 1. The PSNRs of different speckle denoising methods (in dB).

The superiority of the denoising methods acting in the wavelet domain is obvious. The comparison in figure 16 permits the appreciation of the visual aspect of the results of the methods based on 2D DT CWT and HWT for the treatment of the test image in figure 14. These results are compared in figure 16 with the result of another despecklisation method applied in the spatial domain [Walessa&Datcu, 2000]. The superiority of the directional selectivity of the methods based on complex wavelets can be observed comparing the interior of the regions highlighted in yellow. Indeed the methods based on wavelets conserve better the details of the hat. The result with the best visual aspect in figure 16 corresponds to the utilization of the HWT. In figure 17 are compared the results of three different methods based on wavelets. The first two despecklisation methods are those based on 2D DT CWT and HWT and the third one was proposed in [Argenti et al., 2006] and is based on the association of the 2D UDWT with a MAP filter. The 2D UDWT is computed either with the aid of the Daubechies mother wavelets with eight vanishing moments, db8 or with the pair of biorthogonal mother wavelets bior9.7. The first denoising algorithm proposed in [Argenti et al., 2006] performs a local linear minimum mean square error (LLMMSE) filtering in the UDWT domain. The second one uses a MAP filter constructed supposing that the noise-free wavelet coefficients and the wavelet coefficients of the noise are distributed according to Generalized Gaussian Distributions.
The parameters of those pdfs are estimated for each pixel of the input image. The corresponding MAP filter equation is solved with the aid of numerical methods. All the three results in figure 17 have a good visual aspect. The methods based on complex wavelets (2D DT CWT and HWT) prove a better output PSNR, a better directional selectivity, a better treatment of the contours and textures and a better contrast preservation but the method based on 2D UDWT treats better the homogeneous regions. The method's noise corresponding to the use of 2D DT CWT is quite identical with the original noise (compare images (b) and (d)) some differences appearing in the dark regions of the image Lena.

Fig. 16. A comparison of the visual aspect of the results of three despecklisation methods; From up to bottom: noisy sub-images, results obtained applying the method in [Walessa&Datcu, 2000], results obtained applying the method in [Ifremer, 2010] and results obtained applying the method based on the HWT.

Fig. 17. A comparison of three despecklisation methods acting in the wavelets domain; (a) noisy image (PSNR=21.4 dB), (b) speckle noise distributed following a Rayleigh law with unitary mean, (c) result obtained with the method based on 2D DT CWT (PSNR=31.4 dB), (d) 2D DT CWT method's noise, (e) result obtained with the method based on HWT (PSNR=32.5 dB), (f) result obtained with the method in [Argenti et al., 2006] (PSNR=31.1 dB).

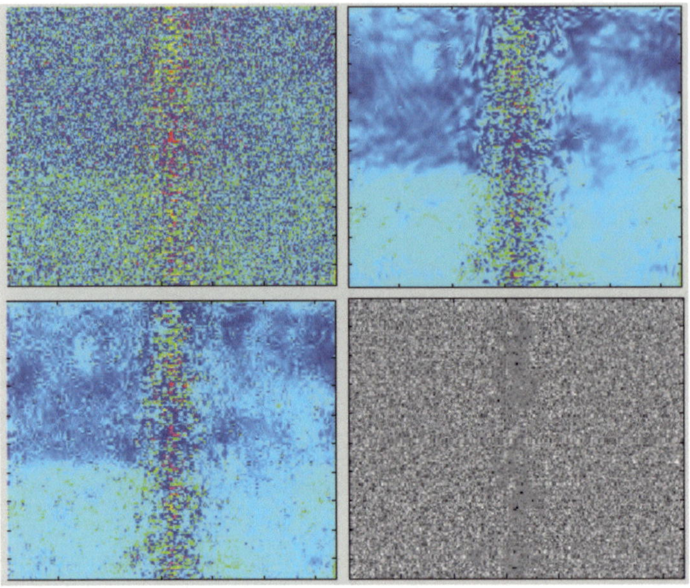

Fig. 18. A comparison of the despecklisation methods based on 2D DT CWT and HWT. From left to right and up to bottom; acquired SONAR image, result of the despecklisation method based on 2D DT CWT, result of the despecklisation method based on HWT, 2D DT CWT method's noise.

### 2.6.2 Real images

In figure 18 is made a comparison of the despecklisation methods based on 2D DT CWT and HWT applied to a very difficult SONAR image. The first picture in figure 18 represents the acquired SONAR image. It is very difficult because it has small size (200×320) and because the relief of the sea floor is flat in the region considered. So, all the pixels have intensities of similar values. The image was artificially colored, to see better the small intensity differences. Even the specular region can be observed with difficulty in this picture. This image is very noisy. It can be observed, analyzing the results of denoising, a demarcation line between two regions (the blue one and the green one) with pixels having different intensities. The difference between the intensities of the pixels belonging to the two regions is smaller than 1 dB. So, the proposed denoising method increases the discrimination capability of the analyst. Another consequence of the denoising based on wavelets is the better separation of the specular region from the rest of the scene. It can be also observed that a great amount of noise was eliminated by the proposed denoising methods. Comparing the second and the third pictures in figure 18, it can be noticed that the result of the despecklisation method based on the 2D DT CWT is over smoothed. So, the despecklisation method based on HWT seems to be the best choice. The comparison of the despecklisation methods based on 2D DT CWT and HWT is continued in figure 19, where other two SONAR images are considered. On the first column are represented the row SONAR images, on the second column are represented the results obtained applying the despecklisation method based on 2D DT CWT and on the last column are represented the results obtained applying the despecklisation method based on HWT. Analyzing figure 19, it can be observed that the two despecklisation methods based on wavelets are equivalent from the visual aspect point of view. Finally, we compare in figure 20 the two denoising methods based on 2D DT CWT and HWT respectively using a SONAR image containing a higher amount of information. The performance obtained by the two despecklisation methods based on wavelets for homogeneous regions is certificated by the important enhancements of ENL obtained considering a region of 120×1000 pixels. The gain in performance of the method based on HWT can be explained by the superiority of HWT versus 2D DT CWT in despecklization applications.

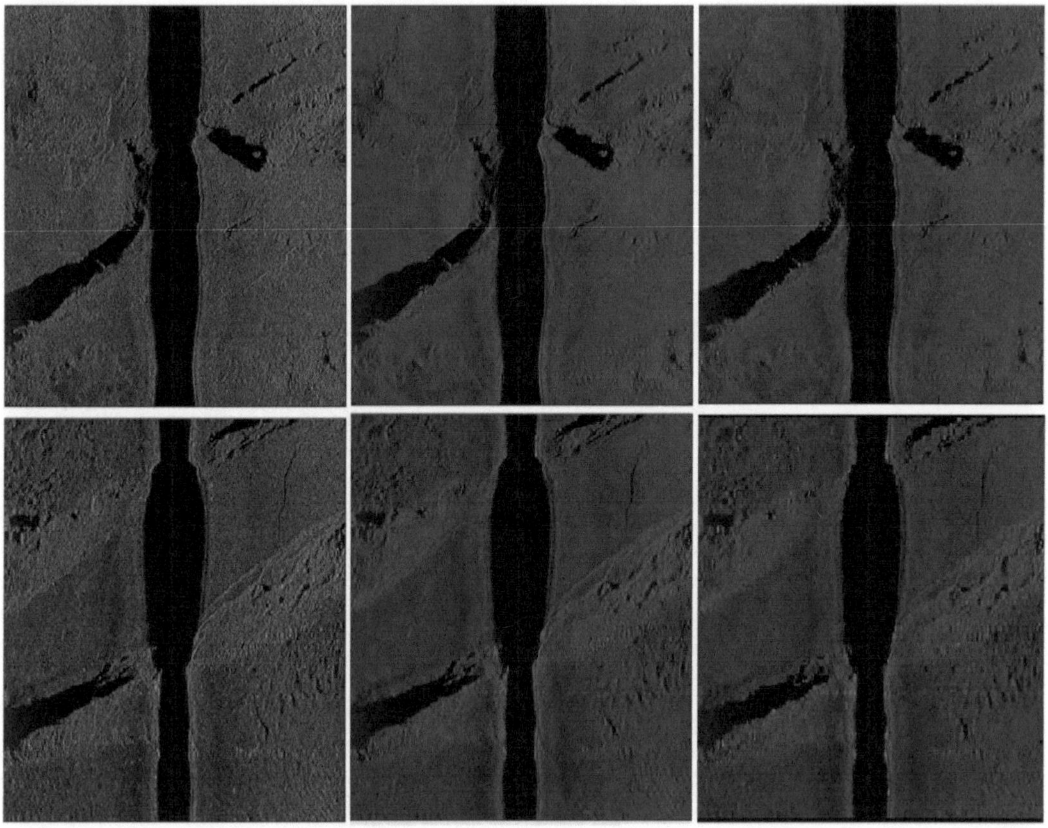

Fig. 19. A comparison of two despecklisation methods based on wavelets; First column: raw images, Second column: results obtained applying the despecklisation method based on 2D DT CWT, Third column: results obtained applying the despecklisation method based on HWT.

## 3. Conclusion

This book is dedicated to SONAR systems. Despite the actual proliferation of this type of images, there are not numerous publications dealing with their denoising. In this chapter was presented the particular case of SONAR images starting with an overview of speckle removal techniques both in the spatial domain and in the wavelet domain. We have proved by simulations the superiority of the methods applied in the wavelet domain. These methods were classified according to the type of the wavelet transform used: 2D DWT, 2D

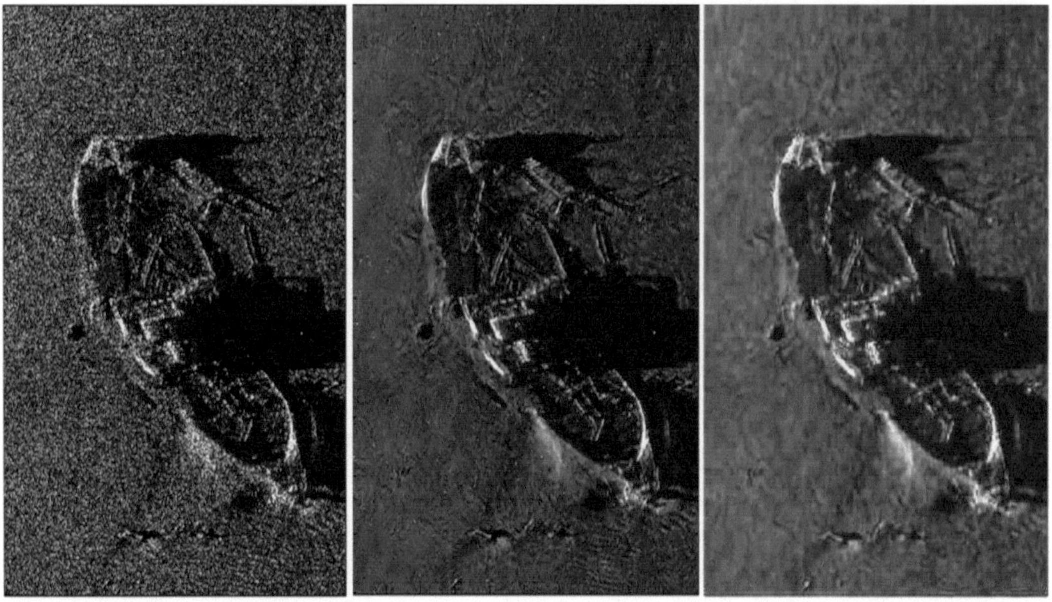

Fig. 20. Left: sea-bed sonar Swansea image (we are thankful to GESMA for providing this image, ENL=3.4), Middle: result of the method based on 2D DT CWT (ENL=102), Right: result of the method based on HWT (ENL=150).

UDWT, 2D DT CWT and HWT. In the case of the 2D DT CWT and HWT, we have used homomorphic filtering techniques and applied the bishrink filter in the 'additive noise denoising kernel' block, in figure 12. The proposed algorithms use two of the best WTs, the 2D DT CWT and the HWT and a very good MAP filter which can be associated for despecklisation purposes. The undesirable bias introduced by the homomorphic methods is corrected using a clever combination of two expectation computation blocks.

The results obtained were satisfactory, especially those in which HWT with Biorthogonal 9/7 as mother wavelet was associated with the bishrink filter, followed by a supplementary correction applied to homogeneous areas. We have proved by simulation that the HWT is a better choice than the DWT, the UWT or the DT CWT for SONAR images despecklisation. The denoising method based on HWT outperforms other remote sensing images denoising methods from the visual aspect (it preserves better the contrast of the noiseless component of the image which must be denoised), the PSNR enhancement and the enhancement of ENL points of view. It is faster than the other despecklisation methods as well.

## 4. Acknowledgment

The research reported in this chapter was developed in the framework of a grant funded by the Romanian Research Council (CNCSIS) with the title "Using wavelets theory for decision making" no. 349/13.01.09. We have established collaboration with the specialists from the French Sea Institute, IFREMER, from Brest; Xavier Lurton and Jean-Marie Augustin, with respect to the denoising of SONAR images. We also acknowledge the contribution of Professor Jean-Marc Boucher, the Ph. D. Advisor of Ioana Firoiu.

## 5. References

[Ifremer, 2010] http://www.ifremer.fr/fleet/acous_sism/sonarscope/index.html.

[Abry, 1994] Abry P. (1994), *Transformées en ondelettes-Analyses multirésolution et signaux de pression en turbulence.* PhD thesis, Université Claude Bernard, Lyon, France.

[Achim&Kuruoglu, 2005] Achim A. and Kuruoglu E. E. (2005), Image Denoising Using Bivariate α-Stable Distributions in the Complex Wavelet Domain, *IEEE Sig. Proc. Let.*, 12 (1), January 2005, pp. 17-20, ISSN 1070-9908.

[Argenti et al., 2006] Argenti F., Bianchi T. and Alparone L. (2006), Multiresolution MAP Despeckling of SAR Images Based on Locally Adaptive Generalized Gaussian PDF Modeling, *IEEE Trans. on Image Proc.*, vol. 15, no. 11, Nov. 2006, pp. 3385-3399, ISSN 1057-7149.

[Argenti et al., 2009] Argenti F., Bianchi T., di Scarfizzi G.M., Alparone L. (2009), LMMSE and MAP Estimators for Reduction of Multiplicative Noise in the Nonsubsampled Contourlet Domain, *Signal Processing*, vol. 89, no. 10, Oct. 2009, pp. 1891–1901, ISSN 0165-1684.

[Buades, 2007] Buades A. (2007), *Image and film denoising by non-local means*, Ph. D. Thesis, Universitat de les Iles Baleares.

[Davenport, 2010] Davenport C., *Commutative Hypercomplex Mathematics*, December 17, 2010, http://home.comcast.net/ cmdaven/hypercplx.htm.

[Donoho&Johnstone, 1994] Donoho D. L. and Johnstone I. M. (1994), Ideal spatial adaptation by wavelet shrinkage, *Biometrika*, 81(3), March 1994, pp. 425-455, ISSN 0006-3444.

[Frost et al., 1982] Frost V. S., Stiles J. A., Shanmugan J. A. and Holtzman J. C. (1982), A Model for Radar Images and Its Application to Adaptive Digital Filtering of Multiplicative Noise, *IEEE Transactions on Pattern Analysis and Machine Intelligence*, Vol. PAMI-4, No. 2, March 1982, pp. 157 - 166 , ISSN 0162-8828.

[Foucher et al., 2001] Foucher S., Benie G. B. and Boucher J.-M., (2001), Multiscale MAP Filtering of SAR images, *IEEE Trans. Ima. Proc.*, vol. 10, no.1, January 2001, pp. 49-60, ISSN 1057-7149.

[Holschneider et al., 1989] Holschneider M., Kronland-Martinet R., Morlet J. and Tchamitchian P. (1989), Wavelets. Time-frequency Methods and Phase Space, chapter *A Real-Time Algorithm for Signal Analysis with the Help of the Wavelet Transform*, pages 289–297, Springer-Verlag, ISBN-10: 3540530142, Berlin, Germany.

[Jalobeanu et al., 2003] Jalobeanu A., Blanc-Féraud L. and Zerubia J., (2003), Natural image modeling using complex wavelets". In *Proc. of SPIE, Wavelets X*, Volume: 5207, 2003, pp. 480-494, ISBN: 9780819450807.

[Katkovnik et al., 2006] Katkovnik V., Egiazarian K. and Astola J. (2006), *Local Approximation Techniques in Signal and Image Processing*, SPIE Press, ISBN 0-8194-6092-3, Bellingham, Washington, USA.

[Kingsbury, 1998] Kingsbury N. G. (1998), "The dual-tree complex wavelet transform: a new efficient tool for image restoration and enhancement". In *Proc. European Signal Processing Conference, EUSIPCO 98*, Rhodes, pp. 319–322, ISBN 960-7620-05-4.

[Kingsbury, 1999] Kingsbury N. G., (1999), "Image processing with complex wavelets". *Philosophical Transactions of the Royal Society London A*, vol. 357 no. 1760, September 1999. pp. 2543–2560, Online ISSN: 1471-2962.

[Kingsbury, 2001] Kingsbury N.G. (2001), "Complex wavelets for shift invariant analysis and filtering of signals", *Journal of Applied and Computational Harmonic Analysis*, 10, no.3 , May, 2001, pp. 234–253, ISSN 1063-5203.

[Kuan et al., 1987] Kuan D. T., Sawchuk A. A. and Strand T. C., (1987), Adaptive restoration of images with speckle, *IEEE Trans. ASSP*, Vol. 35, No. 3, Mars 1987, pp. 373-383, ISSN: 00963518.

[Lee, 1981] Lee J. S., (1981), "Speckle Analysis and Smoothing of Synthetic Aperture Radar Images", *Computer Graphics and Image Processing*, Vol. 17, no.1, September, 1981, pp. 24-32, ISSN: 0146-664X.

[Lurton, 2002] Lurton X., (2002), *An Introduction to Underwater Acoustics, Principles and Applications.* Springer, ISBN 3-540-42967-0, Chichester, UK.

[Mallat, 1999] Mallat S. (1999), *A Wavelet Tour of Signal Processing*, Academic Press, ISBN 0-12-466606-X, San Diego, CA, USA.

[Nason, 2002] Nason G. P., (2002), "Choice of wavelet smoothness, primary resolution and threshold in wavelet shrinkage", *Statistics and Computing*, vol. 12, no. 3, Mars, 2002, pp. 219-227, ISSN : 0960-3174.

[Portilla et al., 2003] Portilla J., Strela V., Wainwright M. J. and Simoncelli E. P., (2003), "Image denoising using Gaussian scale mixtures in the wavelet domain," *IEEE Trans. on Image Processing*, vol. 12, no. 11, October 2003, pp. 1338–1351, ISSN: 1057-7149.

[Selesnick, 2001] Selesnick I. W., (2001), "The Characterization and Design of Hilbert Transform Pairs of Wavelet Bases". In *2001 Conference on Information Science and Systems*, The Johns Hopkins University, Mars 21-23, 2001, ISSN 1937-4771.

[Selesnick, 2008] Selesnick I. W., (2008), "The Estimation of Laplace Random Vector in AWGN and the Generalized Incomplete Gamma Function", *IEEE Trans. on Signal Processing*, vol. 56, no. 8, August 2008, pp. 3482-3496, ISSN: 1053-587X .

[Sendur&Selesnick, 2002] Sendur L. and Selesnick I. W., (2002), "Bivariate shrinkage functions for wavelet-based denoising exploiting interscale dependency", *IEEE Trans. on Signal Processing*, vol 50 no. 11, November 2002, pp. 2744-2756, ISSN: 1053-587X.

[Walessa&Datcu, 2000] Walessa M. and Datcu M., (2000), "Model-Based Despeckling and Information Extraction from SAR Images," *IEEE Transactions on Geoscience and Remote Sensing*, vol. 38, no. 5, September 2000, pp. 2258 – 2269, ISSN: 0196-2892.

9

# SONAR Systems and Underwater Signal Processing: Classic and Modern Approaches

Hossein Peyvandi[1], Mehdi Farrokhrooz[2],
Hossein Roufarshbaf[2] and Sung-Joon Park[3]
[1]*Scientific Applied College of Telecommunication, Tehran,*
[2]*Electrical and Computer Engineering Dept., George Mason University,*
[3]*Department of Electrical Engineering, Gangneung-Wonju National University,*
[1] *Iran*
[2]*USA*
[3]*South Korea*

## 1. Introduction

Although SONAR systems have been in practical use since the turn of the 20[th] century, the Titanic tragedy in Apr. 1912 has become an inflection point for SONAR system and development of its derivations in which SONAR system can be considered as one of the most developed engineering systems. SONAR systems development during especially World War II is undeniable in terms of applications of SONAR systems, SONAR signal processing either in active or in passive systems, however, signal processing in the new area of underwater wireless networks are mainly based on academic research around the world. The main research subjects have concerned with detection/ classification, tracking and telecommunication tasks. In this chapter, we discuss research in signal processing for SONAR systems and underwater wireless networks including classic and modern approaches based on intelligent methods. Each method has been briefly discussed as more details can be found through the references at the end.

Detection/Classification task of SONAR signals, especially in the passive SONAR system, is one of the most challenging areas in this field. In passive SONAR system, received noises with SNR as low as -15dB are very common at the receiver (Urick 1983). Added reverberation and scattering, which are similar to phenomena as clutter in RADAR are responsible for complexity in this task of a SONAR system. There are differences between RADAR and SONAR signal processing. The challenge of SONAR signal processing is to detect/classify targets with very weak noises whereas the challenge of RADAR signal processing is to discriminate Target-of-Interest (ToI) among many uninteresting tracks. In signal processing literature, SONAR signal can be classified to wide range of applications according to the SONAR mode, i.e. active or passive SONAR system, and the nature of received signal, i.e. radiated acoustic noise of marine vessels, received active SONAR ping's echo from ships, bats' echo or marine mammals' sound, and so on. The passive SONAR plays an important role in the modern naval battles. In addition to the long range detection, this mode of SONAR systems works covertly so the underwater surveillance systems utilize these two advantages of passive SONAR to stealthily monitor the surface and sub-surface marine vessels. The "class" of marine vessels is defined according to the application of surveillance system for example surface and sub. For decades, the trained people classified and recognized the class of marine vessels by listening to their radiated noise. Substituting these people with intelligent systems for classifying marine vessels based on their acoustic radiated noise is one of the hot topics in signal processing and artificial intelligence. In this chapter, we will present a brief introduction on radiated and background noise in the sea. In addition, we present a literature review and some case studies.

One of the best techniques in the classic era was Matched-Field Processing (MFP), which was originally developed for SONAR signal processing but has subsequently been transitioned to the RADAR applications. Another method based on Hidden Markov Model (HMM), inspired from speech processing (Becchetti 1999) researches, was exploited to recognize the noises of desired targets. The main idea was determining similarity among detected signals, which were enhanced during processing and recoded noises of desired targets as data. Approaches using the classic methods were eligible to cooperate with human experts to detect/ classify the targets. The well-known classic methods are explained in this chapter. Nevertheless, we show that the classic methods did not have capability of consistency especially when the new techniques to silence targets were exploited as used for atomic submarines. On the other hand, Artificial Neural Networks (ANN) attracted several interests again in academic/industrial areas around the world after landmark article of Hopfield (Hopfield 1982), and Rumelhart and McClelland book on Parallel Distributed Processing (Rumelhart 1987). In the decade of $90^{th}$, research on intelligent processing of SONAR signals has been increasingly grown. We will discuss research headlines on intelligent processing of SONAR signals in the past two decades, which we named as modern methods. We considered two different types of neural networks and their expansions for SONAR signal processing.

Although intelligent methods outperformed classic methods in SONAR signal processing using the same data, they need to large amount of real data for learning the artificial neural networks, thus a considerable number of research papers are based on simulations of SONAR signals and environments. In addition, new trends on mixing classic and modern methods have been introduced. In fact, mixed methods strengthen advantages of each approach and reduce their disadvantages. For instance, the hybrid method of HMM/ANN models has produced significant results in SONAR signal processing in which considerable research papers have been published in this area. We will also explain some robust similarity measure named Hausdorff Similarity Measure (HSM), which can be exploited in both classic and modern methods for detection/ classification task.

Target tracking as another important part of SONAR systems will be discussed. In a target tracking problem, the tracker sequentially estimates the target parameters of interest based on the sensor observations. The application of target tracking is in variety of scientific and engineering including SONAR, RADAR, biomedical imaging, computer vision, etc. In an active SONAR system, a transmitter sends signals, or pings, in the surveillance region, and receivers look for the return signals. In a passive SONAR system, the receiver collects the incoming signals with power greater than a preset threshold. In both active and passive

SONAR systems, the measured signals, known as contacts, are reflected either from targets or from other undesired sources. In the latter case, the measured signal is known as a false alarm or clutter as mentioned before. In a SONAR target tracking system, the existence of clutter measures, nonlinear relation between the observation signal and the target parameters, and low detectable target signals add more uncertainty in the tracking system and must be addressed by the tracker.

Finally, in the last section of the chapter, one application of signal processing methods for underwater wireless networks including wireless telecommunication networks have been explained. Since, enhanced Digital Signal Processor (DSP) hardware modules are needed for modems of underwater wireless networks, fast, robust and reliable methods have been exploited to implement such hardware. The new area of security in underwater wireless networks will be discussed at the end of the chapter.

### 1.1 The literature review

Some studies on this special topic of SONAR systems use real data of passive SONAR to evaluate the performance of their suggested system while the SONAR data acquisition is a time and money consuming project, so most of works use the simulated data. The selection of discriminating features and classifiers are two important issues, as well. Implementation issues for HMM can be find in (Becchetti 1999) which implementation issues of Markov Chains have been also considered. To develop a sensor-adaptive Autonomous Underwater Vehicle (AUV) technology specifically directed toward rapid environmental assessment and mine countermeasures in coastal environments a low-frequency SONAR system has been introduced in Generic Ocean Array Technology Sonar (GOATS) joint research program at MIT (Eickstedt 2003).

The discriminating features introduced in (Lourens 1988) are the locations of poles of the second order Autoregressive (AR) model. These are used for classifying the noises of three propulsion systems (High speed diesel, Low speed diesel and Turbine). The authors in (Rajagopal 1990) defined four classes of marine vessels. These classes can be recognized by the vessels speed, the blade rate of propeller, the location of tonal components of machinery, the gear noise, the injector noise and the low frequency radiation from hull of marine vessels. All of these features could be extracted from Power Spectral Density (PSD) (Stoica 2005) of radiated noise.

The algorithm suggested in (Li 1995) uses six parameters extracted from power spectrum. In this method, a standard feature vector is obtained for each category of ships. Then, the weighted distance between each standard vector and the extracted feature vector from test data is calculated. This distance determines the category to which the test data belongs. There are nine unknown parameters in the feature extraction process and the weighted distance calculation that their values must be selected. Nothing is suggested about selection of these parameters in (Li 1995). So, the implementation of this method is difficult or even impossible.

The discriminating features suggested in (Soares-Filho 2000) are based on power spectral density of radiated noise of four different classes of ships. In this paper the background noise has been estimated using Two-Pass Split-Window (TPSW) (Nielsen 1991) and the tonal components were extracted by subtracting the ambient noise from total power spectral density.

The neural network is the suggested classifier in this paper. The suggested method in (Ward 2000) is based on the Short Time Fourier Transform (STFT) as features and Finite Impulse Response Neural Network (FIRNN) as classifier. For performance evaluation, the authors have utilized the recorded data by Defense Research Establishment Atlantic using underwater sonobuoys in the Bedford Basin off Nova Scotia, Canada. The authors of (Farrokhrooz 2005) represented the acoustic radiated noise of ships by an AR model with appropriate order and coefficients of this model are used for classification of ships. A Probabilistic Neural Network (PNN) (Duda 2000) is used as the classifier and the AR model coefficients are used as the feature vector to this classifier. The performance of this method is examined by using a bank of real data files.

The authors of (Xi-ying 2010) have analyzed the advantages and disadvantages between discriminating features extracted from power spectral density and higher order spectrum, and then combined the power spectrum density estimation and higher order spectrum to extract the distinguishable characteristics synthetically. The proposed classifier is a kind of Back Propagation (BP) neural network (Duda 2000) with some modifications. Two sets of discriminating features proposed in (Farrokhrooz 2011). The first set of features is extracted from AR model of radiated noise and the other is directly extracted from power spectral density of radiated noise. The proposed classifier is the modified probabilistic neural network, which is referred to Multi-Spread PNN (MS-PNN) and a method for estimating the parameters of classifier.

Advanced technologies in SONAR systems have been thoroughly presented in (Silva 2009), the underwater wireless telecommunication networks challenges, signal processing techniques and current studies have been discussed in (Stojanovic 2006, Stojanovic 2008), State-of-the-Art in underwater acoustic sensor networks have been studied in (Akyildiz 2006) and a survey on underwater networks can be found in (Peterson 2006). In addition, hardware issues have been considered in (Benson 2008) as pioneer works of underwater acoustic networks researches and developments. A secure technique for underwater wireless networks has been recently introduced by (Peyvandi 2010).

## 2. Classic approaches

### 2.1. Basic Concepts

SONAR systems are divided into two main categories - active and passive. Passive SONAR system (figure 1: (a)) provides monitoring the undersea environment without sending energy through the water. On the other hand, active SONAR system can act same as RADAR using responses from signals sent towards targets.

Underwater signals obtained from passive SONAR contain valuable clues for source identification even in high noisy environments. Attempts on detection/classification of acoustic signals based on spectral characteristics have met little success in early era of SONAR system development (Urick 1963).

In addition, finding the rules to classify objects in underwater is more difficult than those for surface vessels.

Understanding the nature and characteristics of ambient noise and the acoustic radiated noise of vessels is of great importance in selecting the discriminating features and the classification algorithms.

Therefore, we present a brief explanation about sources of ambient and radiated noises and their spectral characteristics.

**The ambient noise**

The major sources of ambient noise are as follows (Urick 1963, Burdic 1991):

- Seismic disturbances
- Biological organisms' activities
- Oceanic turbulence
- Distant shipping
- Wind
- Thermal noise

The distant shipping and wind are usually the main sources of ambient noise in the passive SONAR system. The level of these two kinds of noise depends on the distance from shipping lines and the sea state, respectively. In deep waters the distant shipping noise is dominant in frequencies from about 10-20 Hz to about 200-300 Hz, and wind noise is dominant in frequencies from about 200-300 Hz to several tens of kHz. The power spectral density of ambient noise is relatively smooth in the acoustic region. Some studies show a Gaussian distribution for ambient noise but other studies claim that it has a non-Gaussian distribution (Wenz 1962).

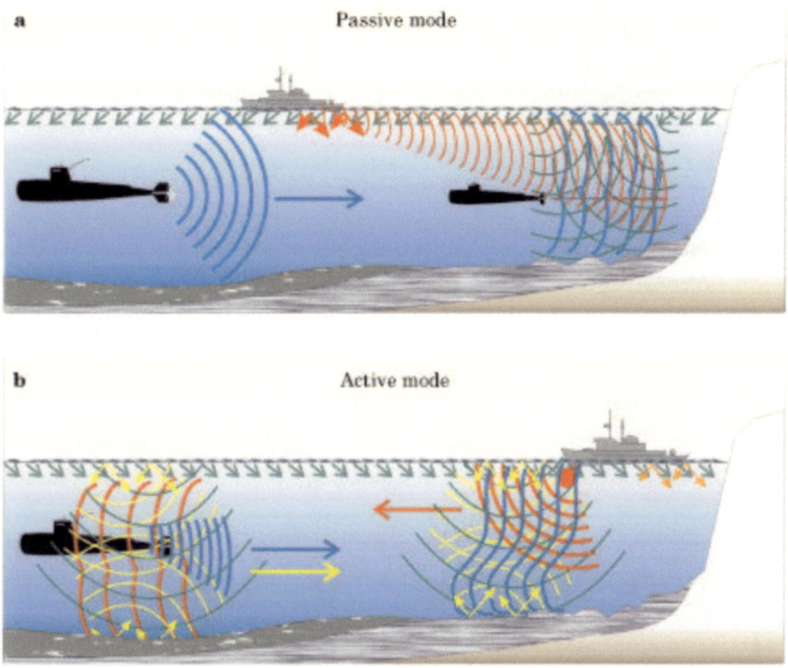


Fig. 1. (a) Passive and (b) Active modes of SONAR system (Physics Online)

**The acoustic radiated noise**

The acoustic radiated noise produced by the vessel's machinery and its motion in sea, consists of broadband and narrowband components. The propeller and the hydrodynamic turbulence produce a broadband noise. The narrowband components are produced by the propeller, propulsion system and auxiliary machinery. Generally, the spectrum of acoustic radiated noise varies with the change of speed (Urick 1963).

Indeed, two main reasons for making the research in this area, challenging back to the nature of acoustic radiated noises which have:

- Rapid changes of characteristics over the time in both frequency and time
- Variations in Signal to Noise ratio (SNR) due to multipath propagation

## 2.2 Hidden markov model

The Hidden Markov Model (HMM) approach to classify the SONAR signals is a promising technique, which has proven capability of detection/classification as well as applications in speech recognition (Rabiner 1989).

In fact, the HMM is a stochastic model for speech applications as classifier of spoken words. Most current speech recognition algorithms are mainly based on HMM. In these models, speech signals are modeled as sequences of states with probabilities assigned to state transitions.

Probability densities are associated with the observation of acoustic parameters. These parameters are considered from spectra or cepstra corresponding to the states or state transitions. The states are identified with the phonemes. A left-to-right HMM topology is mainly used for speech recognition. Nevertheless, a fully connected HMM topology which the transition from any state to other states is possible, can be more useful in our application of SONAR signal detection/classification. The left-to-right topology of HMM is known as Bakis model and the fully connected topology is known as Ergodic, as well.

Exploiting the HMM approach to classify the signals consists of two main problems:

- Feature extraction of signals for training the models
- Preparing each model with training to classify the signals

**Feature extraction**

AR and cepstral coefficients are sensitive to the SNR of signals but the spectra components are more robust to variation of signals over time. However, AR model parameters can be used in stationery conditions, or even in short period SONAR signal processing. Therefore, they are preferred spectral features as input data to train the models. Using the spectra features, derived algorithms based on HMM are often robust. For an example, a typical spectrum of a diesel submarine noise is illustrated in figure 2.

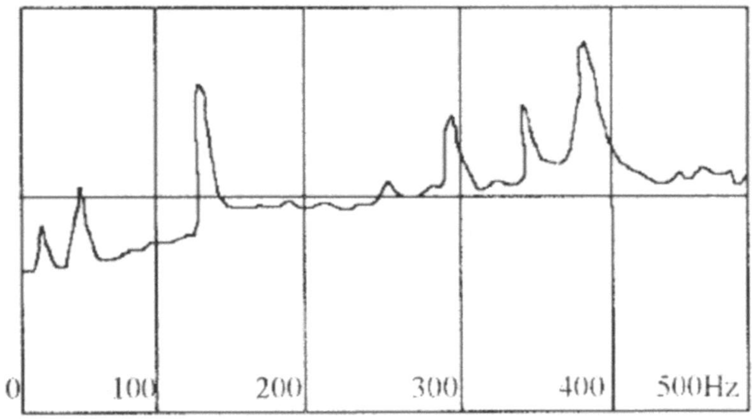


Fig. 2. A typical spectrum of a diesel submarine

Though the SONAR signal received from such targets may be varied in time with respect to their SNR, interferences or even reverberation, but the spectrum features are mainly remained unchanged in which they can be used as main input data for training of HMM models. However, in severe conditions of ambient noise, detecting the main features is not so easy and may produce wrong amplitude or positions of main features.

**HMM Training**

Considering a first order, N-state Markov model (Rabiner 1989) in which state sequence $Q$ is a realization of Markov model with probability:

$$\Pr(Q\,|A,B) = b_{q_1}\prod_{t=2}^{T} a_{q_{t-1}q_t} \tag{1}$$

where;

$$\begin{aligned} &A = \{a_{ij}\}; \quad a_{ij} = \Pr(q_{t+1} = j|q_t = i),\ i,j = 1,2,\ldots,N \\ &B = \{b_{ij}\}; \quad b_{ij} = \Pr(q_1 = i),\ i,j = 1,2,\ldots,N \\ &\text{and} \\ &Q = \{q_t\}; \quad q_t \in \{1,2,\ldots,N\}, t = 1,2,\ldots,T \end{aligned} \tag{2}$$

As a property of HMM, states of model cannot be directly observed, but can be estimated through a sequence of observations. In our purpose, it should be considered observations with continuous probability density. Thereafter, observation density vector will be:

$$\begin{aligned} &C = \{c_j(o_t)\};\ c_j(o_t) = \text{observation density of } o_t \text{ at} \\ &q_t = j,\quad O = \{o_t\} \text{ observations};\ t = 1,2,\ldots,T \end{aligned} \tag{3}$$

where;

$\lambda = (A,B,C)$ specifies an HMM, completely.

By a number of observations of SONAR signals, we are able to train an HMM as optimal model. A density function of O is given by:

$$P(O|\lambda) = \sum_{Q}[b_{q_1}c_{q_1}(o_1)\prod_{t=2}^{T} a_{q_{t-1}q_t}c_{q_t}(o_t) \tag{4}$$

By using above formula we can compute density function of any observation and discriminate among related classes. The procedure of above-mentioned description is:

- Given a set of acoustic training data that is phonetically labeled (possibly with errors); train an estimator to generate the data density for any hypothesized state: i.e. train an estimator for the emission probability density.
- Given a probability estimator for the data likelihood of each state, use the dynamic programming to find the most likely sequence through the states in all possible sequences. This step is sometimes called forced Viterbi algorithm (Viterbi 1967).

The latter procedure can be proved that converges to a local optimum, however, in practice the stopping criterion can be modified by a threshold level. Thereafter, when an $O$ of unknown class is given, we compute:

$$p^* = \arg\max_p \ (O, Q^* | \lambda_p) \tag{5}$$

where;
$p^*$ is optimal model and $Q^*$ is optimal state sequence and we can classify the source signal as belonging to the class $p^*$.
For instance, we considered simulated data of SONAR and applied previously mentioned model to evaluate the capability of HMM in detection/classification of SONAR signals. However, lack of real data has prevented to evaluate the different models. We considered two classes of submarines and simulated their noises.
Figure 3 illustrates two examples of related spectra. We used feature vectors extracted from spectra for training two HMMs, nevertheless above vectors might be noisy in real SONAR systems. As simulation stage, we produced data with different SNR ranged from -20dB to 20dB in order to check robustness of algorithm. Therefore, we had at least two HMMs with optimal number of states.

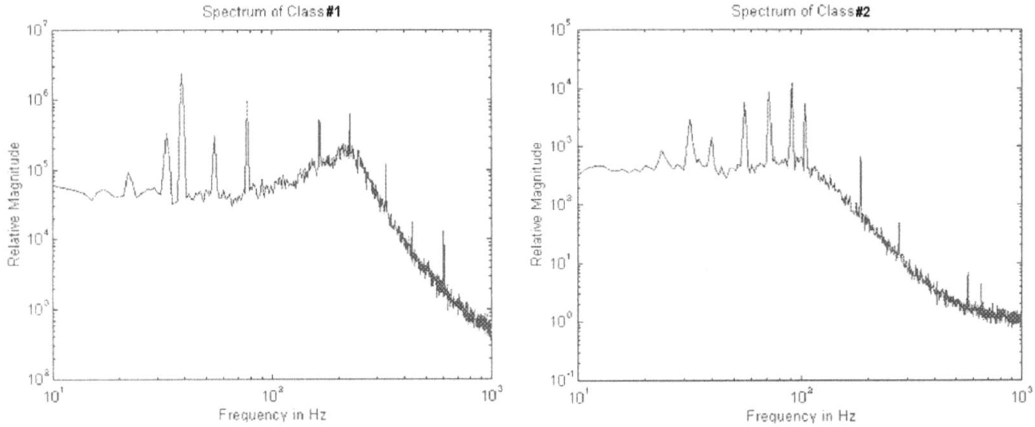


(a) (b)

Fig. 3. Two spectra of simulated noises in frequency band of [10Hz, 1000Hz]: (a) Spectrum of simulated data in class#1, (b) Spectrum of simulated data in class#2

Now, for evaluation of method we consider another set of simulated data in order to provide test data. This procedure is iteratively continued until statistical results are provided. These statistical results are shown in figure 4. In this figure, all 24 results of classification scores have been graphically illustrated. The scores have been converted to grayscales, i.e., bright color means high score of classification.
As shown in this figure, the class two has higher scores (brighter parts) then, as per majority vote to make a decision for received data over time, class #2 will be selected. In the first image, we considered simulated data from class #1. After preparing model (both stages of training test have been completed), a procedure similar to the test stage (named: evaluation stage) is iterated for 12 times. Therefore, we have 24 probability scores, which have been finally produced by two HMMs. The argument of the highest score determines class of input data.

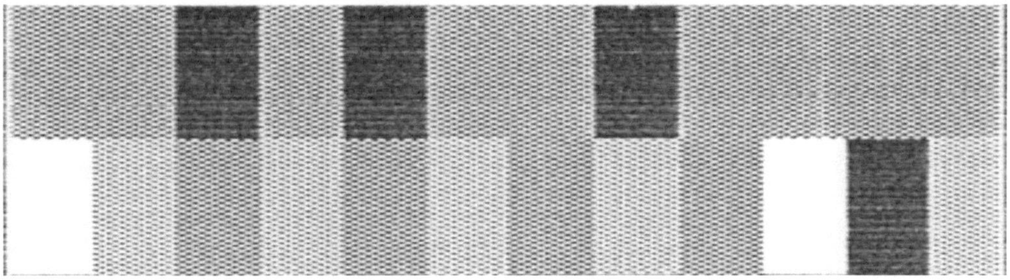

HMM Classification Results: Black: Low Score, Gray: Mediocre, White: High Score

Fig. 4. The simulation result for classification of SONAR signals based on HMM method: {White: High, Grays: Medium, Black: Low} scores as model outputs

Some promising results:

- Using HMM as a new method increases robustness of algorithm against fluctuations of SNR in received signals.
- At low and fixed SNR (< 0dB) has a better performance in recognition of sounds; for instance, we obtained 5.6% increase in performance at SNR=-12dB. Almost the same performance is obtained at high and fixed SNR in compare with the normal classifier.

Meanwhile, new method needs only more computations and therefore faster processors. We used a new method based on HMM, moreover implemented it in new area, which may increase our insight about underwater signal processing.

## 3. Modern approaches

The neural network method is widely used due to its capability of learning and clustering (Gosh 1992, Kundu 1994). Amongst them, Radial Basis Function Neural Network (RBFNN) has more advantages such as fast convergence in train mode and low computations in both train/ evaluation modes. In regular function of RBFNN kernels, Euclidean measure has been used to discriminate membership of an input feature vector. Euclidean measure is extremely sensitive to magnitudes of input values and ignores their separations, totally. This is why two vectors with the same values but in different locations produce a high Euclidean measure score. In contrast, Hausdorff Similarity Measure (HSM) is able to combat with this separation and discriminate the similarity, as well.

### 3.1 RBF Neural Network (RBFNN)

As depicted in figure 5, RBFNN has two major layers, the processing layer with arbitrary neurons therein and the output layer that weights the outputs of previous layer and represent the final approximated value.

Theoretically, there is no constraint for number of neurons. In normal RBFNN Euclidean measure scores the current input vector similarity respect to the central trained vector of neurons. The radial basis function may be defined as following:

$$\phi_i(x) \triangleq \exp[-0.5(\frac{x - m_i}{\sigma_i})^2] \tag{6}$$

**Hausdorff Similarity Measure (HSM)**

HSM is normally used for 2-D features (Basalto 2007). To replace Euclidean, it is needed to consider another axis of input features such as frequencies bins separations (in our study) that insert information of bins locations to the evaluation process. HSM for two vectors with arbitrary lengths is defined as:

$$HSM\ (\overline{A},\overline{B}) = Max\ \{Sup\ [Inf\ (Dist\ (a_i,\overline{B}), Dist\ (\overline{A},b_j))]\} \tag{7}$$

In (7), $\overline{A}(ai)$ and $\overline{B}(bj)$ are input vectors and $Dist(.)$ is Euclidean measure.

**Methodology**

We modified kernel functions of considered RBFNN from Euclidean to Hausdorff. Then we trained the modified RBFNN over several SONAR signals, including few real data and other simulated ones. While spectral features of mentioned data have been exploited as input vectors of networks, we considered three major networks in order to classify three different classes of gathered data. The final output score of every network has been evaluated in test mode and satisfied scores are considered for classification process.

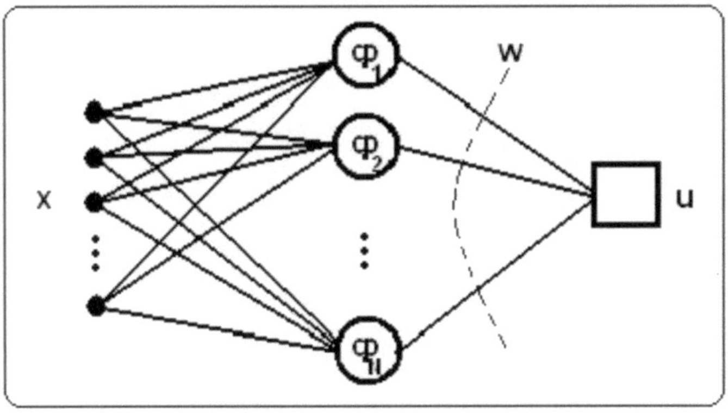


Fig. 5. RBF Neural Network topology with single output

**Feature selection**

The required features for the classifier are extracted from the spectral densities. We noted that spectral densities have more robustness rather than cepstral coefficients and statistical or parametric models. But using a regular approach to the spectral densities can produce incorrect results in front of real data, due to high variations in spectral frequencies within each class of data. These variations may cause high values of Euclidean measures and then incorrect detection/ classification of SONAR signals.

**Simulation results**

Simulation was based on above methodology for both new approach and normal RBFNN. Two separated sets of data as train and test were considered. In train mode, processing was continued for three classes until the fluctuations of learning curves were negligible. Reported results for train mode are based on the final classification rates. Processing in the test mode was done over long term time series including ambient noises and gathered data, randomly. All simulation procedures and algorithms have been written in MATLAB. Table I represents a comparison result between two RBFNN classifier with normal and modified kernels in both train/ test modes.

| Method | RBFNN (NORMAL) | | RBFNN (HSM) | |
|---|---|---|---|---|
| Class No. | TRAIN (%) | TEST (%) | TRAIN (%) | TEST (%) |
| #1 | 93 | 87.4 | 95 | 92.5 |
| #2 | 90 | 83.6 | 93 | 90.4 |
| #3 | 88 | 74.0 | 88 | 83.7 |

Table 1. Classification Results

Although the results in train mode of both normal and modified networks are not distinguishable, but there is a reasonable difference in test mode, (The less numeric precision is due to the lack of real data in the train mode). Sample spectral used in our approach is depicted in figure 6.
The required diversity of each class samples was produced by changing the frequency bins amplitudes, moving them within a random interval and also adding ambient noise ranged from SNR=-20dB to SNR=0dB. As a result, the new method produces a better classification and target detection rates of simulated SONAR signals even in low SNR. In the next section, we present a complimentary study based on PNN and some new results for the classification of Persian Gulf data set.

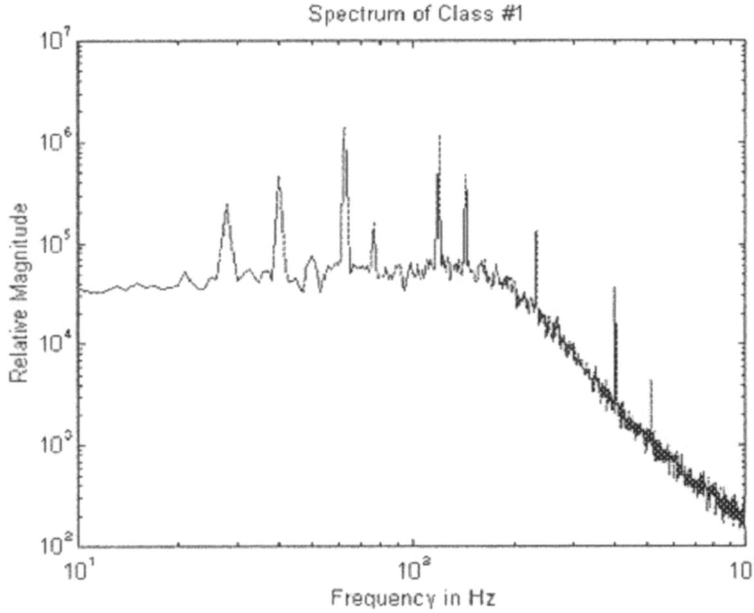


Fig. 6. Spectrum of class #1 in train mode

### 3.2 Probabilistic Neural Network (PNN)

There are three layers in a PNN structure including: input, hidden and output layers. The length of feature vectors determines dimension of input layer. If we use $d$ discriminating features for classification, the input layer dimension will be $d$. The hidden and the output layers consist of n patterns and $c$ classes, respectively. Number of patterns n is chosen such that the convergence is guaranteed. Each pattern is $d$-dimensional and the output of the PNN is chosen from the set of classes $\Omega = \{\omega_1, \omega_2, ..., \omega_c\}$. PNN is based on the Bayes decision rule, i.e. a trained PNN will decide $\omega_i$ if:

$$p(\mathbf{x} | \omega_i) P(\omega_i) > p(\mathbf{x} | \omega_j) P(\omega_j) \quad ; j = 1, ..., c \ \ and \ j \neq i \tag{8}$$

Where, $\mathbf{x} = \{x_1, x_2, \ ... \ , x_d\}$ is the $1 \times d$ feature vector, and $p(\mathbf{x} | \omega_i)$ and $P(\omega_i)$ are the $d$-dimensional conditional Probability Density Function (PDF) and prior probability obtained from the training data for $i$-th class, respectively. $P(\omega_i)$ is estimated by:

$$P(\omega_i) = \frac{number \ of \ training \ data \ for \ i'th \ class}{total \ number \ of \ training \ data} \quad ; i = 1, ..., c \tag{9}$$

$p(\mathbf{x} | \omega_i)$ is estimated using Parzen windows. The Parzen (or Kernel) method is a non-parametric approach for PDF estimation. In this method, a symmetric $d$-dimensional window, which is usually a Gaussian window, is centered on each training sample. The windows corresponding to $i$-th class are superposed and appropriately normalized to give $p(\mathrm{x} | \omega_i)$. The normalization is such that:

$$\int_{-\infty}^{+\infty} \cdots \int_{-\infty}^{+\infty} p(\mathrm{x} | \omega_i) dx_1 ... dx_d = 1 \tag{10}$$

The conventional PNN needs an input parameter which is called the spread parameter. This parameter must be chosen by the user. The standard deviation of the Gaussian Parzen window is proportional to the spread parameter and it is identical for all samples in all classes. The most important disadvantage of the conventional PNN is the difficulty in selecting an appropriate spread value. The classification performance depends significantly on selecting an appropriate spread value. Too small spread values give very spiky PDF's and, too large spread values smooth out the details. To show this disadvantage of conventional PNN, consider a distribution of two dimensional feature vectors as shown in figure 7.

The estimated PDF for set of points shown in figure 7 will be different for different values of spread parameter (the standard deviation of Gaussian windows). For example, using 1, 0.06 and 0.02 as values of the spread parameter, the estimated PDFs will be as shown in figures 8(a)-8(c), respectively.

Although there are some approaches such as cross-validation for estimating an appropriate spread value, but they need a large number of data and high processing time. Thus, we determined the spread parameter by the trail and error. Another disadvantage of the conventional PNN is using the same spread parameter for all classes, which decreases the degree of freedom of the PDF estimator.

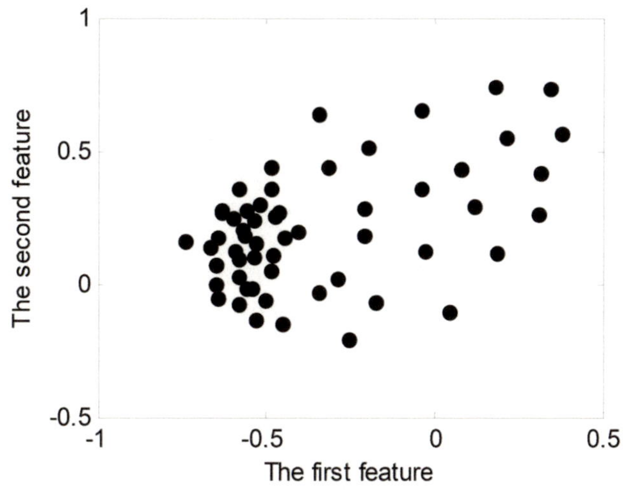


Fig. 7. The scatter plot of an exemplar set of two dimensional feature vectors.

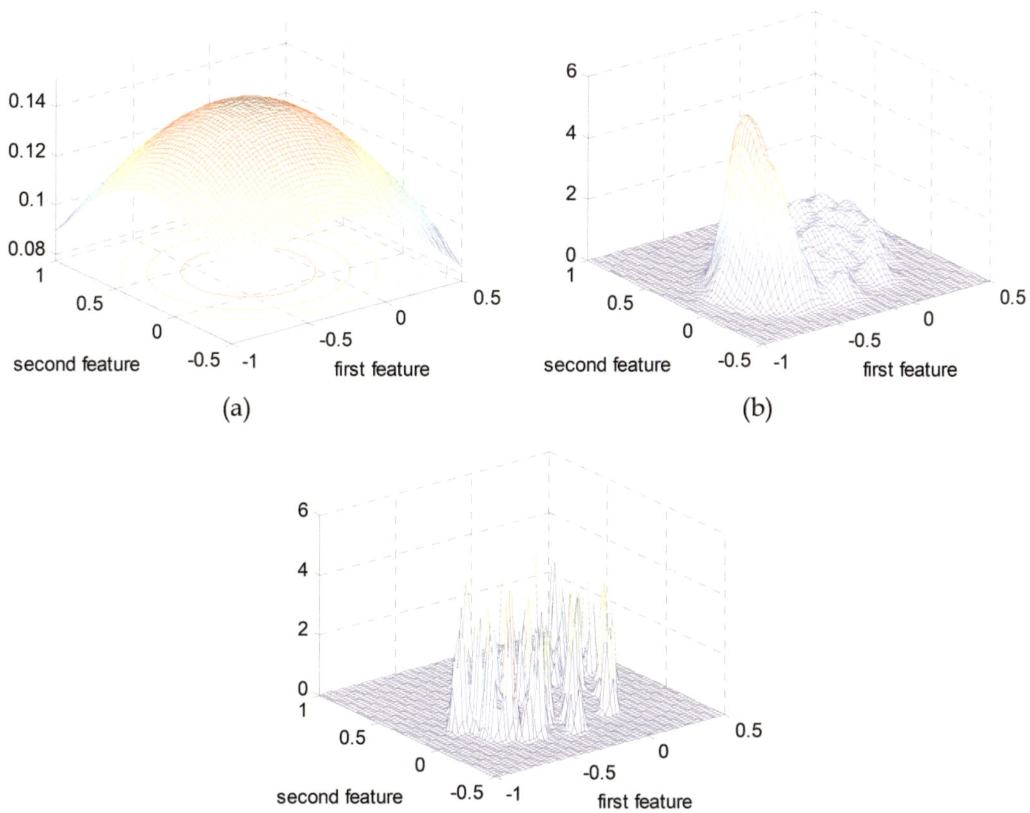


Fig. 8. The estimated PDF for feature distribution of figure 7 using a spread value equal to (a) 1, (b) 0.06, and (c) 0.02.

To overcome these disadvantages, a multi-spread PNN that uses a different spread parameter for each class has been recently introduced (Farrokhrooz 2011). Consider an exemplar classification problem with two discriminating features that is illustrated in figure 9. As it is shown in this figure, the distances between samples of the first class are typically more than those of the second class. Obviously, the appropriate spread parameter for the first class is different from that of the second class. Therefore, it is better to use two different spread values for these two classes of feature vectors. The MSPNN is a Probabilistic Neural Network, which can estimate PDF of the training data using Gaussian Parzen windows with a different spread parameter for each class. Using a different spread parameter for each class causes an increment in degree of freedom of the neural network. Thus, the MSPNN is more flexible than the conventional PNN in the current application.

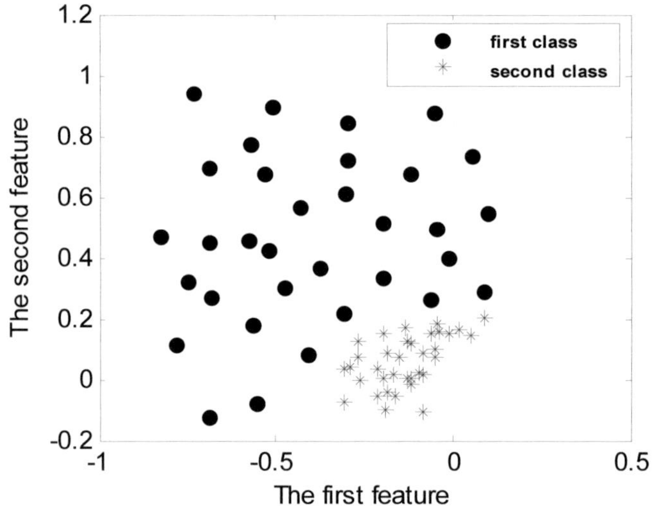


Fig. 9. The scatter plot for an exemplar classification problem.

**Data bank**

The conventional PNN and MSPNN have been used for classification of three classes of marine vessels, i.e., heavy ships, medium ships, and boats. The data set of heavy ships includes 29 radiated noise files from five different merchant ships (supertanker, container ship …). The data set of medium ships consists of 26 files from three different medium ships, and the data set of boats includes 16 files from four boats. The main criteria for dividing vessels into these three categories have been their size and weight. The number of data files in each subset is shown in Table II.

| Subset | 1st | 2nd | 3rd | 4th | 5th |
|---|---|---|---|---|---|
| Heavy | 7 | 4 | 3 | 3 | 12 |
| Medium | 13 | 5 | 8 | - | - |
| Boat | 7 | 2 | 1 | 6 | - |

Table 2. The number of data files in each subset of each class of vessels.

Each subset of files corresponds to one vessel. All files have been recorded in the Persian Gulf at low sea states. The time length of each file is five seconds and the recording sampling frequency equals 2560 samples/second. All data files have contents only in the 125-500Hz frequency band.

**Feature selection**

The set of selected features for classifying the radiated noises of ships can be divided into two categories: the features extracted from AR model of signals and the features directly extracted from the estimated PSD of signals. The first category contains the coefficients of AR model of signals with AR order 15 or 13, and the poles of AR model with order 2. The second category of discriminating features includes 6 features directly extracted from PSD of signals.

**Performance evaluation**

In this section, we evaluate the classification performance of conventional PNN and MSPNN with the method in (Farrokhrooz 2011) for spread estimation by applying them to the bank of real acoustic radiated noises of marine vessels. We define the performance of a classifier for a class as the probability of correct decision for the data of that class.
In order to evaluate the performance of the classifiers, we divided the data bank into two parts, the training data set and the test data set. After feature extraction, the extracted features of training data set are used for classifier training and the features of test data set are used for evaluating the performance of the classifier. We used a training data set as introduced in Table III. In each of the performance evaluations in this section, this data set of Table III is used for training and the rest of data bank files are used as the test data set. In each performance evaluation, the performance of conventional PNN will be shown versus spread parameter values and the performance of MSPNN will be demonstrated versus the multiplier $x$.

| Subset used for training | heavy ship file | medium ship file | boat file |
|---|---|---|---|
| **Training data set** | 7 randomly selected files from all subsets | 6 randomly selected files from all subsets | 4 randomly selected files from all subsets |

Table 3. The number of data files in each subset of each class of vessels

In this subsection the performance of conventional PNN is evaluated using various sets of features. Figure 10 shows the performance evaluation of conventional PNN when the discriminating features are spectral features. As shown in this figure, selecting a proper spread value causes high probability of correct decision for heavy and medium ships. However, the probability of correct decision for boats is not high in these cases. In the next performance evaluation, the six spectral features directly extracted from PSD, and the coefficients of AR(15) model are used as discriminating features. The evaluated performances are shown in figure 11.

It can be seen that the performance of the classifier in this case is approximately similar to the previous case. In the last performance evaluation, the six spectral features directly extracted from PSD, the poles of AR(2) model, and the coefficients of AR(15) model are used as discriminating features. The evaluated performances are shown in figure 12. Again, it can be seen that the performance of the classifier in this case is approximately similar to the previous cases. The performance of Multi-Spread PNN is evaluated using various sets of features. figure 13 shows the performance evaluation of MSPNN when the discriminating features arespectral features. It can be seen from figure 13 that the performance of MSPNN is not so good in this special range.

In the next performance evaluation, the six spectral features directly extracted from PSD, and the coefficients of AR(15) model are used as discriminating features. The evaluated performances are shown in figure 14. It can be seen from this figure that, the performance of MSPNN is good for all three classes of ships in a special range of multiplier values. In the last performance evaluation, the six spectral features directly extracted from PSD, the poles of AR(2) model, and the coefficients of AR(15) model are used as discriminating features. The evaluated performances are shown in figure 15.

It can be seen from this figure that, the performance in most cases is good for all three classes of ships in a special range of multiplier values.

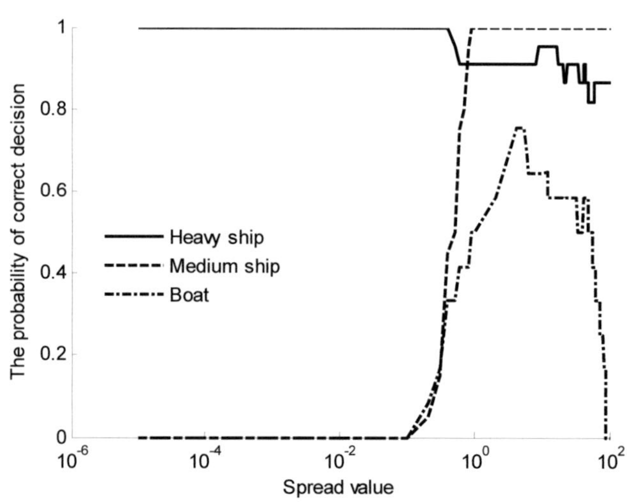


Fig. 10. The performance of conventional PNN in classifying heavy and medium ships and boats using the six features directly extracted from PSD as discriminating features.

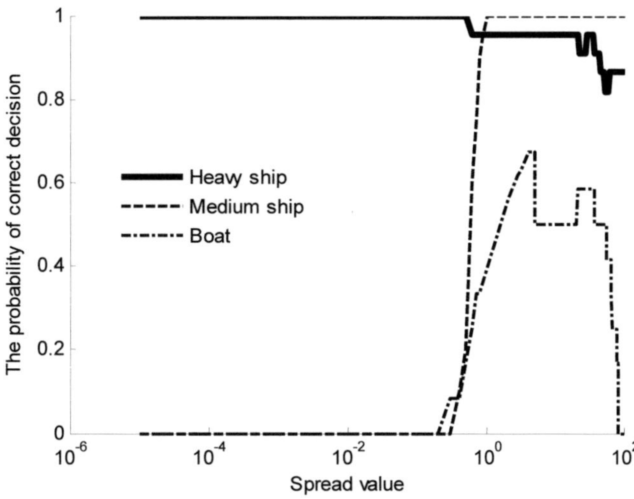


Fig. 11. The performance of conventional PNN in classifying heavy and medium ships and boats using the six features directly extracted from PSD and coefficients of AR model as discriminating features.

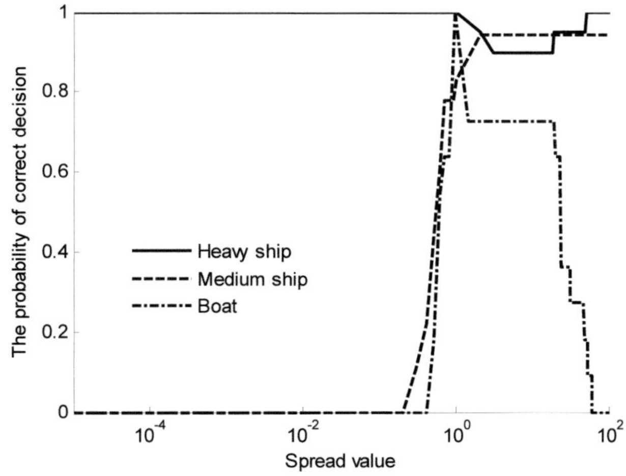


Fig. 12. The performance of conventional PNN in classifying heavy and medium ships and boats using the six features directly extracted from PSD and poles and coefficients of AR model as discriminating features.

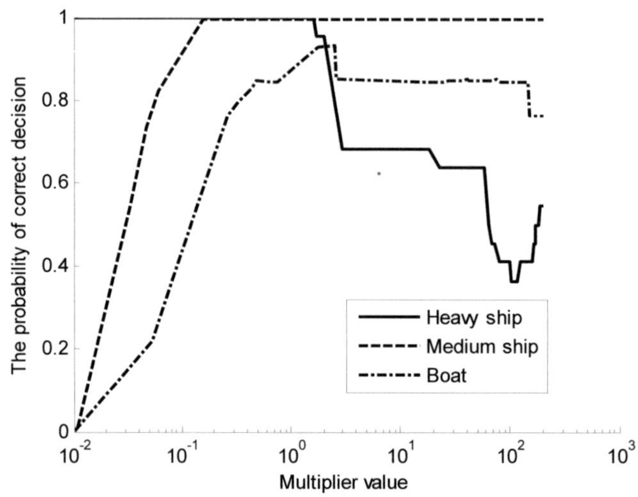


Fig. 13. The performance of MSPNN in classifying heavy and medium ships and boats using the six features directly extracted from PSD as discriminating features.

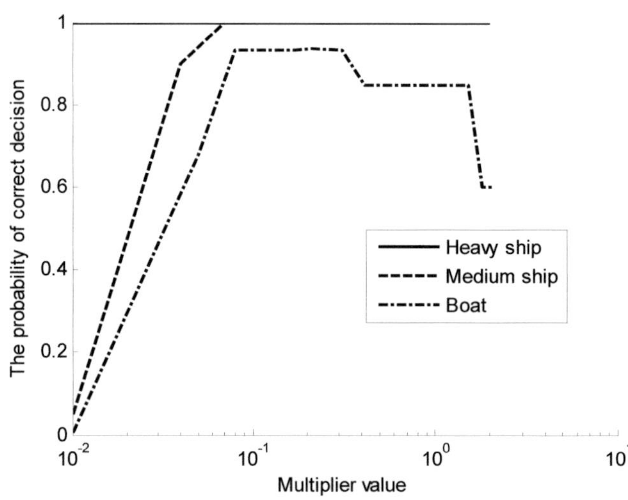


Fig. 14. The performance of MSPNN in classifying heavy and medium ships and boats using the six features directly extracted from PSD and AR coefficients as discriminating features.

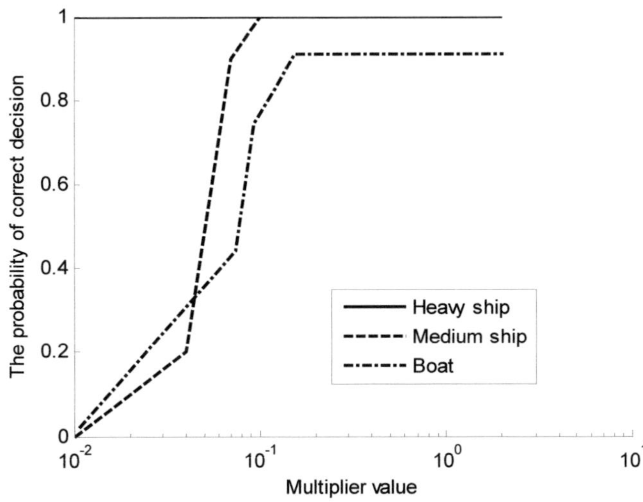


Fig. 15. The performance of MSPNN in classifying heavy and medium ships and boats using the six features directly extracted from PSD and poles and coefficients of AR model as discriminating features.

## 4. Target tracking

The goal in target tracking problem is to sequentially estimate the target parameters of interest from the sensor observations. The target parameters of interest are modeled by the state vector $\mathbf{x}_k$ and the sensor observations are modeled by the observation set $\mathbf{Z}_k = \{\mathbf{z}_k^1, \cdots, \mathbf{z}_k^{m_k}\}$ where; k denotes the time index (or scan index) and $m_k$ denotes the number of sensor observations at the k-th scan. In SONAR target tracking systems, the system state vector may contain information about the location and speed of the target; observations are the signal strength, bearing, and arrival time of the returned signal.

**Dynamic state space model**

Dynamic state space model (DSSM) has been widely used for modeling a target tracking problem. The dynamic state space model formulates variations of the system state and observations using two processes: the system state process and the observation process. In the system state process, variations of the target state vector $\mathbf{x}_k$ in different scans *k* are modeled using a first order Markov process (Rabiner 1989) as follows:

$$\mathbf{x}_k = f_k(\mathbf{x}_{k-1}, \mathbf{v}_k) \tag{11}$$

where;

$f_k(.)$is the known state transition function and $\mathbf{v}_k$ denotes the state transition noise, which is modeled by a random vector with known statistics. The observation model or measurement model relates the observed target signal $\mathbf{z}_k^o$ to the target state vector $\mathbf{x}_k$ by:

$$\mathbf{z}_k^o = h_k(\mathbf{x}_k, \mathbf{w}_k) \tag{12}$$

where;

$h_k(.)$represents the observation function and $\mathbf{w}_k$ denotes the observation noise, assumed to be independent of the state transition noise $\mathbf{v}_k$. The target observation vector $\mathbf{z}_k^o$ may appear in the observation set $\mathbf{Z}_k$ with the probability of $P_D$. While the target observed signal is related to the target state vector by (12), the clutter generated observations are assumed to be uniformly distributed in the surveillance region. The number of clutter generated contacts is modeled using a Poisson distributed random variable with parameter of *CV* where *C* denotes the clutter density or the average number of clutter per unit volume and *V* denotes the volume of the surveillance region.

**Bayesian filtering**

To simplify the tracking problem, it can be assumed that there is no clutter contact and the target is always detected at the receiver, i.e. $\mathbf{Z}_k = \{\mathbf{z}_k^o\}$. The Bayesian filter (Anderson 1979, Stone 1999) provides the optimum solution to the state space model with respect to the Maximum A Posteriori (MAP) criterion. The Bayesian filtering technique implements the Bayes' rule and provides a recursive formula for estimating the posterior probability density of the target state vector given the sensor observations. The estimation of the posterior probability density function is through two steps; the motion update step and the observation update step. The motion update step implements the target motion model in (11) and provides a prediction estimate to the posterior probability density function by:

$$p(\mathbf{x}_k|\mathbf{Z}^{k-1}) = \int p(\mathbf{x}_{k-1}|\mathbf{Z}^{k-1})p(\mathbf{x}_k|\mathbf{x}_{k-1})d\mathbf{x}_{k-1} \tag{13}$$

where;

$\mathbf{Z}^{k-1}$ denotes all the observations up to scan *k*−*1*. The observation update process uses recent sensor observations and the target observation model to refine the predicted posterior probability density function in (13) by:

$$p(\mathbf{x}_k|\mathbf{z}^k) = \frac{p(\mathbf{x}_k|\mathbf{z}^{k-1})p(\mathbf{z}_k|\mathbf{x}_k)}{\int p(\mathbf{x}_k|\mathbf{z}^{k-1})p(\mathbf{z}_k|\mathbf{x}_k)d\mathbf{x}_k} \tag{14}$$

By setting the initial posterior probability distribution of the target state vector $p(\mathbf{x}_0|\mathbf{Z}_0) = p(\mathbf{x}_0)$ and following (13) and (14), the posterior probability density function of the target state can be estimated. Equations (13) and (14) do not require the system state space models provided in (11) and (12) to be linear. Therefore, the Bayesian filter estimates the posterior probability density function in the general case that is known as the general *non-linear tracking*. Kalman filtering technique (Kalman 1960) is one of the most famous special cases in which the system state space models are linear and the noise vectors **v** and **w** are additive and Gaussian. This case is known as the linear-Gaussian state space model and is represented using the following equations:

$$\mathbf{x}_k = F_k\mathbf{x}_{k-1} + G\mathbf{v}_k \tag{15}$$

$$\mathbf{z}_k^o = H_k\mathbf{x}_k + \mathbf{w}_k, \tag{16}$$

where; $F_k$ is the state transition matrix, $\mathbf{v}_k$ is the state process noise with zero mean and covariance matrix of Q, G is a transform matrix, $H_k$ is the transition matrix from the target state space to the observation space, and $\mathbf{w}_k$ is the Gaussian additive observation noise with zero mean and covariance matrix of R. Using the linear-Gaussian model, the target state vector is the linear combination of the Gaussian distributed vector **v** then is Gaussian. This is the Key assumption in deriving the Kalman filter equations. The estimation of the target state vector, denoted by $\hat{\mathbf{x}}_{k|k}$ is computed as follows:

$$\hat{\mathbf{x}}_{k|k-1} = F_{k-1}\hat{\mathbf{x}}_{k-1|k-1} \tag{17}$$

$$\Sigma_{k|k-1} = GQ_{k-1}G^T + F_{k-1}\Sigma_{k-1|k-1}F_{k-1}^T \tag{18}$$

$$\hat{\mathbf{x}}_{k|k} = \hat{\mathbf{x}}_{k|k-1} + K_k(\mathbf{z}_k^o - H_k\hat{\mathbf{x}}_{k|k-1}) \tag{19}$$

$$\Sigma_{k|k} = \Sigma_{k|k-1} - K_k S_k K_k^T \tag{20}$$

where;

$$S_k = H_k\Sigma_{k|k-1}H_k^T + R_k \tag{21}$$

is the covariance matrix of $\mathbf{z}_k^o - H_k\hat{\mathbf{x}}_{k|k-1}$ and,

$$K_k = \Sigma_{k|k-1}H_k^T S_k^{-1} \tag{22}$$

is known as the Kalman gain. The Kalman filter provides the expected value $(\hat{\mathbf{x}}_{k|k})$ and the covariance matrix $(\Sigma_{k|k})$ of the posterior probability density function $p(\mathbf{x}_k|\mathbf{z}^k)$. In a linear-Gaussian state space model, Kalman filter is the optimum solution in the senses of the Minimum Mean Square Error (MMSE), Maximum Likelihood (ML), and Maximum A Posteriori (MAP) criteria (Anderson 1979, Kalman 1960). Because of the efficiency and simple implementation of the Kalman filter, many approaches have been developed that focus on linearization of state space model and applying Kalman filtering to the linear model. The most popular approach in this group is the well known extended Kalman filter (EKF) (Bar-Shalom 1988, Kailath 2000) which makes the state space model linear using Taylor series expansions of the nonlinear functions. While the EKF performs well under mild nonlinearity, it suffers from performance loss under sever nonlinearities. Another approach to linearization of state space model, presented in (Norgaard 2000), implements the polynomial approximation of a nonlinear function. This technique, known in the literatures as the central-difference Kalman filter (CDKF), can replace the EKF in some practical applications in which linearization using Taylor series is not so accurate. For both EKF and CDKF techniques, there is an implicit Gaussian prior assumption on the posterior density function due to applying the Kalman filter. Another group of proposed techniques assumes an explicit prior form (normally Gaussian) for the posterior density function. Perhaps the most famous approach in this group is the unscented Kalman filter (UKF) (Julier 2000, Julier 2004) which uses a novel nonlinear transformation of the mean and covariance matrices. In this approach, the mean and covariance matrices of the assumed Gaussian posterior density are parameterized by a set of samples, and these samples are updated through processing by a nonlinear filter. In the Quadrature Kalman filter (QKF) (Ito 2000, Arasaratnam 2007) the assumed Gaussian posterior density is parameterized through a set of Gauss-Hermite quadrature points, and the nonlinear process and observation functions are linear using statistical linear regression (SLR). The Cubature Kalman Filter (CKF) (Cubature 2009) uses a spherical-radial cubature rule for numerical computation of the multivariate moment integrals that appear in the Bayesian filter formulation. The Gaussian posterior density function is the key assumption in this numerical computation. Some existing approaches do not have explicit assumptions on the posterior distribution. Instead, they estimated a discrete set of parameters and used them to approximate the posterior density function. The Gaussian mixture filter (Alspach 1972), for example, estimates a set of discrete weight parameters and implements these weights to approximate the posterior density function using a weighted sum of Gaussian density functions. It is shown in (Sorenson 1971) that any practical density function can be represented by a weighted sum of Gaussian density functions and as the number of Gaussian densities increases, the approximation will converge uniformly to the desired density function. Recently, particle filtering (PF) (Djuric 2003) has become popular in many applications of statistical signal processing. In particle filtering, a desired posterior density function is estimated through a set of particles and associated weights. At each time update, new particles are generated via sampling from an importance function, and the weight associated with each particle is updated. With fast development of powerful computational systems, the numerical approach to solve the Bayesian inference equation that was first introduced in (Bucy 1971) has renewed its attraction through the point mass approach (Simandl 2006). In the point mass approach, a regular grid discretizes the system state, and the Bayesian filter is evaluated numerically on the discrete points. The tree search target tracking technique, introduced in (Nelson 2009, Roufarshbaf 2009) approximates the Bayesian filtering recursions by generating a tree of possible target state variations and finding the solution by navigating the search tree using the Stack algorithm.

**Data association techniques**

Thus far, we assumed that there are no clutter observations at the receiver and the target is detected in each scan. When clutter observations appear at the receiver, the above techniques cannot be performed directly for tracking the target and the observation contacts must be associated to either the target(s) or clutter. In this section, we discuss the most famous techniques for data association. We also consider the cases in which the target contact is not detected at the receiver. The early work on target tracking that addresses the challenge of data association was introduced in (Sittler 1964). The approach splits a track whenever more than one measurement is present. Then it calculates the likelihood function for each track and the tracks with likelihoods less than a predefined threshold are deleted. In the nearest neighbor Kalman filter (NNKF) approach (Singer 1971, Sea 1971, Singer 1973), the contact that has the smallest statistical distance to the predicted target track is considered as a true target observation and the others are considered as clutter. The probabilistic data association filter (PDAF) introduced in (Bar-Shalom 1975), provides a suboptimal Bayesian approach for data association in the single target tracking problem. In this approach, the target track is updated using the weighted sum of all possible track updates from each data association hypothesis of the current scan. When a single target is present, the optimal Bayesian approach for data association considers all possible association hypotheses from the target initiation time index to the current time and updates the target track using sum of the all possible track updates weighted by evaluated association probabilities (Bar-Shalom 1988). Since the optimal Bayesian method is computationally intractable, suboptimal approaches have been suggested. The joint probabilistic data association filter (JPDAF) (Fortmann 1983) is an extension to the PDAF algorithm for multiple target tracking. In this approach, all possible association hypotheses of the current time are evaluated for all targets, and the tracks are updated by a weighted sum of all possible track updates from the data association hypotheses. Other data association algorithms implements either MAP or ML approaches to find the best sequence of measurements that can be associated to each target. The Viterbi algorithm (Viterbi 1967, Wolf 1989) and the Expectation Maximization (EM) algorithm (Gauvrit 1997) have been widely used for this purposed. Multi-hypothesis tracking (MHT) (Blackman 2004) is a MAP estimator in which at each scan, instead of combining the hypotheses like JPDAF or selecting the best hypothesis like nearest neighbor (NN), the algorithm makes a tree of possible association hypotheses and decides based on the future observations through a MAP estimator. All of the data association approaches discussed above is combined with a tracking technique such as EKF, to estimate the target track within the clutter observations. As an example, we discuss the details of the PDAF algorithm which is one of the most common approaches in SONAR target tracking. In the PDAF technique, the data association probabilities $\beta^j$ and j = $0, \cdots, m_k$ are defined for each scan as the probability that the jth observation of the scan is generated by the target and others are related to the clutter. The special case $\beta^0$ is defined as the probability that all the observations are generated by clutter and the target generated observation is not detected. Using the data association probabilities in the Kalman filter, it is sufficient to change the equations (19) and (20) as follows:

$$\hat{\mathbf{x}}_{k|k} = \hat{\mathbf{x}}_{k|k-1} + K_k \tilde{\mathbf{Z}}_k \tag{23}$$

$$\Sigma_{k|k} = \Sigma_{k|k-1} - (1-\beta^0) K_k S_k K_k^T + \Sigma_k \tag{24}$$

where;

$$\tilde{\mathbf{Z}}_k = \sum_{j=1}^{m_k} \beta^j \tilde{\mathbf{z}}_k^j, \tag{25}$$

$$\tilde{\mathbf{z}}_k^j = (\mathbf{z}_k^j - H_k \hat{\mathbf{x}}_{k|k-1}) \tag{26}$$

$$\Sigma_k = K_k \left[ \sum_{j=1}^{m} \beta^j \tilde{\mathbf{z}}_k^j \tilde{\mathbf{z}}_k^{jT} - \tilde{\mathbf{Z}}_k \tilde{\mathbf{Z}}_k^T \right] K_k^T \tag{27}$$

The association probabilities $\beta^j$ have been calculated in literatures and the results are as following:

$$\beta^j = \frac{\exp\left(-1/2 \times \tilde{\mathbf{z}}_k^j S_k^{-1} \tilde{\mathbf{z}}_k^{jT}\right)}{b + \sum_{i=1}^{m_k} \exp\left(-1/2 \times (\tilde{\mathbf{z}}_k^i) S_k^{-1} \tilde{\mathbf{z}}_k^{iT}\right)} \tag{28}$$

$$\beta^0 = \frac{b}{b + \sum_{i=1}^{m_k} \exp\left(-1/2 \times (\tilde{\mathbf{z}}_k^i) S_k^{-1} \tilde{\mathbf{z}}_k^{iT}\right)} \tag{29}$$

where;

$$b = (2\pi)^{M/2} C |S|^{1/2} (1 - P_D)/P_D \quad (30)$$

and $M$ denotes the dimension of the observation vector $\mathbf{z}_k^j$.

**Recent techniques**

Some recent techniques in target tracking including sequential Monte Carlo techniques and the tree search techniques combine the data association problem and the target tracking problem as a unified problem. Sequential Monte Carlo approaches (Hue 2002) have been introduced in target tracking problems for both track estimation and data association. In the approach presented in (Hue 2002), Gibbs sampling is implemented to solve the data association problem, and particle filtering is used for target track estimation. The algorithm is developed for the cases of single target, multiple targets, single receiver, and multiple receiver systems for both fixed and variable number of targets (Hue 2002, Arulampalam 2002). Recently the stack-based tree search algorithm (Roufarshbaf 2011, Nelson 2009, Roufarshbaf 2009), has been developed for tracking the targets in non-linear environments with high clutter density. The approach is classified as a sub-optimum Bayesian approach for both track estimation and data association and is implemented for linear/non-linear state space models as well as Gaussian/non-Gaussian additive noise. The stack algorithm was originally developed as a sequential method for decoding error-correcting convolutional codes in digital communications (Jelinek 1969). It has more recently been suggested as a lower-complexity alternative to trellis-based techniques, e.g. the Viterbi algorithm (Viterbi 1967), for ML sequence detection of data transmitted over a dispersive channel (Xiong 1990, Nelson 2006). In target tracking applications, the stack algorithm navigates a tree in search of the target path with the largest likelihood, or metric. A set of possible paths and their associated metrics are stored in a list (or stack), and at each iteration, the algorithm extends the path with the largest metric. A simple example of the stack algorithm for target tracking is given in figure 16. The algorithm starts from an initial target estimate and makes a search tree of possible target movements from its currently approximated locations. At the end of the observation, the target path that has the highest probability among the extended paths of the search tree is selected as the estimated target track.

As a single example, consider a target that follows a linear motion model. The target state vector is defined by $\mathbf{x} = [x \; y \; \dot{x} \; \dot{y}]^T$, where; $(x, y)$ denotes the position of the target in Cartesian coordinates, and $(\dot{x}, \dot{y})$ denotes the target velocity in Cartesian coordinates. The parameters of the target motion model (15) are set as follows:

$$F = \begin{bmatrix} 1 & 0 & \Delta_t & 0 \\ 0 & 1 & 0 & \Delta_t \\ 0 & 0 & 1 & 0 \\ 0 & 0 & 0 & 1 \end{bmatrix}, \quad G = \begin{bmatrix} \frac{\Delta_t^2}{2} & 0 \\ 0 & \frac{\Delta_t^2}{2} \\ \Delta_t & 0 \\ 0 & \Delta_t \end{bmatrix} \quad (31)$$

where; $\Delta_t = 1$ is the time between scans. In this model, the state transition noise vector $\mathbf{v}_k$ determines the acceleration (in the x and y directions) over the time interval and the covariance matrix of $\mathbf{v}_k$ is set to $Q = 4 * 10^{-4} I_2$ where I denotes the identity matrix. The system is assumed to be an active SONAR and a single transmitter and receiver is placed at (0, 0) in Cartesian coordinates. The receiver measures the angle of arrival (bearing angle) and the range of reflecting subjects. Therefore, the observation transform function in (12) will be:

$$h(\mathrm{x}) = \begin{bmatrix} \tan^{-1}(y/x) \\ \sqrt{x^2+y^2} \end{bmatrix} \tag{32}$$

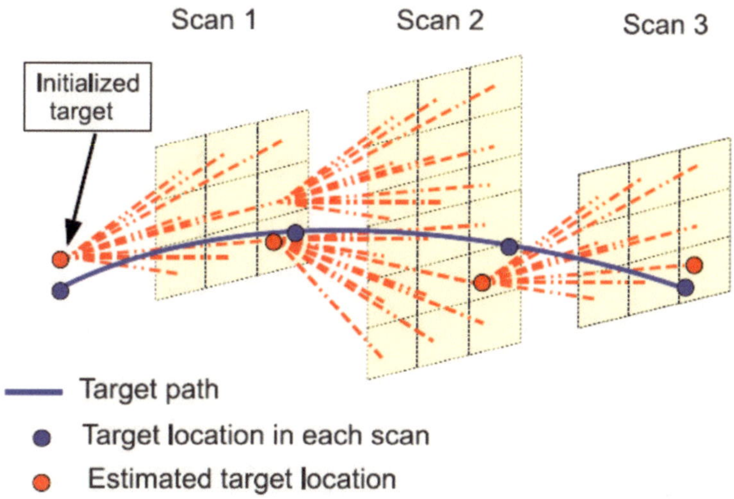


Fig. 16. A tree search target tracking structure

We assume that the measurement noise covariance matrix is given by:

$$\Sigma_w = \begin{bmatrix} \left(\frac{0.15\pi}{180}\right)^2 & 0 \\ 0 & 0.15^2 \end{bmatrix} \tag{33}$$

We also assume that the target generated contact is detected at the receiver with the probability of $P_D = 0.9$. Figure 17 shows an ensemble of the generated target and the estimated path using the EKF-PDAF technique and the tree search technique. The target initially starts its path from (10, 10) and follows its linear motion model. The clutter observations are uniformly distributed in the surveillance region and the average number of clutter observations per unit area (*C*) is set to 1. We can see that both algorithms are able to track the target in this special case. In the above example, the target motion model is linear and the non-linear transition function in the observation model is well approximated by its Taylor series expansion. Therefore, the EKF algorithm performs well in this scenario. However, in non-linear target motion models, the performance of the EKF technique decreases drastically while the other recent techniques such as tree search technique is able to follow the target track (Roufarshbaf 2010, Roufarshbaf 2011).

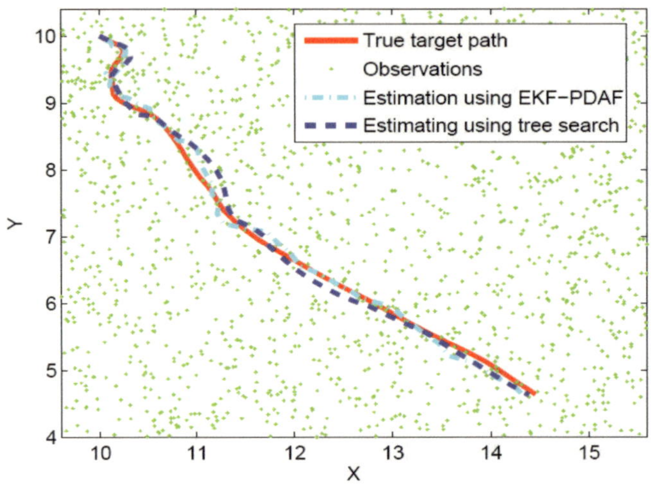


Fig. 17. Example of a target tracking problem. The target starts at initial point and follows according to its motion model.

## 5. Underwater wireless networks

Underwater wireless telecommunication, in general, and security and privacy of conversations, in particular, are new challenging areas and have recently attracted many interests. In spite of the advances in underwater wireless telecommunications (Stojanovic 2006, Benson 2008), the research on security and privacy of such systems has not been conducted. Indeed, despite achievements in navigation, ranging and telecommunication; a few researches on security aspects have yet been completed. Nevertheless, continuing study on SONAR systems and underwater telecommunication networks, plenty of research reports have been published (Peterson 2006, Stojanovic 2008).

### 5.1 Security and privacy

In this section, we concern with security aspect of an underwater telecommunication network and a secure scheme. The main idea stems from the fact that all telecommunication networks, including underwater networks, have been substantially prepared to transfer speech and voice (Kondoz 2005, Peyvandi 2009). As figure 18 depicts, the input voice is transformed to the bit stream using a low bit rate encoder. Then, the whole bits are mapped to the predefined symbols. Using the secure scheme, the overall bit error rates become noticeably low in which there has no any major effect on quality of speech. On the other hand, the noises are quite unintelligible for intruders who try to access to the conversations through the channel. Produced noises are signals including random scrambled speech-based symbols, which make no any sense to the listener. The entire network has been simulated to transfer speech to obtain results in an experimental state.

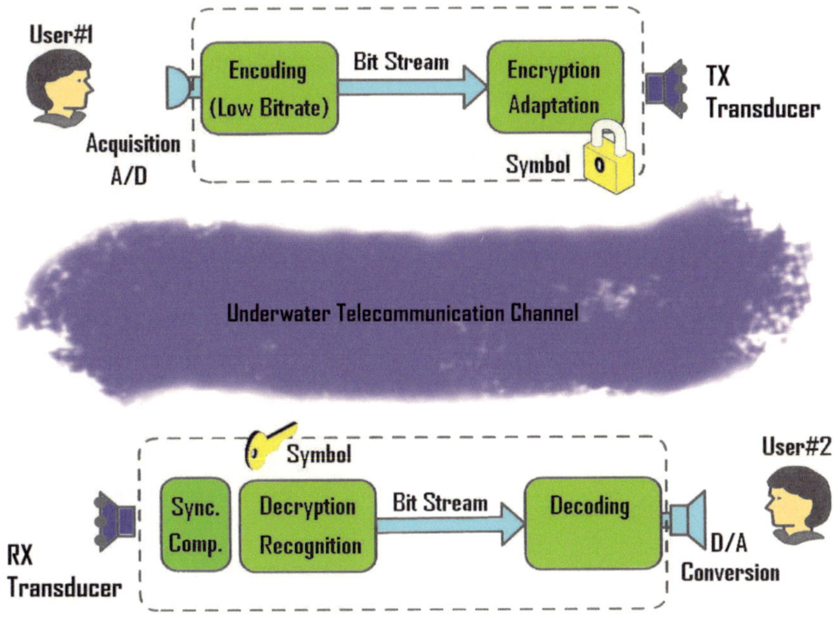


Fig. 18. Half-Duplex of underwater telecommunication network

**Methodology**

Figure 18 shows the modular illustration of designed underwater telecommunication network including the encryption module, which uses the secure scheme.

In this figure, at transmitter side, a low bit rate speech encoder is used to produce bit stream to the second module. The encryption module encrypts the bit stream in which each frame is mapped to a predefined symbol that has been originally designed using hi-fi speech records (Peyvandi 2009). In fact, for each frame, 7 bits in this study, a unique symbol is produced. Therefore, 128 symbols are needed to be stored in both sides of telecommunication network as lookup table. The consequent symbols, produced by encryption module, are windowed, filtered and shaped in order to transmit over channel. No more compression or coding is used. At receiver, synchronization/ compensation module provides the symbols from received samples after preprocessing. Due to severe distortion of signals in underwater channels, a Decision-Feedback Equalizer (DFE) at receiver is requisite for compensation. The final symbols, with 40 samples in length for 8 KHz, are represented to decryption module, which uses pattern recognition (Theodoridis 2006) to measure similarity among final and lookup table symbols. Final bit stream are represented to decoding module in order to produce speech.

**Simulation results**

All modules of figure 18 have been exploited in simulation of underwater telecommunication network and simulated environment for underwater channel as per real acoustic data (Shahbazi 2008). In a real system, other modules in order to modulate signals, at transmitter, and to demodulate, at receiver, are required; however, they have been omitted in the simulation. Using secure scheme over simulated underwater channel

Bit Error Rates (BER) has been obtained. The BER results with transmission of three different speeches were as low as 1%. In another experiments, the number of bit for each underlying frame (BPF) has been changed whilst the number of samples per frame was constant.
Figure 19 shows the BER results, which for 6- and 7-bit are less than 1% and for 8- and 9-bit are unacceptable for speech encoding. Thus, 7-bit is considered as optimum BPF.

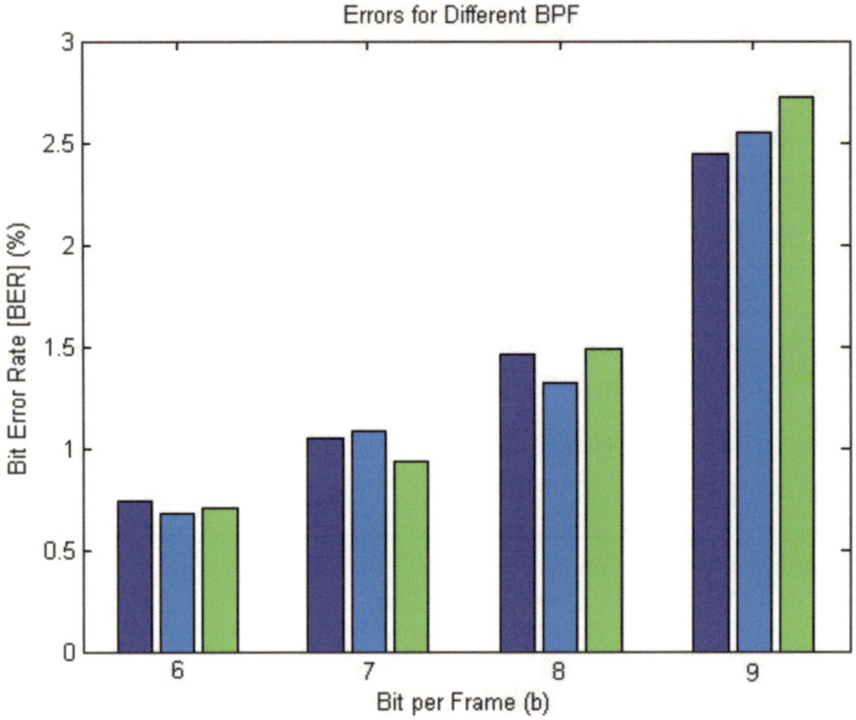


Fig. 19. Bit error rates (BER) for different experiments (BPF)

Table IV shows the statistical results for several transmission experiments for three different speeches with 7-bit BPF. The means and variances are separately shown for three speeches. In addition, results of Symbol Error Rates (SER), an important rate in speech communication, show that most errors occur non-uniformly.
Although there are bit errors in transmission of a longtime speech but they have no major effect on quality of speeches. Particularly, objective test of quality using NTIMIT dataset shows non-meaningful differences in final speeches. TIMIT (Becchetti 1999) dataset are for quality test of speeches and NTIMIT data are among the network datasets. Figures 20 and 21 depict the waveform of two different NTIMIT samples used in quality test.
The NTIMIT dataset is used for test of speech quality in telecommunication network. However, there are differences among original and decrypted samples but these differences make no meaningful distortion from auditory point of view. Adding to low bit error rates, the other reason is the non-uniform distribution of errors. To evaluate secure scheme under real conditions the test bed of figure 22 is considered.

| Exp. No. | BER (%) | | SER (%) | |
|---|---|---|---|---|
| | Mean | Variance | Mean | Variance |
| #1 | 0.96 | 0.05 | 0.55 | 0.09 |
| #2 | 1.07 | 0.07 | 0.60 | 0.12 |
| #3 | 0.93 | 0.06 | 0.52 | 0.10 |

Table 4. Simulation Results

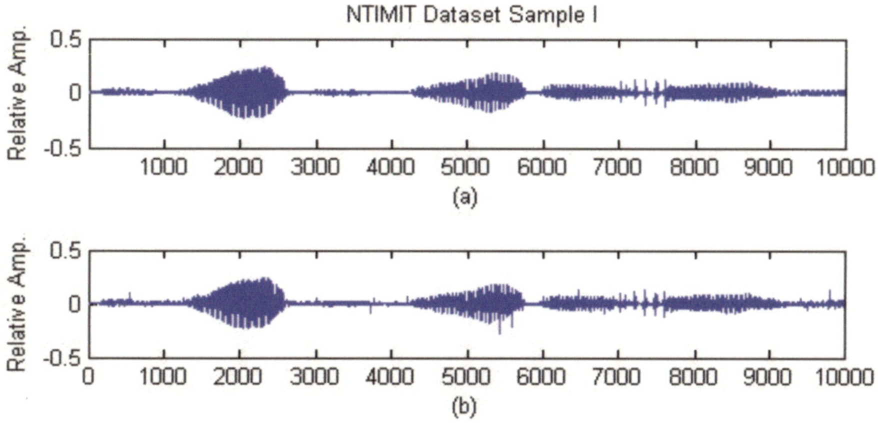


Fig. 20. NTIMIT dataset sample I (a) Original (b) After decoding

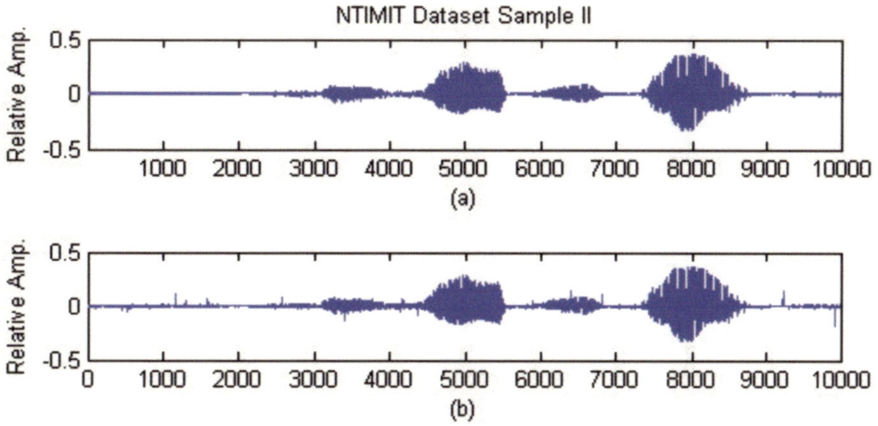


Fig. 21. NTIMIT dataset sample II (a) Original (b) After decoding

In the figure 22, the intruder can access to the underwater telecommunication link, but the voices are unintelligible even they are decoded by intruder. As of terrestrial wireless telecommunication networks, the public/private keys have been considered in coding/decoding modules as a fundamental part of them. Similarly, we should use the public/private keys in the security module of underwater wireless networks, as well. The private keys can be considered among different underwater networks (Peyvandi 2010).

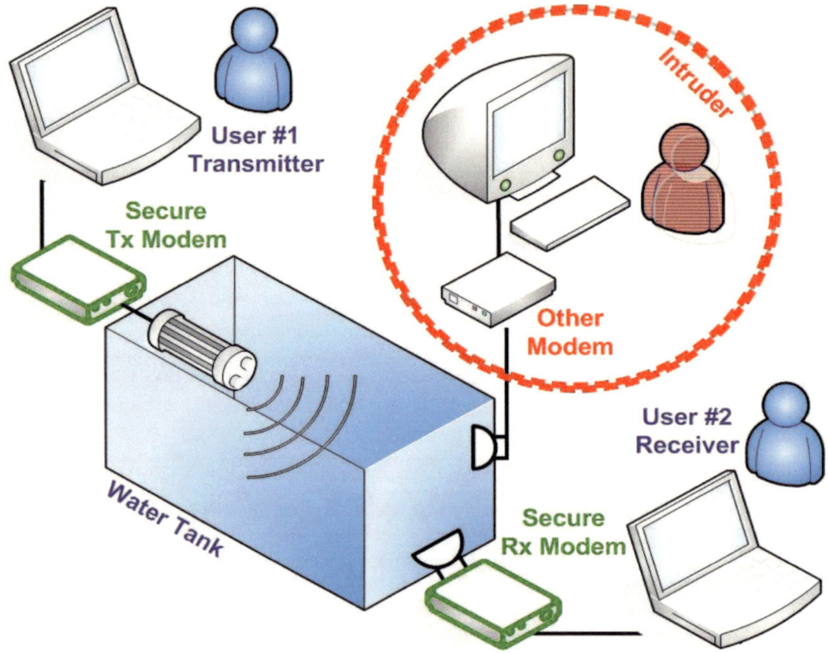


Fig. 22. Schematic diagram of test bed for secure scheme in underwater acoustic network

## 6. Conclusion

In this chapter, we introduced methods for underwater signal processing which are used for SONAR systems, either passive or active system, and underwater wireless network as an emerging area in the field. Method based on Hidden Markov Model (HMM) has been investigated and the simulation results have been released with examples. The capability of HMM in modelling different classes of underwater noises is notable even in low SNR, which is common in most SONAR systems. Hausdorff Similarity Measure (HSM) as a new measure for discriminating different features of classes has been studied in this chapter. HSM outperforms the other one-dimentioan measures, i.e, Minkovsky based, in low SNR environments when the spectral components are used as features.

In addition, we explained the necessity of a multi-spread probabilistic neural network. We used the method for estimating the spread value for each class in the multi-spread PNN. The multi-spread PNN is more flexible than the conventional PNN. This increased flexibility is obtained by using a different spread value for each class. Using the proposed method for estimating the spread value reduces the need for trial and error approach for spread parameter determination. The conventional PNN and the MS-PNN with the proposed method for estimating the spread parameter were applied to the radiated ship noise classification problem. A bank of real radiated noise data were used to evaluate the performance of the conventional and MS-PNN. In all performance evaluations, which is used for conventional PNN as the classifier, selecting a proper spread value caused high probability of correct decision for heavy and medium ships. In all performance evaluations, which is used for the MS-PNN as the classifier, there is a relatively wide range of multipliers in which the best performance occurs.

We also considered target tracking problem in sonar systems. A target tracking problem was modeled using the conventional DSSM. Then, we reviewed the classic Bayesian target tracking techniques and its closed form Kalman filtering approach that is optimum for linear-Gaussian models. Then, we studied tracking of low observable targets through clutter contacts and we reviewed the recent approaches in target tracking that are more efficient in non-linear scenarios. Through an example, we show how a classic EKF technique and a recent tree search technique are able to track a single target through clutter contacts.

In the last section in this chapter, we have proposed a new scheme for providing secure conversation in underwater telecommunication systems. Although, there are bit errors in near all communications, low bit errors with non-uniform distribution have unintelligible effect on quality of speeches. The quality test has been provided using NTIMIT dataset samples. The results showed that the proposed scheme for underwater wireless telecommunication systems is reliable in an experimental state. The environmental test of proposed scheme is ongoing.

## 7. Acknowledgment

We would like to acknowledge people who have commented and made suggestions on this work. Without their help, there would be fewer results and the existing descriptions would be more difficult to read. In particular, we would like to express our thanks to Dr. M. Karimi for his valuable advices, Dr. A. Ebrahimi for discussion on early version of section 5, Dr. M. Dianati for helpful discussion and Mr A. Maleki, H. Kamalpour for their help to provide additional analysis in datasets. At finally yet importantly, the first author would like to thank Dr. Hassan Peyvandi for his support and help during tough time in the last year.

## 8. References

Abrahat, R. J.; Kneale, P. E & See, L. M. (2004). *Neural Networks for Hydrological Modelling*, A. A. Balkema Publishers, London.

Alspach, D. L. & Sorenson, H. W. (1972). "Nonlinear Bayesian estimation using Gaussian sum approximations," *IEEE Trans. On Automatic Control*, vol. 17, no. 4, pp. 439-448.

Anderson, B. D. O. & Moore, J. B. (1979). *Optimal Filtering*. Dover Publications.

Arasaratnam, I. & Haykin, S. (2007). "Discrete-time nonlinear filtering algorithms using Gauss-Hermite quadrature," *Proceedings of the IEEE*, vol. 95, no. 5, pp. 953-977.

Arasaratnam, I. & Haykin, S. (2009). "Cubature Kalman Filter (CKF)," *IEEE Trans. on Automatic Control*, vol. 54, pp. 1254-1269.

Arulampalam, S.; Maskell, S.; Gordon, N. & Clapp, T. (2002). "A tutorial on particle filters for online non-linear/non- Gaussian Bayesian tracking," *IEEE Trans. on Signal Processing*, vol. 50, no. 2, pp. 174-188.

Bar-Shalom, Y. & Tse, E. (1975). "Tracking in a cluttered environment with probabilistic data association," *Automatica*, vol. 11, pp. 451-460.

Bar-Shalom, Y. & Fortmann, T. E. (1988). *Tracking and Data Association*. Academic Press, Inc.

Basalto, N.; et al, (2007). "Hausdorff Clustering of Financial Time Series," *Physica A, 379*, pp. 636-644.

Becchetti, C. & Ricotti, L. P. (1999). *Speech Recognition*, John Wiley, ISBN: 0-471-97730-6.

Blackman, S. (2004). "Multiple hypothesis tracking for multiple target tracking," *IEEE Aerospace and Electronic Systems Magazine*, vol. 19, no. 1, pp. 5-18.

Jelinek,F. (1969), "Fast sequential decoding algorithm using a stack," IBM journal of research and development, vol. 13, no. 6, pp. 675–685

Bucy, R. & Senne, K. (1971). "Digital synthesis of nonlinear filters," *Automatica*, vol. 7, pp. 287-298.

Burdic, W. S. (1991). *Underwater Acoustic System Analysis*, ISBN: 0139476075, Prentice-Hall, 2nd edition.

Djuric, P. M.; Kotecha, J. H.; Zhang, J.; Huang, Y.; Ghirmai, T.; Bugallo, M. F. & Miguez, J. (2003). "Particle filtering," *IEEE Signal Processing Magazine*, vol. 20, pp. 19-38.

Duda, R. O.; Hart, P. E. & Stork, D. G. (2000). *Pattern Classification*, ISBN: 0471056693, John Wiley & Sons, Inc., 2nd edition.

Eickstedt, D. P. & Schmidt, H. (2003). "A low-frequency SONAR for sensor-adaptive, multi-static, detection and classification of underwater targets with AUVs", *IEEE Oceans Conference*, USA.

Farrokhrooz, M. & Karimi, M. (2005). "Ship noise classification using Probabilistic Neural Network and AR model coefficients", *IEEE OCEANS 2005 Europe.*

Farrokhrooz, M. & Karimi M. (2011). "Marine Vessels Acoustic Radiated Noise Classification in Passive SONAR using Probabilistic Neural Network and Special Features", *J .Intelligent Automation and Soft Computing, Vol. 17, NO. 4.*

Fortmann, Y. B.-S. T.E. & Scheffe, M. (1983). "Sonar tracking of multiple targets using joint probabilistic data association," *IEEE Journal of Oceanic Engineering*, vol. OE-8, pp. 173-184.

Gauvrit, H.; Cadre, J. P. L. & Jau_ret, C. (1997). "A formulation of multitarget tracking as an incomplete data problem," *IEEE transactions on Aerospace and Electronic Systems*, vol. 33, no. 4, pp. 1242-1257.

Ghosh, J.; Deuser, L. M. & Beck S. D. (1992). "A Neural Network Based Hybrid System for Detection, Characterization and Classification of Short-Duration Oceanic Signals," *IEEE J. Ocean Eng.*, vol. 17, no. 4.

Hopfiled, J. (1982). "Neural networks and physical systems with emergent collective computational abilities", *PNAS April,vol. 79 no. 8 2554-2558.*

Hue, C.; LeCadre, J. P. & Perez, P. (2002). "Sequential Monte Carlo methods for multiple target tracking and data fusion," *IEEE Transactions on Signal Processing*, vol. 50, pp. 309-325.

Ito, K. & Xiong, K. (2000)."Gaussian filters for nonlinear filtering problems," *IEEE Trans. on Aut. Control*, vol. 45, pp. 910-927.

Julier, S. J. & Uhlmann, J. K. (2004). "Unscented filtering and nonlinear estimation," *Proc. of the IEEE*, vol. 92, pp. 401-422.

Julier, S. J.; Uhlmann, J. K. & Durrant-Whyte, H. F. (2000). "A new method for the nonlinear transformation of means and covariances in filters and estimators," *IEEE Trans. on Automatic Control*, vol. 45, pp. 477-482.

Kalman, R. E. (1960). "A new approach to linear filtering and prediction problems," *Trans. ASME*, vol. 82, pp. 34-45.

Kondoz, A.; Katugampala, N. N.; Al-Naimi, K. T.; Villette, S. (2008). "Data Transmission", US Patent 2008/0165885 A1.

Kundu, A.; Chen, G. C. and Persons, C. E. (1994). "Transient SONAR Signal Classification using Hidden Markov Models and Neural Nets," *IEEE J. Ocean. Eng.*, vol. 19, no. 1.

Li, Q. H. & Wei, W. (1995). "An application of expert system in recognition of radiated noise of underwater targets"*Proceedings of OCEANS'95 Conference*, vol. 1, pp. 404-408.

Lourens, J. G. (1988). "Classification of ships using underwater radiated noise", *Proc. of COMSIG 88 Conference*, pp. 130-134.

Nelson, J. K. & Roufarshbaf, H. (2009). "A tree search approach to target tracking in clutter", *Proceedings of the 12th International Conference on Information Fusion (FUSION).*

Nelson, J. K. & Singer, A. C. (2006). "Bayesian ML sequence detection for ISI channels," *Proc. of the IEEE Conf. on Info. Sciences and Systems*, pp. 693-698.

Nielsen, R. O. (1991). *SONAR Signal Analysis*, ISBN: 0890064539, Boston, MA: Artech House.

Norgaard, M.; Poulsen, N. K. & Ravn, O. (2000). "New developments in state estimation of nonlinear systems", *Automatica*, vol. 36, p.1627.

Peyvandi, H.; Park, S.-J. & Lucas, C. (2010). "An Efficient Technique for Secure Communication in Underwater Wireless Networks", 5th ACM *Wireless UnderWater Networks (WUWNet'10)*, USA.

Peyvandi, H. (2009). "A Novel Approach for Data Transmission over Voice Dedicated Channel of Worldwide Wireless Telecommunication Networks", 8th IEEE *Wireless Telecommunication Symposium (WTS'09)*, Czech Rep.

Physics Today, Online, *http://misclab.umeoce.maine.edu.*

Rabiner, L. R. (1989). "A tutorial on hidden Markov models and selected applications in speech recognition", *Proceedings of the IEEE.* pp. 257-286.

Rajagopal, R.; Sankaranarayanan, B. & Rao, R. R. (1990). "Target classification in a passive SONAR an expert system approach", *Proceedings of IEEE ICASSP Conference.*

Ross, D. (1976). *Mechanics of underwater noise*, ISBN: 0-932-14616-3, Pergamon press.

Roufarshbaf, H. (2011). "A tree search approach to detection and estimation with application to communications and tracking," *PhD Thesis*, George Mason University.

Roufarshbaf, H. & Nelson, J. K. (2009). "Target tracking via a sampling stack-based approach", *Proceedings of the 2009 Asilomar Conference on Signals, Systems, and Computers.*

Roufarshbaf, H. & Nelson, J. K. (2010). "Evaluation of multistatic tree-search based tracking on the SEABAR dataset," in *Proceedings of the 13th International Conference on Information Fusion (FUSION).*

Rumelhart, & McClealan, (1987). *Parallel Distributed Processing*, ISBN-10: 0-262-68053-X, MIT Press, Vol. I, II.

Sea, R. G. (1971). "An efficient suboptimal decision procedure for associating sensor data with stored tracks in real-time surveillance systems," in *Proceedings of the IEEE Conference on Decision and Control.*

Shahbazi, H. & Karimifard, M. (2008). "Design and analysis of low frequency communication system in Persian Gulf", *MTS/IEEE Oceans Conference,* Canada.

Silva, S. R. *editor* (2009). *Advances in SONAR Technology,* InTech Publishing Inc., ISBN 978-3-902613-48-6, Vienna, Austria.

Simandl, M.; Kralovec, J. & Soderstrom, T. (2006). "Advanced point-mass method for nonlinear state estimation," *Automatica,* vol. 42, pp. 1133-1145.

Sittler, R. W. (1964). "An optimal data association problem in surveillance theory,"*IEEE Trans. on Military Electronics,* vol. 8.

Singer, R. A. & Stein, J. J. (1971). "An optimal tracking filter for processing sensor data of imprecisely determined origin in surveillance systems," in *Proceedings of the IEEE Conference on Decision and Control,* pp. 171-175.

Singer, R. A. & Sea, R. G. (1973). "New results in optimizing surveillance system tracking and data correlation performance in dense multitarget environments,"*IEEE Transactions on Automatic Control,* vol. 18, pp. 571-581.

Soares-Filho, W.; Seixas, J. M. & Caloba L. P. (2000). "Averaging spectra to improve the classification of the noise radiated by ships using neural networks", *Proceedings of Sixth Brazilian Symposium on Neural Networks.*

Sorenson, H. W. & Alspach, D. L. (1971). "Recursive Bayesian estimation using Gaussian sums," *Automatica,* vol. 7, pp. 465- 479.

Stoica, P. & Moses, R. (2005). *Spectral Analysis of signals,* ISBN: 0131139568, Pearson Education, Prentice Hall.

Stone, L.; Corwin, T. & Barlow, C. (1999). *Bayesian Multiple Target Tracking.* Artech House.

Theodoridis; S & Kourtroumbas, K. (2006). *Pattern Recognition,* 3rd. Ed., ISBN: 0-12-369531-7, Academic Press.

Urick, R. J. (1983). "*Principles of Underwater Sound*", 3rd Ed., ISBN 0-932146-62-7, McGraw-Hill.

Viterbi, A. (1967). "Error bounds for convolutional codes and an asymptotically optimum decoding algorithm", *IEEE Trans. Information Theory, PP. 260 – 269.*

Ward, M. K. & M. Stevenson (2000). "SONAR signal detection and classification using artificial neural networks", *Canadian Conference on Electrical and Computer Engineering.*

Wenz, G. M. (1962). "Acoustic Ambient Noise in the Ocean: Spectra and Sources," *J. Acoustical Society of America,* Vol. 34, pp. 1936-1956.

Wolf, J. K.; Viterbi, A. J. & Dixon, G. S. (1989). "Finding the best set of K paths through a trellis with applications to multitarget tracking," *IEEE transactions on Aerospace and Electronic Systems,* vol. AES-25, no. 2, pp. 287-296.

Xi-ying, H.; Jin-fang, C.; Guang-jin, H. & Nan L. (2010). "Application of BP neural network and higher order spectrum for ship-radiated noise classification", *International Conference on Future Computer and Communication (ICFCC).*

Xiong, F.; Zerik, A. & Shwedyk, E. (1990). "Sequential sequence estimation for channels with intersymbol interference of finite or infinite length," *IEEE Trans. on Communication magazine,* vol. 38, no. 6, pp. 795-804.

# Part 3

# Biology and Marine Science Sonar Applications

# 10

# Bats Sonar Calls and Its Application in Sonar Systems

Tadeusz Gudra[1], Joanna Furmankiewicz[2] and Krzysztof Herman[1]
[1]*Wroclaw University of Technology*
[2]*University of Wroclaw*
*Poland*

## 1. Introduction

Sonar system operating in gas medium (air) are based on echolocation phenomenon (active sonar). Many scientist and specialists in the field of air coupled ultrasound localisation systems have been inspired by echolocation mechanism of bats. Those mammals mastered echolocation perfectly using ultrasound waves. A resolution of so called real time biosonar is out of range of human made equivalent due to scientific and technological limitations. The main aim of the chapter presented below is to introduce book readers with the echolocation mechanism, evolved in bats, which are an example of animals that have achieved perfection in biosonar usage. The next problem that authors cover is an evolution of sonar systems from simple one beam ranging echolocation devices to advanced multibeam array beamformers operating with digital signal as serviceable form of information representation. Some properties of ultrasound waves in gas medium and physical phenomena involved with air coupled ultrasound waves generation, transmission and detection are described in the chapter as follows: (1) generation of ultrasound wave in gas medium, (2) a short review of air operating transducers including piezoceramics, sandwich, electrostatic, EMFi and MEMS, (3) problem of ultrasound wave transmission in gas medium (the influence of temperature and static pressure on speed of sound, attenuation in air) and range equation, (4) reception and detection of ultrasound waves, target strength. Considering the fact that the detection of echo signal does not give us an information about target a review of signal processing methods is also presented. Taking into account global trend in applying digital signal processing methods to sonar application authors describe some of DSP solutions which are adaptation of echo signal processing by bats at higher level of their auditory system. In order to explain some important issues from bat echo analysis, authors to present some information about bat auditory system and the role of its particular organs and auditory pathways in spatial information processing.

## 2. The bases of echo imaging in echolocating bats

Echolocation plays the most important role in the environment perception in the echolocating bats and odontocete cetaceans. The sonar system of bats and dolphins evolved however in a different way, because of differences in animal size, environment, prey and sound propagation in the air and water Au (2004). Most of the over 1200 bat species echolocate. Echolocation is used for spatial orientation and prey pursuing. Bats ordinarily emit echolocation calls when moving in light and dark, suggesting that echolocation is a very important component of their sensorimotor system. Nonetheless, echolocation is not the only sensory system of bats, since they probably make use of all senses (including vision, olfaction and kinesthetic sense) to integrate them for cognitive processing Masters & Harley (2004).

In the echolocating sensing system of bats, ultrasound vocalisation is actively propagated over the air and returning echoes are used for determination of the position, size and structure of sonar target. A bat can determine between its own echo and echoes coming from other individuals. Sometimes bats can even eavesdrop on emitted sound and its echo of conspecifics and other species to find good foraging sites and forage with members of social group Balcombe & Fenton (1988), Dechmann et al. (2009). These findings and other researches show that echolocation pulses serve as acoustic cues in communication of bats (see review in Jones (2008)).

### 2.1 Sound emission

Sound emission in bats is produced in the larynx, a common sound generator in all mammals. Bats emit calls through the mouth or the nostrils. Nostril emission have advantage when foraging, because a bat can fly and echolocate with the prey in mouth. Moreover, the emission of sound through two nostrils results in interference of emitted sound beams. The interference pattern depends on the distance between nostrils. If it's half of the wavelength of the emitted frequency, the sound energy straight ahead will be intensified, and attenuated on either side, resulting in a narrowing of the sound beam. Experiments with a leaf-nose bat *Carollia perspicillata* (family Phyllostomidae) showed that when one of the two nostrils is blocked the extend of the sound beam doubles in the horizontal plane Fig. (1) Hartley & Suthers (1987).

Bats using the nostrils often have complex noseleaves composed of fold of skin and cartilage. The noseleaves have varying complexity and shape: e.g. shape of horseshoes in horseshoe bats Rhinolohphidae or vertical blade-like noseleaf in leaf-nosed bats Phyllostomidae. Noseleaf acts as acoustic lens, focusing the sound into a narrow beam, which cause increasing of directionality of emitted calls and enhanced echolocation performance. The experiments with *Carollia perspicillata* indicated that if the noseleaf is removed, the vertical beam width increased from 60 to 120 [*deg*] Hartley & Suthers (1987).

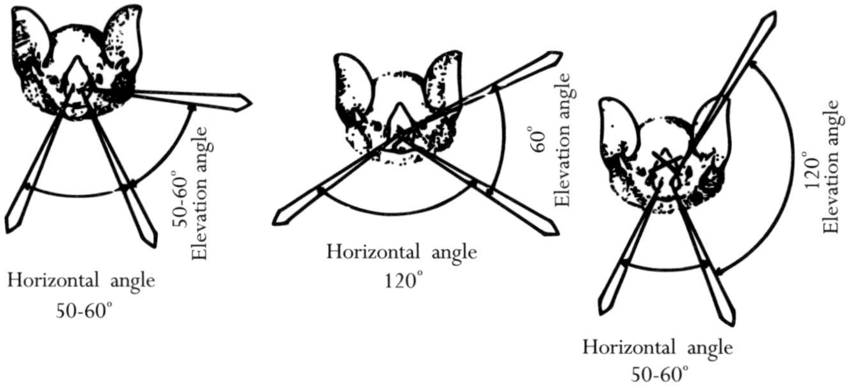


Fig. 1. The relation between the width of the sound beam and nose-leaf and nostril emission in *Carollia perspicillata* (adapted from Hartley & Suthers (1987)).

### 2.2 How bats avoid deafening?

The intensity of the emitted echolocation pulses may be up to maximum level above 140 $[dB]$ of sound pressure level (SPL) Surlykke & Kalko (2008). This is loud enough to cause hard pain and hardly damage the eardrum and cochlea. Bats posses special adaptations to avoid deafening. Echolocation calls are emitted in short bursts. During the burst, bats make themselves temporarily deaf. The deafening is caused by the contraction of the muscle stapedius, which is attached to stapes (a small ossicle in the middle ear, see the chapter *'Auditory adaptations to sound perception')*. Just before a bat emits an echolocation pulse, the stapedius contracts and pulls the stapes away from the oval window Henson JR (1965). The sound cannot be passed to the cochlea, which prevent the damage of this delicate structure. In the final stage of pursuing for an insect, a bat's sonar pulse repetition rate may exceed 200 $[Hz]$, and its stapedius muscle may operate at the same frequency - one of the highest rates recorded in vertebrate muscle Altringham (1999).

### 2.3 The diversity of bat echolocation calls

The diversity of echolocating bats and its habitat leads to diversity of acoustic signals used for echolocation and therefore acoustic information received from echoes. Bats' sonar pulses are usually described as narrowband constant frequency calls (CF) and broadband frequency modulated pulses (FM), however there are many combinations of those two call types Fig. (2) Neuweiler (2000). FM pulses starts at high frequency and sweeps downward to progressively lower frequency. They last for a short period of time, usually up to several ms. There is interspecies and between habitat variation in the slope of the pulse. CF signal is a pure tone or only slightly modulated in frequency at the beginning. In horseshoe bats (Rhinolophidae), hipposiderid bats (Hipposideridae) and mormoopid *Pteronotus parnellii* the CF tone ends and/or starts with short FM part. CF pulse usually last for $10 - 100$ $[ms]$.
Many FM bats used open, semi-open, edges of forest, separating pulse from echo in the time and discriminating delay in echo when estimating target distance, whereas CF bats usually forage in clutter environment and emit long-duration calls separating pulse and echo in frequency Neuweiler (2000).

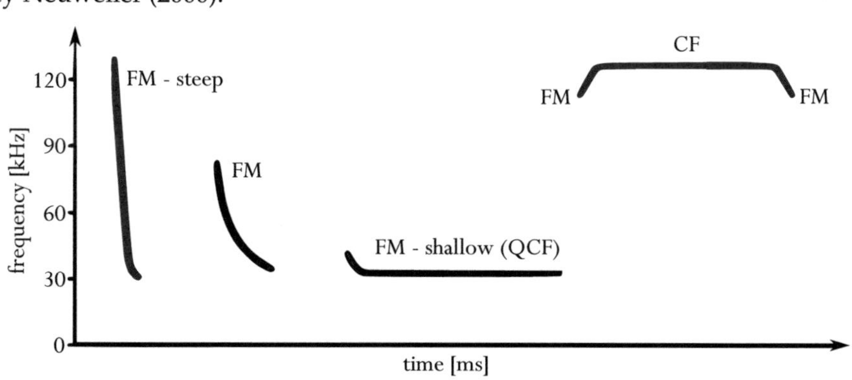


Fig. 2. The main categories of bat echolocation pulses: broadband FM sweeps, narrowband quasiconstant frequency pulse with short FM element (FM-QCF) and constant frequency (CF) call.

An echolocating bat emits pulses with irregular time intervals. During pauses between pulses no information is available. The time without information varied between species, habitats and pulse type. The ratio of the time during which a signal is present to the entire duration of the time from one call onset to the next is called duty cycle. FM bat echolocate on low-duty cycle, i.e. most of the echolocation sequence is occupied by interpulse interval (gaps between pulses), and the pulses covered usually less than 20 [%] of the sequence. CF bat produces high-duty cycle sequence, where signals occupied more than 50 [%] of the sequence time Jones (1999).

The echolocation pulse sequence of a flying bat foraging for insects consists usually of three phases. When a bat is searching for an insect it emits search pulses (search phase). As soon as the bat detect target, it begins to pursuit it, emitting signals faster, progressively decreasing intervals between them in order to avoid overlap between the outgoing signal and the returning echo. This is called approach phase and a bat can emit calls with the signal repetition rate up to 40-50/s. Just before the bat captures an insect, it emits a final buzz, called also terminal buzz or feeding buzz Fig. (3). During the final buzz, the duty cycle increases to about 90 [%].

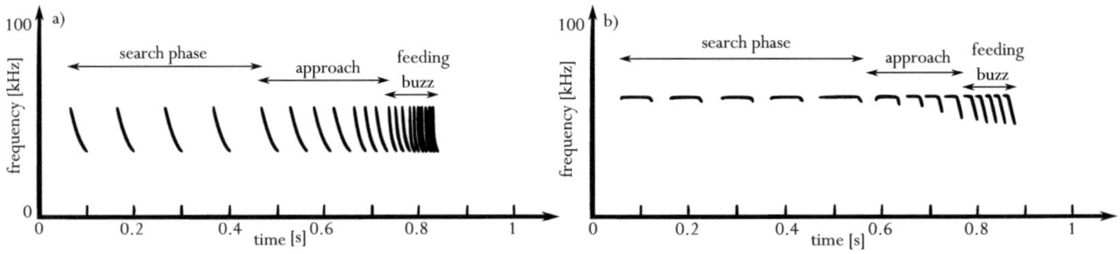


Fig. 3. Echolocation sequence of FM (a) and CF (b) bats emitted during prey pursuit. The pulse repetition rate increases when bats detect a prey reaching the maximum rate just before a catch.

### 2.4 How bats detect faint echoes?

The detection of echoes encounter the problems related to the echo strength and the interference of echoes of different individuals. Echolocation calls of most bats are very loud at the source but they are very quickly attenuated in the air, because most of the bats use ultrasound calls of high frequencies. So the returning echo is very weak. Moreover, bat's auditory system is exposed to extraneous sounds, e.g. sonar of other bats, which may interfere with bat's own echo processing. Bat overcome these problems by opening neural gate of short duration. The neural gate is triggered by the bat's own echolocation pulses, when they are emitted. This process briefly opens time windows enabling the bat's auditory system to receive and interpret bat's own echoes Roverud & Grinnell (1985), Altringham (1999), Neuweiler (2000). The detection of weak pulses may be additionally enhanced by the process called forward masking. During forward masking hearing thresholds increase when acoustic stimuli are presented shortly after a sound of similar spectral composition, i.e. weak echoes returning shortly after the emission of their own comparatively intense sonar calls Siewert I. (2004).

### 2.5 Determination of sonar target features

The echo imagination of an object is affected by the structure of the echolocation pulse and the physical parameters of an object. FM and CF calls are adapted in a different way to intercept the objects. CF bats concentrate all or most of the energy into one frequency what carry the call further. CF signals are adapted to detect wing beats of fluttering insect in clutter environment, whereas FM pulses are well suited for three dimensional target localization, giving broadband echospectra. FM bats spread the energy over a broadband FM pulse with low frequency components. Some FM bats use also shallow FM pulses with low frequency CF tail. Some of them, e.g. noctule *Nyctalus noctulu*, also alternate CF with FM pulses. This strategy is used by bats foraging in the open environments and helps bats to detect prey simply by maintaining the intensity of the pulse, and hence that of the echo, over greater distance Neuweiler (2000).

### 2.5.1 Target location in 3-dimensional space

FM bats discriminate an object's location in horizontal and vertical planes, the distance to the object and its relative velocity and direction of movement. For an analysis of the distance to a target an echolocating FM bat uses the time interval (delay) between their emitted sonar vocalization and a returning echo to obtain target distance information Simmons (1973). An FM bat appears to determine prey velocity primarily by measuring the changes in distance between it and its prey from one echo to the next.

The horizontal position of an object in space is determined by differences in the intensity and/or time of arrival of echoes at the two ears Fig. (4). Tragus plays probably an important role in the vertical positioning, since it is responsible for generating multiple reflections in the external ear as echoes travel to the eardrum Lawrence & Simmons (1982). The reflections of sound waves from tragus and the wall of the pinna create interference patterns, which change according to the vertical direction of the sound Fig. (4). Therefore the vertical positioning may be resolved by analyzing secondary echoes which follow different paths through the inner ear and around the tragus, depending upon their direction of origin. The big brown bat *Eptesicus fuscus* can resolve vertical angles down to 3.5 $[deg]$, and horizontal angles of 1.5 $[deg]$. Bats that lack tragus (horseshoe bats and hipposiderids) move their pinnae in a pattern correlated with the emission of their echolocation calls: one ear moves upward and downward, the second ear moves forward and back. This pinnae movements play a role in sound localization a the vertical plane Neuweiler (2000).

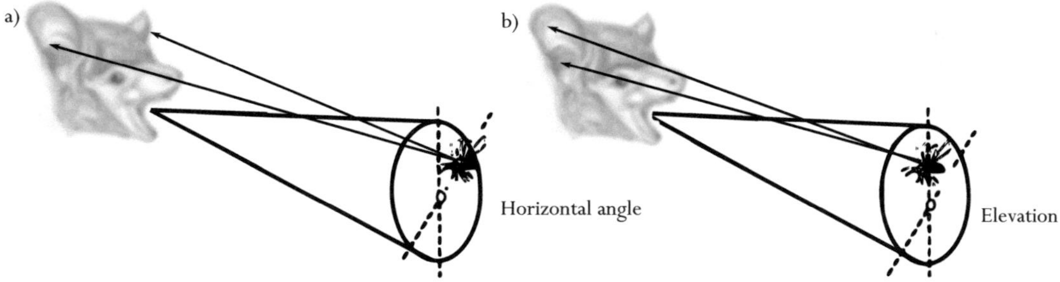


Fig. 4. The schematic imagination of the vertical and horizontal angles determination by an echolocating bat.

### 2.5.2 Target range differences discrimination

FM bats well resolve the discrimination of small differences in the distance to two objects, utilizing matched-filter and the envelope of autocorrelation function. The bat cross-correlates the emitted pulses with returning echo. The ACR describes the average period $T$ of the signal (the reciprocal of the fundamental frequency, which decreases with increasing frequency) and $t$, the width of the envelope at the half of its maximum. This is approximately the reciprocal of the signal's bandwidth. The introduction of FM higher harmonics decreases the height of the side peaks, reducing $t$. Bat neural system can determine unambiguously the timing (i.e.

position) of the peak of the ACR. The narrower the central peak, and the lower the side peaks, the more acute the bat's target ranging. It is achieved with the FM pulses of large bandwidth (small $t$), high fundamental frequency (small $T$) and with additional harmonics (the side peaks are of low amplitude). The CF bat (constant frequency call with short, narrow bandwidth FM tail) can only distinguish between two objects if one is more than 12 $[mm]$ closer from the bat than the other. The FM bat can distinguish distance of 4 $[mm]$, which means that it is able to differentiate between echoes arriving as little as 20 $[\mu s]$ apart Fig. (5) Simmons (1973).
Important role in target ranging plays the neural gate. It acts as a distance filter for bat's own sound. In *Noctilio albiventris, Eptesicus fuscus* and *Rhinolophus rouxi* the time window is open for around 30 $[ms]$, allowing the bats to target range over distances up to 5 $[m]$. The bat can ignore echoes from more distant objects Altringham (1999).

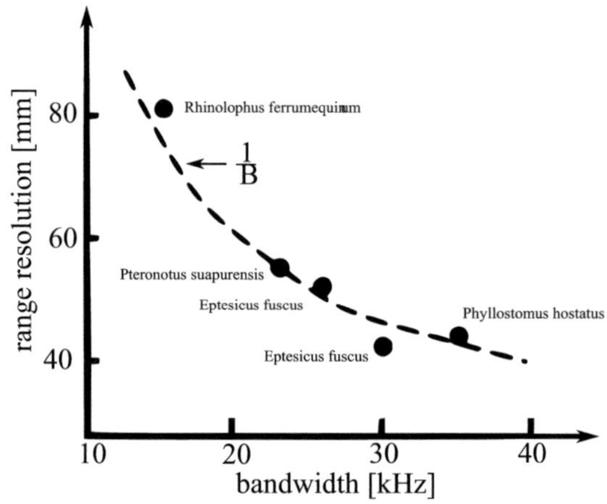


Fig. 5. Relationship between call bandwidth $(B)$ and target ranging accuracy in several species of bat (*Rhinolophus* and *Pteronotus* are known to used CF, the remaining species are FM bats) (from Altringham (1999) based on Simmons (1973))

#### 2.5.3 Determining object size

Bats determine target size using amplitude modulation, i.e. the intensity of the echo. The bigger an object reflecting echolocation pulses, the smaller amplitude modulation. By small objects the amplitude of returning echo is decreasing. Different species of bats can detect and avoid wires with a diameter of $0.06 - 0.1$ $[mm]$.

#### 2.5.4 Determining object structure

Most of the objects perceived by bats is very complex and have many reflective surfaces that result in a chaotic acoustic image of the target without systematic spectral interference pattern. The echo perceived by a bat consists of its ultrasonic emission convolved with the acoustic impulse response (IR) of the ensonified object. The IR is the sum of reflections from the complex target. Its extraction implies great echolocation capabilities of bats. The experiment with *Phyllostomus discolor* (Phyllostomidae) showed that bats can classified echoes and evaluate IR roughness. Hence bats can evaluate complex texture, such as e.g. foliage type Grunwald et al. (2004). Thus the structure of more complex objects, ensonified by ultrasounds, can be probably detected with high precision by echolocating bats. This capability allow bats to recognize not only environment features but also find and classify prey or food source inaccessible for other (bat) species. This ability is especially important in FM bats, that forage in a clutter environment and detect insects against background or feed on nectar and pollen of flower. The precision in structure classification and insect detection in a clutter leads to niche partitioning in FM clutter forager bat species, because prey detection in clutter increases as the width of the FM call band broadens Siemers & Schnitzler (2004)

Some of the flower-visiting bats (glossophagines from the family Phyllostomidae) can probably estimate the stage of flower inflorescense using echolocation. *Mucuna holtoni*, a bat-pollinated flower attracts bats by conspicuous echoes. The plant open new flowers successively from dusk to about midnight, so the bat must search for newly open flowers which serve with pollen. When a bud becomes mature, its vexillum (upper petal) is raised. The experiments showed that the vexillum indicate the presence of mature flowers to the bats. Based on this feature, exploded flowers can be distinguished from virgin flowers Helversen & Helversen (1999). When the flowers of *Mucona holtoni* are exposed to artificial sounds (imitating echolocation calls of bats), they give different echoes (amplitude modulations) in bud and open flower. The raised vexillum strongly reflect sounds so the echo of the whole flower is strongly dominated by echo of the vexillum Helversen O. von (2003).

The best performance in determining target structure is achieved by FM bats. As opposed to CF calls, FM signals contain more information about spectral changes (modulation of the frequency components) and amplitude changes (modulation of the strength of the echo). The very broadband FM call ($100 - 20$ $[kHz]$) used by *Megaderma lyra* enables it to analyse textural depths in the range $0.9 - 4.2$ $[mm]$. The narrowband FM tail to the call of greater horseshoe bats is limited to $1.1 - 1.5$ $[mm]$ Altringham (1999). Experiments with *Megaderma lyra* showed that the bat can detect the difference in depth of only 0.2 $[mm]$ Schmidt (1992). The bat can do that not only using echolocation but also by passive listening to the virtual echoes.

Bats can perceived sound structure by analyzing the interference patterns of echos. Signals reflecting from the rough structure contains peaks and nulls at different frequencies, depending on the depth of the textural surface of the target. The interference between sound waves of identical frequency and amplitude causes addition or cancellation of the energy. The addition of the energy is maximal when the waves are in phase, whereas maximal cancellation occurs when the echo and emitted signal are 180 $[deg]$ out of phase. Such cancellation within a spectrum is called a notch. A notch occurs when the distance between the high and low points on a target is $\lambda/4$ or an uneven multiple of this relationship. The addition of the energy occurs when the differences in depth between the target surface is $\lambda/2$ or even multiple thereof Fig. (6). The position of the peaks and nulls within the spectrum provides a measure of the surface texture of an object. The smoother target the higher the frequency of the maximal notch Neuweiler (2000).

#### 2.5.5 CF calls and prey detection in CF bats

CF bats intercept the space and prey in a different way as FM bats. The CF pulses has poor range resolution, although they also need determine the range for interception or avoidance. The CF bats probably therefore introduce FM component to their signals. These bats detect moving prey using the Doppler shift (Doppler effect) of the constant frequency or by detecting the changes in echoes of a flying insect caused by its wing beats Schnitzler et al. (1983), Schnitzler (1987). The echo reflected from the fluttering insect contains momentary glint-like energy maxima. Such glints consists of amplitude and frequency modulations. Changes in wing profile causes variation in the intensity of the echo amplitude modulation. The wing movement towards or away from the direction of the bat broadens sound spectrum due to

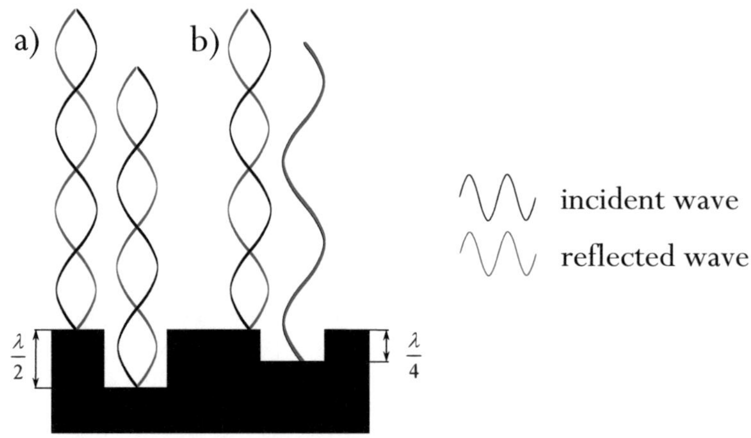


Fig. 6. The interference pattern of sounds reflecting from the object if rough structure.

Doppler effects. Echoes received by a bat approaching a stationary insect will be increased in frequency, and increased further if the insect is approaching the bat Schnitzler et al. (1983), Schnitzler (1987). Similarly, echospectra of fluttering insect oriented at three different angles demonstrates that the echo will be amplitude modulated within a pulse, and there will be also fluctuations in echoes from pulse to pulse Schnitzler et al. (1983). Insects of different size and from different species/families can therefore give different amplitude modulation, which may be an useful information for foraging bats.
The Doppler shift arises from the differences in velocity between bat and insect and the CF bats can compensate for this effect Schnitzler (1968). FM pulses are Doppler tolerant, so they are not affected significantly by either their own motion or that of the prey. CF bats compensate for Doppler shift induced by bat's flight speed by lowering the frequency of emitted pulses when flying. Doppler shift compensation causes that emitted pulse and returning echo differ in frequency, however the frequency of returning echo always falls within the range of the frequency emitted when resting, which is the frequency of best hearing '(BF)'. CF bats are very sensitive to a very narrow frequency band '(BF)', actually one frequency of CF signal, which is called auditory fovea (Neuweiler (2000), see the next chapter). Therefore the bat alters its emission frequency to keep an echo at a strictly defined frequency. The shift in emitted frequency tells the bat how fast it is moving, and in what direction, relative to the object. As a target moves and the echo frequency changes, the bat shifts its emitted frequency to keep the echo frequency constant Neuweiler (2000).

### 2.6 Auditory adaptations to sound perception

Bats auditory system is adapted to special tasks of the spatial orientation using sound (pulse and its echo) to perceive the environment, but it is built and operate in the similar manner as in other mammals, i.e. pinna with tragus, external ear, middle ear and inner ear, auditory nerve and auditory centers in brain Neuweiler (2000).

#### 2.6.1 Functional anatomy of the ear

Echolocating bats have relatively large pinnae. The pinna acts as movable directional antenna. The directional hearing and sound picking by the pinna depends on the size and geometry of the pinna as well as the wavelength of sound. The pinna functions as a directional antenna when $ka > 1.25$, that is, when the wavelength is smaller than 4-5 times the radius of the

pinna opening where $a$ - radius of the pinna opening. The larger the pinna, the better low frequencies will be transmitted to the eardrum. In *Plecotus auritus*, a bat with relatively big pinnae, the cut-off frequency is $6-8$ $[kHz]$. The pinna can not exert its directionality when the wavelength of sound exeeds the dimensions of the pinna. The higher the frequency, the more pronounced the directional characteristic of the pinna Neuweiler (2000).
The highly variable pinna form and size in different families of bats affect their acoustic properties. The pinna in most echolocating bats is rather small, because it is tuned to ultrasound frequencies with short wavelengths between 2 and 16 mm. In contrast, in gleaning bats (taking prey from the surface) which listen to the low frequency rustling noise made by prey, the pinna is extraordinarly large. The greatest pinna amplification is in the range of $6-10$ $[kHz]$.
Unique to most echolocating bats is the tragus, cartilaginous projection at the base of the pinna. Tragus enhances discrimination of the elevation of the target Lawrence & Simmons (1982). Bats that do not posses tragus have antitragus, a horizontal fold of the skin along the opening of the ear canal, and frequently move their pinnae.
Sound perceived by pinna passes to middle ear and causes vibration of eardrum (tympanic membrane), which is only $2-11$ $[\mu m]$ thick in echolocating bats. The vibrations are passed to the oval window along 3 ears ossicles: malleus, incus and stapes. Those media acts as filters, because of the vibration capabilities. The higher frequencies bats produce and perceive the thinner the eardrum is and smaller and lighter middle ear ossicles, because they vibrate more rapidly. Vibrations of the oval window are transmitted along the spiral canal of the cochlea. Cochlea is a bone with three canals spiral in parallel with the organ of Corti between them. It is highly complicated sound receiver and frequency analyzer, which categorizes the sound by frequency. Echolocating bats have $2.5-3.5$ complete turns, relative to the 1.75 of non-echolocating bats and primates Altringham (1999).
The inner canal of the cochlea is a tube, scala media, filled with fluid. It is covered by a second tube, which upper part is called scala vestibuli and the lower part scala tympani. Both of those tubes are separated from middle ear by membranes of the oval and round windows Fig. (7). The floor of the scala media is formed by the basilar membrane with sensory hair cells. Each hair cell posses bundle of stereocilia. The sterocilia tips are covered by the tectorial membrane. Vibration of the basilar membrane against the tectorial membrane causes shearing of the hair cell sterocilia and consequently, oscillations in the receptor potential that follow the rhythm of basilar membrane movement. This movement is caused by the momentary pressure differences between the scala vestibuli and scala tympani, which arises when the stapes moves and press against the oval window. The basilar membrane moves up and down at the same frequency as the stimulus. The oscillation progresses along the length of the basilar membrane, decreasing its speed. When the stimulus is a high frequency sound, the travelling wave moves only a short distance along the basilar membrane. The lower the frequency, the farther the traveling wave moves within the cochlea. Thus, high frequencies activate only the most basal hair cells, and lower frequencies activate the apical hair cells most strongly Fig. (7) Neuweiler (2000). Each hair cells is activated by specific frequency, its best frequency ($BF$). The frequency map of the basilar membrane varied between CF and FM bats. The CF bats have 'personal' constant frequency which show small variation in a resting specimen ($\pm 50$ $[Hz]$, Schnitzler (1968)). Therefore there is a mechanical filter in cochlea tuned to very narrow frequency band of about several kHz. The cochlea of CF bats has the expanded representation of a narrow frequency band of only 6 $[kHz]$ around the individual's pure-tone echo frequency Fig. (8). The hair cells sensitive to that narrow band occupy the entire second turn of the

cochlea, what correspond to one-fourth of the length of the basilar membrane. In FM bats this region contains the representation of about one octave and the frequency progression of the hair cells id approximately logarithmic Vater et al. (1985), Neuweiler (2000). This very narrow filter in CF bats is called auditory fovea. The auditory fovea of each individual bat is precisely tuned to its own emitted frequency Vater et al. (1985). The corresponding fitter exists in the in consecutive stages of sound processing in the brain (see next subchapter).

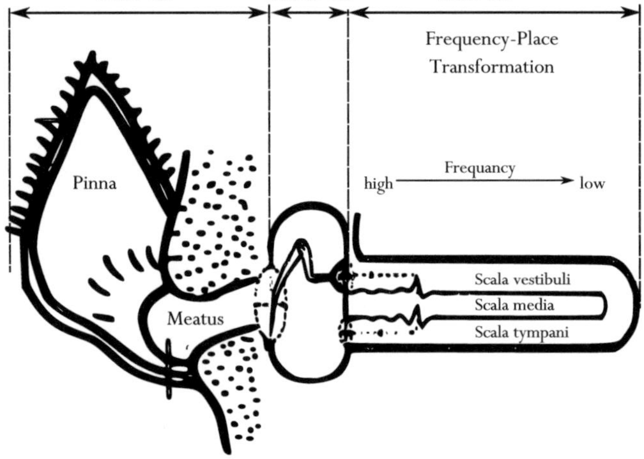


Fig. 7. The schematic plan of the mammalian ear and sound wave flow from the pinna to the inner ear (adapted from Neuweiler (2000)).

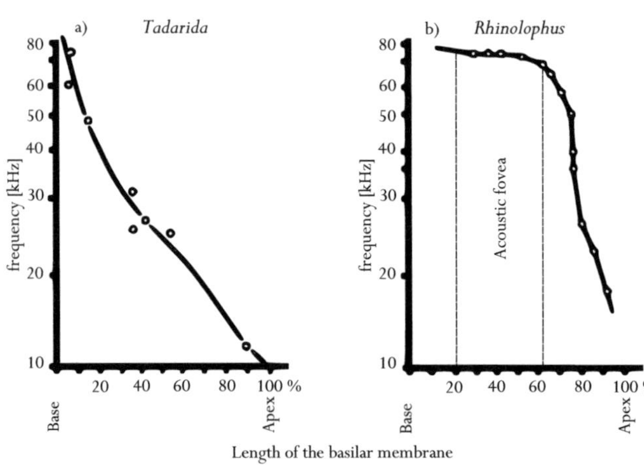


Fig. 8. Frequency representation on the basilar membrane in the inner ear of FM *(Tadarida brasiliensis)* and CF bats *(Rhinolophus rouxi)*. CF bat show specific auditory fovea (adapted from Vater et al. (1985) and Neuweiler (2000).

### 2.6.2 Auditory pathways in the central nervous system

The central auditory system of bats consists of the same nuclei as in other mammals, however some of these nuclei are relatively larger than they are in other mammals. The impulses from the hair cells in the inner ear are transfer to the auditory nerve fibers, which further transmit the impulse to the midbrain auditory centre, called *inferior colliculus*. In echolocating bats the *inferior colliculus* is large. It gather all pathways together and transmits information to the *medial geniculate* body of the thalamus. From thalamus the information is proceeds to auditory cortex in forebrain where the final processing of the sound is happened and sound is translated into images of the environment Neuweiler (2000). The output of the bank of

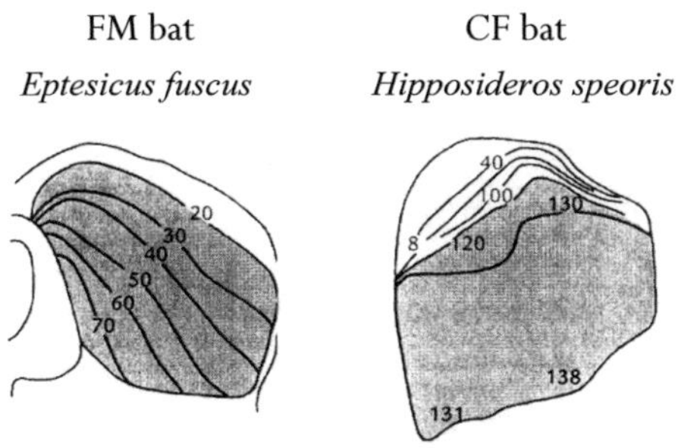


Fig. 9. Tonotopic frequency representation in the inferior colliculus in FM bat *(Eptesicus fuscus)* and CF bat *(Hipposideros speoris)* (from Ostwald (1984))

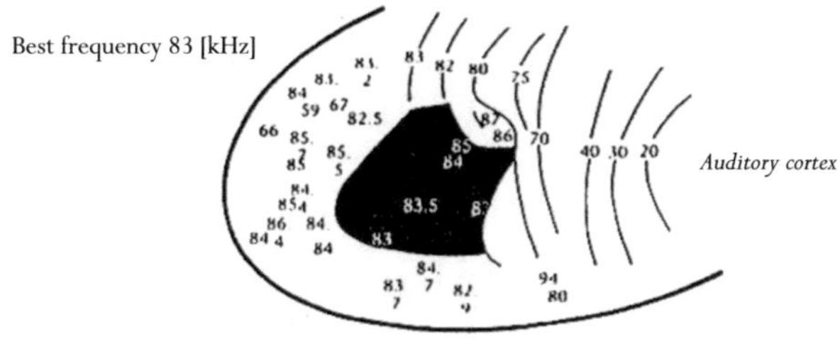


Fig. 10. Tonotopic frequency representation in the auditory cortex of CF bat *(Rhinolophus ferrumequinum)* with narrow frequency band of auditory fovea (from Neuweiler (2000).

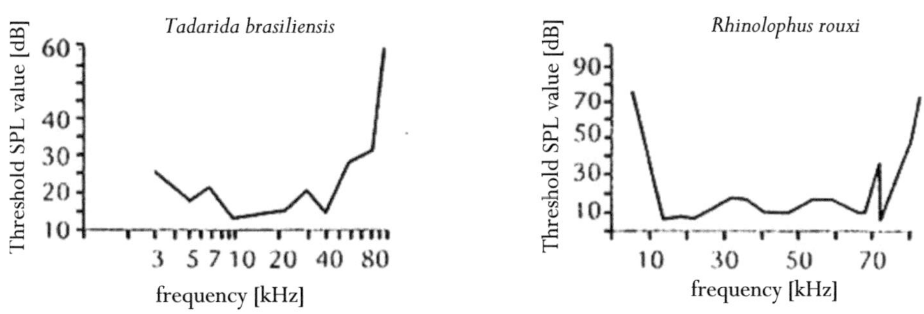


Fig. 11. Audiograms of FM *(Tadarida brasiliensis)* and CF bats *(Rhinolophus rouxi)* obtained from the measurements of the neurons in the inferior colliculus. CF bat shows fine tuning to a narrow frequency band (at about 73 [$kHz$]) of the auditory fovea (adapted from Vater et al. (1985), Neuweiler (2000).

spectral filters in the inner ear is organized into a tonotopic (cochlear) map with a laminar arrangement of neurons of the *inferior colliculus*. The cells within a given lamina all have a similar best frequency Fig. (9). Thus, each lamina is responsible for processing information from one specific frequency band. In FM bat frequency bands are distributed less or more evenly, whereas in CF bat, the lamina responsible for one narrow frequency band is relatively large. The auditory cortex of CF bat has adequate an overrepresentation of the neurons sensitive to the frequencies of only 1.5 $[kHz]$ band Fig. (10). These narrow filters correspond to the auditory fovea and is seen in the auditory tuning curves of CF bats, as a region of very low threshold for the frequencies of CF portion of the echolocation pulse Schuller & Pollak (1979) Fig. (11). In FM bats the neurons are almost always either broadly tuned or only fairly sharp Neuweiler (2000) Fig. (11).

## 3. Ultrasound waves in gas medium

The nature of a gas medium, which is determined by its parameters, is extremely ultrasound unfriendly in relation to ultarasound waves generation transmission and detection. The main parameter of the gas medium is sound velocity $c$. It varies depending on such parameters as static pressure and temperature of gas. A theoretical dependence between sound velocity and gas properties is given by Eq. (1)

$$c = \sqrt{\frac{\kappa (T) \cdot p_0 \cdot T}{\rho_0 \cdot T_0}} \tag{1}$$

where $\kappa (T)$ – ratio of specific heats ${}^{c_p}/_{c_v}$, $p_0$ – static pressure in normal conditions, $\rho_0$ – gas density in normal conditions, $T_0$ – 293 K. The main problems in case of an air coupled sonar system based on TOF are temperature and pressure changes, which have to be compensated using external sensors or reference channels. The second important parameter which is critical in sonar systems is sound attenuation in a given medium. In gases sound attenuation is described by clasicall Stokes-Kirchhoff Eq. (2)

$$\alpha_{cl} = \alpha_\eta + \alpha_\sigma = \frac{\omega^2}{2\rho c^3}\left[\frac{4}{3}\eta' + \sigma\left(\frac{1}{c_v} - \frac{1}{c_p}\right)\right] \tag{2}$$

where $\omega$ – circular frequency, $\eta'$ – shear viscosity, $c_v$ – specific heat at a constant volume, $c_p$ – specific heat at a constant pressure, $\sigma$ – heat conductivity. The most significant information contained in Eq. (2) is that the attenuation factor is proportional to the square of circular frequency $\alpha \approx \omega^2$. As a result of the above the range of air operating sonar system is significantly limited to a few centimeters for frequencies above 100 $[kHz]$. Additionally, echolocating bats use band limited signals from the range of $20 - 200$ $[kHz]$. This varies between species, but in some cases also depends on hunting strategy and the type of surrounding obstacles. A precise expression of attenuation in air Eq. (3) was formulated by Evans et al. (1972).

$$\alpha = \frac{2\pi^2 f^2}{\kappa c_0 p_0}\left(\frac{4}{3}\eta' + \frac{\kappa - 1}{c_p}\sigma + \kappa a_d D_{12}\rho\right) \tag{3}$$

The model of sound absorbtion in air takes into account both the classical processes (viscous losses, heat conduction) and relaxation. The two new variables $a_d$ and $D_{12}$ stand for molecular diffusion constant, and diffusion coefficient for the mixture of oxygen and nitrogen repectievly. The relation between attenuation and frequency for different humidity is shown in Fig. (12)

Considering the simplest case of an acoustic TRX (Transmitting - Receiving) system made of a coaxially situated single transmitter and receiver shown in Fig. (13) it is possible to define

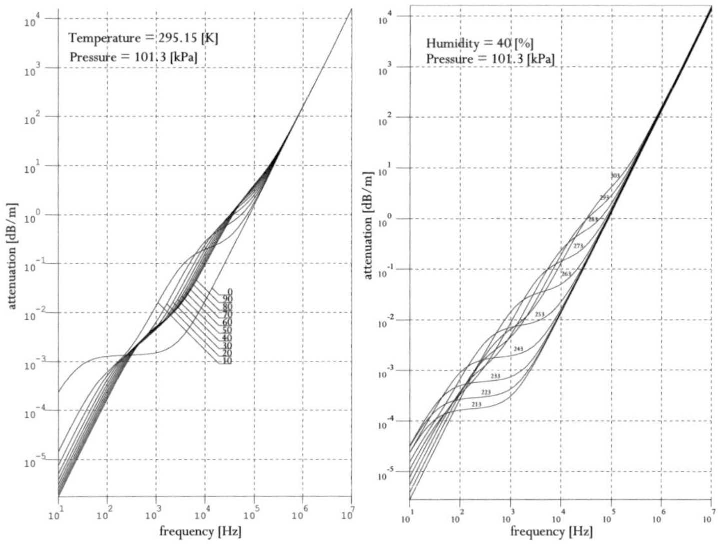


Fig. 12. Attenuation coefficient as a function of frequency with various parameters: humidity in a range of 0 − 90 [%] with 10 percent resolution (left), temperature in a range of 213 − 303 [$K$] with 10 [$K$] resolution (right). According to Evans et al. (1972) and Bass et al. (1995).

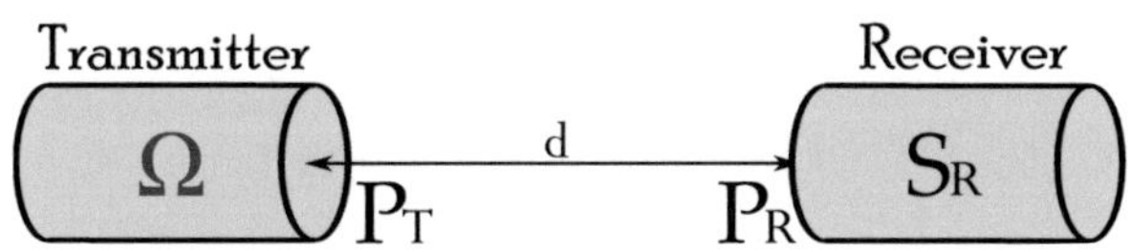


Fig. 13. An illustration of a simple acoustic link

acoustic link range equation (4), widely described by Gudra & Opielinski (2006).

$$\frac{P_T}{P_R} = \frac{4\pi d^2 e^{2\alpha d}}{\Omega \cdot S_R} \tag{4}$$

The equation defines a relation between both transmitted and received acoustic power as a function of distance $d$, directivity of the transmitter $\Omega = \pi^2 f^2 D_T^2 / 2c^2$ for $\pi f D_T / 2c \rangle\rangle 1$, receiver's surface $S_R$ and attenuation factor $\alpha$.

The third important limitation, especially important in air coupled sonar systems, is acoustic impedance mismatch. According to transmission line theory the energy transmission coefficient equals to Eq. (5)

$$T = 2\frac{\sqrt{Z_1 Z_2}}{Z_1 + Z_2} \tag{5}$$

In most ultrasonic applications the acoustic impedance of the transmitter or receiver is a few order higher than the impedance of the medium. As a result of this phenomena most acoustic energy is reflected from impedance discontinuity. Some methods of reducing the impedance mismatch will be described in section (4).

Target strength $TS$ is another very important parameter in the context of object detection using a sonar system. This parameter characterises the reflecting properties of an object and is defined by Eq. 6)

$$TS = 10 \log_{10}\left(\frac{I_r}{I_I}\right) \tag{6}$$

where $I_r$ and $I_t$ indicate intensities of reflected and incident waves respectively Because of the fact that target strength depends on objects' geometry and their reflecting capabilities it can be calculated for simple solids like sphere, cylinder etc. More information about target strength measurements for different objects can be found in the papers Gudra et al. (2010) and Norman & Jones (2000). To evaluate the performance of an active sonar operating in air we include target strength *TS* Eq. (7)

$$SL - 2TL + 2TS - NL + DI + PT > RT \tag{7}$$

where: *SL* - source level, *TL* - transmission losses, *NL* - noise level, *DI* - directivity index, *PG* - processing gain, *RT* - reception threshold.

## 4. Air coupled ultrasound transducers

Since air coupled acoustic signal is used in passive and active sonar systems, ultrasound transducers have to be applied in these solutions. Due to problems related to ultrasound signals transmission described in section (3) signal transmitters and receivers have to guarantee some critical parameters in order to achieve eligible results. Efficiency $S_n$ and sensitivity $S_0$ are very basic parameters of acoustic transmitters/receivers Due to the transducer dimensions comparable with an acoustic wave length the ultrasonic sensors are usually highly directional. The directional parameters of transducers are described by directivity coefficient $\Omega$ given by Eq. (8)

$$\Omega = \frac{2}{\int\limits_0^\pi K^2(\theta)\sin(\theta)\,d\theta} \tag{8}$$

The directivity characteristic for a circular piston vibrating in an infinite baffle is described by Eq. (9)

$$K(\theta) = \frac{2J_1\left(\frac{\pi f D_T \cdot \sin(\theta)}{c}\right)}{\frac{\pi f D_T \cdot \sin(\theta)}{c}} \tag{9}$$

where $D_T$ – piston diameter, $J_1$ – Bessel function of first order. The problem of simplifying the Eq. (9) is widely described by Gudra & Opielinski (2004). In practice for the directivity coefficient calculations of ultrasonic transducers working in the air it can be assumed that:

$$\Omega = \begin{cases} \dfrac{\left(\pi f D_T/_c\right)^2}{2\left[1 - \dfrac{J_1\left(2\pi f D_T/_c\right)}{\pi f D_T/_c}\right]} & for\ \pi f D_T/_c \leq 10 \\ \frac{(\pi f D_T)^2}{2c^2} \ for\ \pi f D_T/_c > 10 \end{cases} \tag{10}$$

There is a variety of transducers that are used to generate or receive ultrasounds in a gas medium. Some selected properties of the transducers are presented in a table (1)
A dynamic transducer may be used to produce low frequency ultrasound. The basic limitation of cut-off frequency is high impedance of vibrating coil and decrease of effective radiating surface of the loudspeaker membrane. An electro-mechano-acoustical model of dynamic loudspeaker is shown in Fig. (14). The parameters of analog circuit and it analysis are widely

| Type | Frequency range | Transmitter | Receiver |
|---|---|---|---|
| Dynamic | broadband | Yes | No |
| Condenser | broadband | Yes | Yes |
| Piezocermic | resonant | Yes | Yes |
| PVDF | resonant/boadband | Yes | Yes |
| EMFi | broadband | Yes | Yes |
| Electret | broadband | Yes | Yes |

Table 1. Comparison of ultrasound transducers' basic properties.

described in a book Borwick (2001). The most important conclusion is that the SPL produced by this type of transducer decreases 12 [dB/oct] because of voice coil impedance $L_c$ and air mass load $M_{md}$.

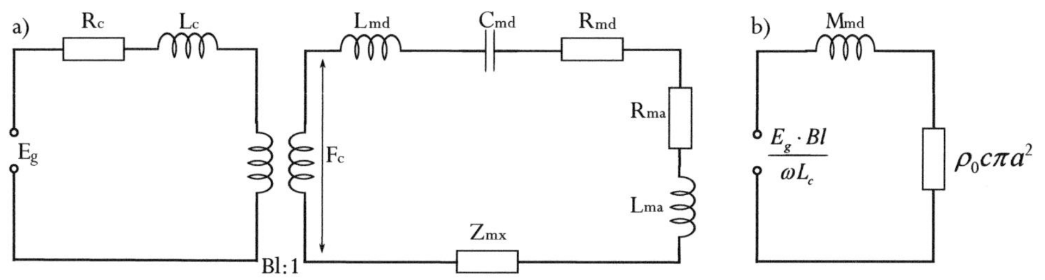


Fig. 14. Dynamic transmitter model a) in a full range of frequencies, b) at very high frequencies (above 2-nd resonance).

The next group of transducers associated with ultrasound wave generation are condenser microphones and transmitters. Most of ultrasound microphones are based on condenser receiver principle. The common known limitation of such a device is its small membrane surface in comparison to the wave length, which causes low sensitivity. The microphones are also very susceptible to humidity and dirt. Condenser transmitters are not as popular as the receivers due to low effectiveness, but in a range of frequencies above 50 $[kHz]$ condenser construction is, next to EMFi film, the only broad band alternative. The most common material in ultrasound transducers technology is piezoceramics. This is due to the resonant character of generated waves and its highest effectiveness/sensitivity. Since Langevine developed a sandwich type transducer in 1927, it has been the most known type of transducer, next the one that uses piezoceramic rings vibrating in radial symmetrical mode. The main advantage of sandwich type transducer coupled with vibrating plate is its high effectiveness and ability to work continuously and generate high acoustic power. Additionally, a surface of vibrating plate improves energy transmission coefficient Eq. (5). A typical construction of a sandwich transducer is shown in Fig. (15).

Sandwich transducer elements' dimensions are described by Eq. (11).

$$tg\left(\frac{2\pi d f_0}{c_p}\right) \cdot tg\left(\frac{2\pi b f_0}{c_b}\right) = \frac{Z_p}{Z_b} \tag{11}$$

where:

$Z_p$ - impedance of piezoceramics material,

$Z_b$ - impedance of loading mass material,

$c_b$ - speed of sound in loading mass material,

$c_b$ - speed of sound in piezoceramics material,

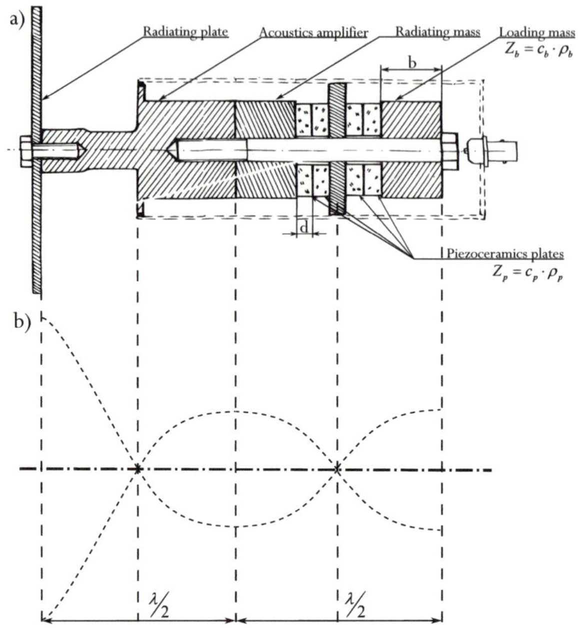


Fig. 15. Sandwich transducer a) structure of a transducer with vibrating plate, b) acoustic velocity distribution.

$d$ - thickness of piezoceramic plate,
$b$ - thickness of loading mass,
$f_0$ - resonance frequency.
The length requirements for acoustic velocity transformer is presented in Fig. (15 b) and equals $\lambda/2$. Radiating mass length has to fulfill requirements $\lambda/4$. The maximum radiating acoustic power can be achieved by applying stepped plate, which vibrates flexurally in its axial mode. The major problem of the presented construction is the high mechanical $Q$-factor of the plate and mechanical part of the transducer and electrical $Q$-factor of piezoceramics.
Radial and thicness vibration of piezoceramic disc are commonly used as a source of acoustic wave in many sensors. An piezoelectric disc and its electrical model are shown in Fig. (16). Radial $u_r$ and thickness $u_z$ displacement of radiating surfaces are given by Eq. (12).

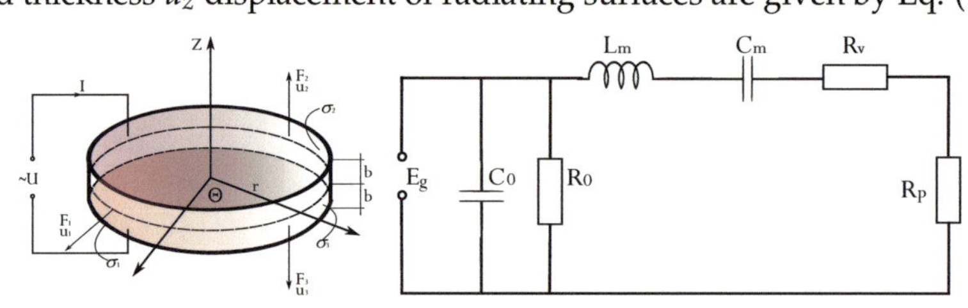


Fig. 16. Coordinate system, force distribution, dimensions of piezoelectric disc and its electrical equivalent circuit at resonance frequency.

$$\begin{aligned} u_r &= -\frac{u_1}{j\omega J_1(k_1 a)} J_1(k_1 r) \\ u_z &= \frac{1}{j\omega}\left[-\frac{(u_2+u_3)}{2\sin(k_3 b)}\sin(k_3 z) + \frac{(u_3-u_2)}{2\cos(k_3 b)}\cos(k_3 z)\right] \end{aligned} \tag{12}$$

where:

$$k_1 = \frac{\omega}{c_1}, c_1 = \sqrt{\frac{c_{11}^D}{\rho}}, k_2 = \frac{\omega}{c_3}, c_3 = \sqrt{\frac{c_{33}^D}{\rho}}, \tag{13}$$

The thickness vibrations of disc with matching layer generate quasi-plane wave while the radial ones are the source of cylindrical wave. To produce a wave with reduced near field and side lobes it is possible to use radial vibrations of the piezoceramic disc to actuate the transducer mechanical enclosure and radiate the wave in $z$ direction.

Maximum displacement across $x$ direction in thickness vibration mode occures at $r = 0.$ . Radial vibration displacement reaches its maximum at $r = a$ across $r$ direction. Normalised displacement across both directions as a function of radius is shown in Fig. (17)

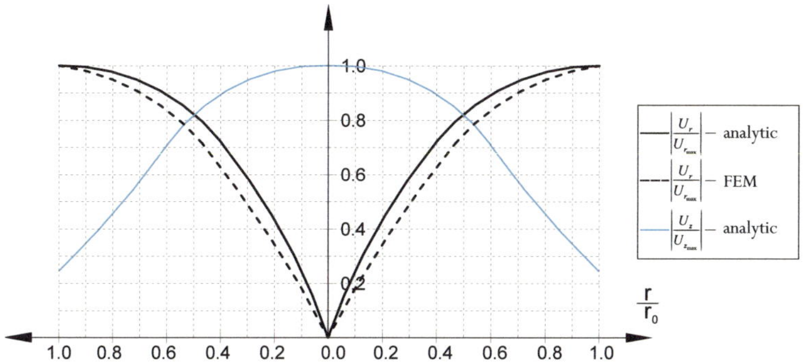


Fig. 17. Distribution of radial and thickness components of displacements of piezoelectric disk calculated using analytical method and FEM (according to Gudra (2008))

The acoustic energy in air operating active sonar systems is transmitted from transducer to a gas medium. The impedance of the piezoceramics PZT equals $Z_p = 29.6 \cdot 10^6$ while the air acoustic impedance equals $Z_{air} = 427$. One of the methods to reduce the adverse energy transmission coefficient Eq. (5) is to use impedance matching layer as it is shown in Fig. (18). Using one-quarter of the wave length matching layer made of material the acoustic impedance of which is $Z_p = 1.12 \cdot 10^5$ guarantees maximum energy transmission. The matching layer material is usually made of composite porous materials. Another way to improve impedance matching is applying multilayered matching structure, where $Z_k << Z_{k+1}$. Sound attenuation in particular layers is a limitation of this method.

The problem of impedance mismatch is less significant for EMFi (Electromechanical Film) based transducers. The physical and acoustics parameters of foil are presented in papers by Paajanen et al. (2000) and Paajanen et al. (2001). The most important advantages of EMFi film is high value of longitudinal piezoelectric constant $d_{33}$, which varies in the range of $130 - 450[pC/N]$ and low value of acoustic impedance, which equals $2.6 \times 10^4 kg/(m^2 \cdot s)$ EMFi film could be used as a broadband transmitter/receiver and can produce bats-like echolocation pulses. This attribute was used in some robotic applications described in Jiménez et al. (2008), Steckel et al. (2008) and Schillebeeckx et al. (2008). Since the bats echolocation mechanism is the most sophisticated biosonar, engeeniers have attempted to use it as a pattern and inspiration in their echolocation devices. Target position estimation is made on the basis of echo signal processing. Target position is specified by range $d$, azimuth $\phi$ and elevation $\theta$ angles.

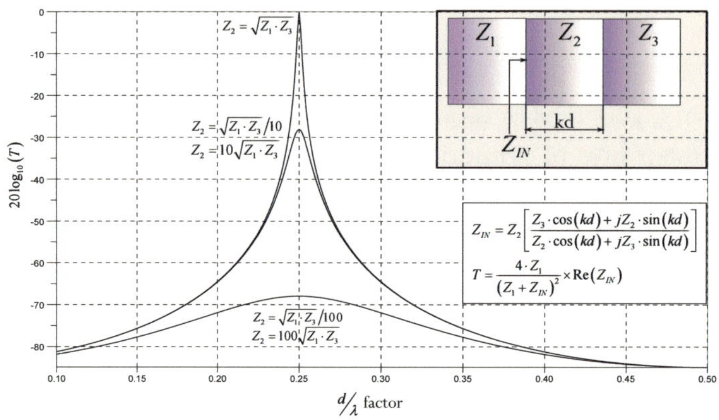


Fig. 18. Impedance matching using one layer and transmission line theory according to Lynnworth (1965).

## 5. Signal processing in sonar systems

Nowadays, ultrasound signal processing in air-coupled sonar systems is dominated by digital technology. Digital signal processing techniques give engineers a whole spectrum of possibilities and enables them to adapt the equipment to specified solutions. On the other side, every ultrasound device consists of an analog signal processing block. A generalised ultrasound localisation system is shown in Fig. (19). Each echolocation system can be configured using the proposed generalisation. The extreme simplification of presented system could be realised as one TRX transducer with analog filter, an amplifier and a fullwave rectifier in the receiving circuit plus an analog switching circuit and a step-in transformer in the transmitting circuit. The ADC block is based on an analog comparator and DAC block consists of an analog switching circuit driver. Everything is controlled by microcontroller which drives DAC and measures the time it takes comparator signal to reach the unit. Much more interesting echo analyzing could be applied using full analog to digital conversion and DSP algorithms. One of the most popular examples in ultrasound ranging system is based on autocorrelation of digitised $x(k)$ echo signal Eq. (14).

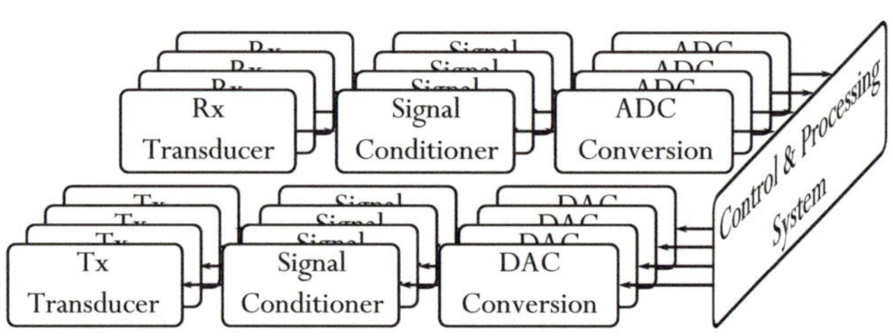


Fig. 19. Generalised block scheme of air operating echolocation system.

$$y(n) = \sum_{k=0}^{N-1} x(k)\, x(k+n) \tag{14}$$

A signal processing simulation of an echo from a two different targets using CF *(Constant Frequency)* Eq. (15) and LFM *(Linear Frequency Modulation)* Eq. (16) signals is shown in Fig.

(20).

$$s_{CF}(n) = A \cdot \sin\left(2\pi \frac{f_0}{f_s} n\right) \tag{15}$$

$$s_{LFM}(n) = A \cdot \sin\left[2\pi\left(f_0 + \frac{m}{2} \cdot \frac{n}{f_s}\right) \cdot \frac{n}{f_s}\right] \tag{16}$$

Analysis of the results obtained using different signals proves that the modulation of the

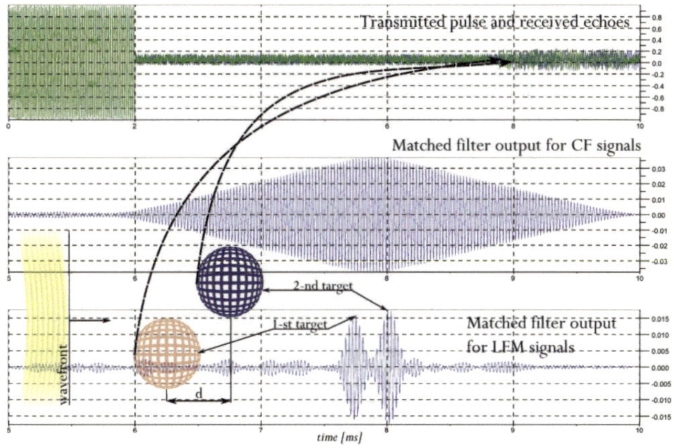


Fig. 20. Simulation of target ranging capabilities using pulse compression technique. Echolocation system performance - $f_s$ - 400 $[kHz]$, $f_0$ - 40 $[kHz]$ (for CF wave) and $f_0$ - 40 $[kHz]$ and $m$ - 10 (for LFM wave). Signal to noise ratio is set to 6 $[dB]$ and the target separation distance $d$ equals 8.5 $[cm]$.

transmitted signal ensures better target discrimination and range resolution. This phenomena is used by bats as it shown in section (2). Bats also reproduce direction of arrival DOA of the echo signals. The ILD Interaural Level Difference and ITD Interaural Time Difference are a source of information about DOA. In air coupled ultrasound signal processing systems the ITD parameter is commonly used due to attenuation changes across parameters described in section (3) and different target strengths for different objects. Some interesting information about ILD processing by bats and modeling of the bats' hearing mechanism using VLSI elements could be found in the paper by Shi & Horiuchi (2007). Interaural time differences follows from the spatial grid of receiving sensors and the sound speed in a gas medium. Since the assumption is that wavefront returning echo signal is plane, we can evaluate horizontal and elevation angle using sensor arrays and beamforming algorithms. Additionally, there are methods allowing beam steering in a particular direction. This enables "looking" using ultrasounds in a selected direction. Using multiple receivers (an array of transducers in different configurations see. Fig. (21) it is possible to modify the directional characteristics of the array. This technique is commonly used in underwater hydrolocation systems and radars. There were some attempts of developing an ultrasonic air-operating beamformers for echolocation systems described in papers by Strakowski et al. (2006), Kleeman & Kuc (1995), Medina & Wykes (2001), Harput & Bozkurt (2008) and Webb & Wykes (1996), Wykes et al. (1993). Most of the solutions cited above are based on a linear array and the signal processing method used is time-delay beamforming see Johnson & Dudgeon (1993.). Additionally, the authors developed an ultrasonic array for bats inspired echolocation study presented in Fig. (22) and described in Herman & Stach (2011). In the proposed solution time-delay algorithm

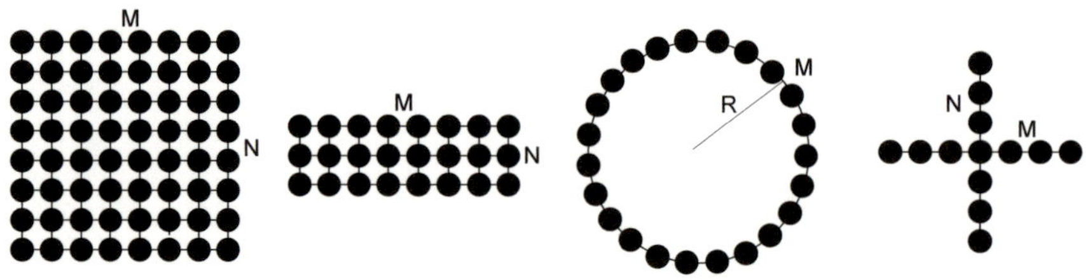


Fig. 21. Some regular grids of sensor arrays, based on Johnson & Dudgeon (1993.)

Fig. 22. A MEMS based linear array developed by authors. Horizontal intersensor spacing is equal to 3.8 $[mm]$ and array size is $32 \times 2$.

was used for a 32 x 2 MEMS sensor array. Due to the fact that signal from each microphone is digitized with sampling frequency $f_s$, the minimal phase shift equals sampling interval. When using digital beamforming algorithms it is critical to delay a beam by a number of samples that corresponds to a given beam deflection angle. To ensure proper angular resolution, the sampling frequency is given by modified Eq. (17):

$$f_s = \frac{c}{d} \cdot \frac{180}{\pi \cdot \Phi_{\min}} \tag{17}$$

where $c$ - velocity of sound, $d$ - intersensor spacing in linear array, $\Phi_{\min}$ - minimum phase shift angle. To improve horizontal angle estimation performance, the time series was upsampled using interpolation algorithm for each channel. The minimum $\Phi_{\min}$ could be obtained assuming that it equals half of the parabolic width $PW$ (at 0 steering angle) of the linear array given by Eq. (18):

$$PW = \frac{4 \cdot \sqrt{3}}{k \cdot d \cdot \sqrt{M^2 - 1}} \tag{18}$$

where: $M$ - number of sensors. Considering the fact that pulse compression method combined with bats-like downsweept chirps, it is possible to process array signals using pathways shown in Fig. (23): The proposed solution was simulated using similar methodology as it is shown in Fig. ( (20), where the difference is that the acoustic echo signal is processed by a linear array of 32 MEMS sensors. The results of the simulation are presented in the Fig. (24).
Unfortunately, increasing number of channels and sampling frequency takes results in the system not being able to operate in a real time. The system processing speed may be increased by paralleling channel information processing, reducing the number of sensors, applying so called redundant-array solution by Johnson & Dudgeon (1993.) or combining differences at respective receivers with information determined from evaluation directional

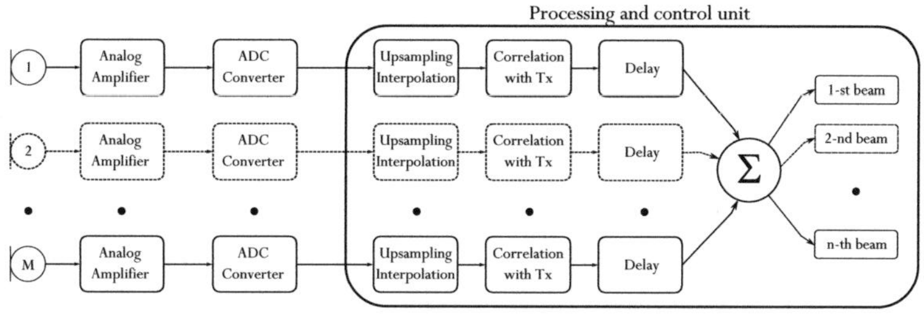


Fig. 23. Digital interpolation delay-sum beamforming scheme based on correlation of sensor data with transmitted signal Tx.

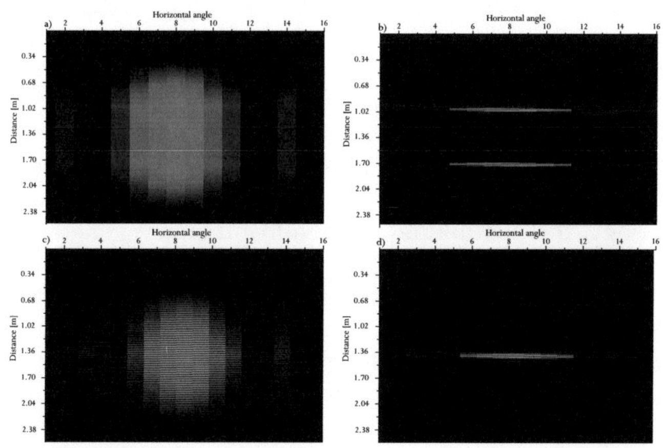


Fig. 24. A simulation of a target position estimation: a) multiple target with usage of CF signal, b) multiple target with usage of LFM signal, c) single target with usage of CF signal, d) single target with usage of LFM signal.

characteristics of the ultra sound transducers described in Georg Kaniak (2010). Due to the fact that processing speed of present-day signal technology is not increasing as rapidly as a few years before, we can observe that signal processing methods are migrating to multi-threaded calculations on multiprocessor platforms. An excellent illustration of the situation are the recent news in sonar technology samples Nilsen & Hafizovic (2009), Coiras et al. (2009), where parallel computing of sensor data are being preformed on GPU targets.

## 6. Conclusions

The chapter describes selected matters associated with a variety of science fields such as: acoustics, electronics, signal processing and biology or more precisely chiropterology. It presents, in a concise manner, the properties of air as a gas milieu in the context of using it as a medium transmitting information encoded in ultrasonic waves. The performed analyses suggest that from the point of view of ultrasound technology air is a very unfavourable medium. This is evidenced by characteristics of gases and the methods of ultrasonic wave generation and reception. On the other side, this medium is used by bats, which can reproduce an image of the surrounding space using ultrasounds. Although current technology is far behind bats in the field of information processing, researchers are relentlessly trying to catch up and judging by the number of scientific publications interest in bats' echolocation is still high. The selected issues related to the bats' biology and presented in section 2 prove that bats can reproduce images of its surrounding with resolution exceeding the physical limitations arising from wave length. It is even possible to formulate a hypothesis stating that a bat is from a technological viewpoint "an intelligent multisensor", which not only processes acoustic information but also environmental conditions (temperature, humidity, flight speed). Some of the bat specific echolocation methods are applied in ultrasonic technology. These include simple distance (angle) meters based on processing information related to time period between signal transmission and echo reception but also advanced beamforming systems. The very structure of the signal used by bats is, however, a much more interesting subject. It appears that better space imaging is possible when using modulated signals, which, as was found, compensate Doppler effect and increase both range and angular resolution. As a direct application of this echolocation strategy, the authors present a solution based on the use of pulse compression method in digital beamforming. It was shown, by means of simulation, that such a system is characterised by a higher range resolution and is capable of more accurate discrimination of targets located at the same angle. As a next step in verifying simulation results, the developed solution will be tested on a platform specifically constructed for that purpose. The primary limitation of pulse compression method is the assumption that acoustic waves with frequency modulation will be generated. With present-day advancement of ultrasonic technology, the possibilities of broadband signal generation are very limited. Even if such a signal can be generated, its SPL level is significantly lower than in case of widely used and well developed resonance transducers. Since space imaging methods and models if reality prepared for air are often a modified transposition of the models used in hydrolocation, it is possible to use solutions and development directions from this field in the area of object localisation in air. Due to required resources and processor capacity it is likely that multibeam systems will turn to distributed computing on multiprocessor platforms. As mentioned in the work, one of possible applications, which has already been the subject of research of several scientific groups, is to construct a device that would help visually handicapped people to ambulate Abdul et al. (2008), Borenstein & Ulrich (1997), Mihajlik et al. (2001), Shoval et al. (1994), Ifukabe et al. (1991). Considering the detail of the problem presented in this chapter and using bat signals and DSP algorithms utilised on an efficient computing platform, construction of such a device seems possible but will require an effort of an interdisciplinary research team.

## 7. References


Abdul, A. I., Sajee, M., Mazlina, M. & Nazleeni, S. (2008). Blind echolocation using ultrasonic sensors, *Information Technology, 2008* 4: 1–7.

Altes, R. (1995). Signal processing for target recognition in biosonar, *Neural Networks* 8: 1275–1295.

Altringham, J. (1999). *Bats. Biology and Behaviour*, Oxford University Press Inc.

Au, W. W. L. (2004). *Echolocation in Bats and Dolphins*, The University of Chicago Press, Chicago and London, chapter A comparison of the sonar capabilities of bats and dolphins, pp. 12–27.

Balcombe, J. P. & Fenton, M. B. (1988). Eavesdropping by bats: The influence of echolocation call design and foraging strategy, *Ethology* 79(2): 158–166.

Barclay, R. M. R. (1982). Interindividual use of echolocation calls: Eavesdropping by bats, *Behavioral Ecology and Sociobiology* 10: 271–275.

Bass, H. E., Sutherland, L. C., Zuckerwar, A. J., Blackstock, D. T. & Hester, D. M. (1995). Atmospheric absorption of sound: Further developments, *The Journal of the Acoustical Society of America* 97(1): 680–683.

Borenstein, J. & Ulrich, I. (1997). The guidecane - a computerized travel aid for the active guidance of blind pedestrians, *IEEE Int. Conf. on Robotics and Automation*, pp. 1283–1288.

Borwick, J. (2001). *Loudspeaker and Headphone Handbook*, Focal Press.

Carmena, J. & Hallam, J. (2004a). Narrowband target tracking using a biomimetic sonarhead, *Robotics and autonomous systems* 46(4): 247–259.

Carmena, J. & Hallam, J. (2004b). The use of doppler in sonar-based mobile robot navigation: inspirations from biology, *Information Sciences* 161(1-2): 71–94.

Chou, T. N. & Wykes, C. (1997). An integrated vision/ultrasonic sensor for 3d target recognition and measurement, *Proc. Sixth International Conference on Image Processing and Its Applications*, Vol. 1, pp. 189–193.

Coiras, E., Ramirez-Montesinos, A. & Groen, J. (2009). Gpu-based simulation of side-looking sonar images, *OCEANS 2009 - EUROPE*, pp. 1 –6.

Dechmann, D. K., Heucke, S. L., Giuggioli, L., Safi, K., Voigt, C. C. & Wikelski, M. (2009). Experimental evidence for group hunting via eavesdropping in echolocating bats, *Proceedings of the Royal Society B: Biological Sciences* 276 B: 2721–2728.

Ealo, J., Seco, F. & Jimenez, A. (2008). Broadband emfi-based transducers for ultrasonic air applications, *Ultrasonics, Ferroelectrics and Frequency Control, IEEE Transactions on* 55(4): 919 –929.

Evans, L. B., Bass, H. E. & Sutherland, L. C. (1972). Atmospheric absorption of sound: Theoretical predictions, *The Journal of the Acoustical Society of America* 51(5B): 1565–1575.

Georg Kaniak, H. F. S. (2010). Angle determination, scene analysis, ultrasound, *Key Engineering Materials (* 437: 258–262.

Griffin, D. (1958). *Listening in the dark*, New Haven: Yale University Press.

Grunwald, J.-E., Schörnich, S. & Wiegrebe, L. (2004). Classification of natural textures in echolocation, *Proceedings of the National Academy of Sciences of the United States of America* 101(15): 5670–5674.

Gudra, T. (2008). Ultrasounds in gas media: generation, transmission, aplications, *Archives of Acousics* 33(4): 581–592.

Gudra, T. & Opielinski, K. (2004). Ultrasonic transducers working in the air with the continuous wave within the 50-500 khz frequency range, *ULTRASONICS* 42(1-9): 453–458.

Gudra, T. & Opielinski, K. J. (2002). Influence of acoustic impedance of multilayer acoustic systems on the transfer function of ultrasonic airborne transducers, *Ultrasonics* 40(1-8): 457 – 463.

Gudra, T. & Opielinski, K. J. (2006). The range equation of the ultrasonic link in gas media, *ULTRASONICS* 44(Suppl. 1): E1423–E1428.

Gudra, T., Opielinski, K. J. & Jankowski, J. (2010). Estimation of the variation in target strength of objects in the air, *Physics Procedia* 3(1): 209 – 215. International Congress on Ultrasonics, Santiago de Chile.

Harput, S. & Bozkurt, A. (2008). Ultrasonic phased array device for acoustic imaging in air, *Sensors Journal, IEEE* 8(11): 1755 –1762.

Hartley, D. J. & Suthers, R. A. (1987). The sound emission pattern and the acoustical role of the noseleaf in the echolocating bat, carollia perspicillata, *The Journal of the Acoustical Society of America* 82(6): 1892–1900.

Helversen, D. V. & Helversen, O. V. (1999). Acoustic guide in bat-pollinated flower, *Nature* 759-760: 759–760.

Helversen O. von, W. Y. (2003). *Bat ecology*, University of Chicago Press. Chicago and London, chapter Glossophaginae bats and their flowers: costs and benefits for plants and pollinators, pp. 346–397.

Henson JR, O. W. (1965). The activity and function of the middle ear muscles in echolocating bats, *J Physiol* 180: 871–887.

Herman, K. & Stach, L. (2011). Analysis and elaboration of an air-coupled ultra- sound wideband sensor array, *Acta Physica Polonica A* In Press.

Holderied, M. W., Baker, C. J., Vespe, M. & Jones, G. (2008). Understanding signal design during the pursuit of aerial insects by echolocating bats: tools and applications, *Integrative and Comparative Biology* 48(1): 74–84.

Ifukabe, T., Sasaki, T. & Peng, C. (1991). A blind mobility aid modeled after echolocation of bats, *IEEE Transactions on Bio-Medical Engineering* 38(5): 461–465.

Jiménez, A., Álvaro Hernández, Urena, J., Pérez, M. C., Álvarez, F. J., Marziani, C. D., García, J. J. & Villadangos, J. M. (2008). Emfi-based ultrasonic transducer for robotics applications, *Sensors and Actuators A: Physical* 148(1): 342 – 349.

Johnson, D. & Dudgeon, D. (1993.). *Array Signal Processing*, Prentice-Hall, Englewood Cliffs.

Jones, G. (1999). Scaling of echolocation call parameters in bats, *J Exp Biol* 202(23): 3359–3367.

Jones, G. (2008). Sensory ecology: echolocation calls are used for communication., *Current Biology* 18: R34–R3.

Kamata, E., Inoue, S., Zheng, M., Kashimori, Y. & Kambara, T. (2004). A neural mechanism for detecting the distance of a selected target by modulating the fm sweep rate of biosonar in echolocation of bat, *Biosystems* 76(1-3): 55 – 64.

Kay, L. (1964). An ultrasonic sensing probe as a mobility aid for the blind, *Ultrasonics* 2(2): 53 – 59.

Kleeman, L. & Kuc, R. (1995). Mobile robot sonar for target localization and classification, *International Journal of Robotics Research* 14: 295–318.

Lawrence, B. & Simmons, J. (1982). Echolocation in bats: the external ear and perception of the vertical positions of targets, *Science* 218(4571): 481–483.

Lynnworth, L. (1965). Ultrasonic impedance matching from solids to gases, *Sonics and Ultrasonics, IEEE Transactions on* 12(2): 37 – 48.

Masters, W. M. & Harley, H. E. (2004). *Echolocation in Bats and Dolphins*, The University of Chicago Press, Chicago and London., chapter Performance and Cognition in Echolocating Mammals, pp. 249–259.

Medina, L. & Wykes, C. (2001). Multiple target 3d location airborne ultrasonic system, *Ultrasonics* 39(1): 19–25.

Mihajlik, P., Guttermuth, M., Seres, K. & Tatai, P. (2001). Dsp-based ultrasonic navigation aid for the blind, *Procceedings of the 18-th IEEE Instrumentation and measurement technology conference*, Vol. 1-3, pp. 1535–1540.

Moss, C. & Sinha, S. (2003). Neurobiology of echolocation in bats, *CURRENT OPINION IN NEUROBIOLOGY* 13(6): 751–758.

Neuweiler, G. (2000). *The biology of bats*, Oxford University Press Inc. Oxford.

Nilsen, C.-I. & Hafizovic, I. (2009). Digital beamforming using a gpu, *Acoustics, Speech and Signal Processing, 2009. ICASSP 2009. IEEE International Conference on*, pp. 609 –612.

Norman, A. P. & Jones, G. (2000). Size, peripheral auditory tuning and target strength in noctuid moths, *Physiological Entomology* 25(4): 346–353.

Ostwald, J. (1984). Tonotopical organization and pure tone response characteristics of single units in the auditory cortex of the greater horseshoe bat, *Journal of Comparative Physiology A: Neuroethology, Sensory, Neural, and Behavioral Physiology* **155**: 821–834.

Paajanen, M., Lekkala, J. & Kirjavainen, K. (2000). Electromechanical film (emfi) – a new multipurpose electret material, *Sensors and Actuators A: Physical* 84(1-2): 95 – 102.

Paajanen, M., Lekkala, J. & Valimaki, H. (2001). Electromechanical modeling and properties of the electret film emfi, *Dielectrics and Electrical Insulation, IEEE Transactions on* 8(4): 629 –636.

Peremans, H. & Van Campenhout, J. (1993). Tri-aural perception on a mobile robot, *Proc. IEEE International Conference on Robotics and Automation*, pp. 265–270.

Peremans, H., Walker, A. & Hallam, J. C. T. (1998). 3d object localisation with a binaural sonarhead, inspirations from biology, *Proc. IEEE International Conference on Robotics and Automation*, Vol. 4, pp. 2795–2800.

Roverud, R. C. & Grinnell, A. D. (1985). Echolocation sound features processed to provide distance information in the cf/fm bat noctilio albiventris: evidence for a gated time window utilizing both cf and fm components, *Journal of Comparative Physiology A: Neuroethology, Sensory, Neural, and Behavioral Physiology* 156: 457–469.

Schillebeeckx, F., De Mey, F. & Peremans, H. (2008). Bio-inspired sonar antennae: Enhancing directivity patterns for localization, *Proc. 2nd IEEE RAS & EMBS International Conference on Biomedical Robotics and Biomechatronics BioRob 2008*, pp. 366–371.

Schmidt, S. (1992). Perception of structured phantom targets in the echolocating bat, megaderma lyra, *The Journal of the Acoustical Society of America* 91(4): 2203–2223.

Schnitzler, H., D., M., R., K. & K, H. (1983). The acoustic image of fluttering insects in echolocating bats, *Neuroethology and Behavioural Physiology* 3: 236–250.

Schnitzler, H. U. (1968). Die ultraschall-ortungslaute der hufeisen-fledermaeuse (chiroptera: Rhinolophidae) in verschiedene orientierungssituationen, *Journal of Comparative Physiology A: Neuroethology, Sensory, Neural, and Behavioral Physiology* 57: 376–408.

Schnitzler, H. U. (1987). *Recent advances in the study of bats*, Cambridge University Press, Cambridge., chapter Echoes of fluttering insects: information for echolocating bats, pp. 226–243.

Schuller, G. & Pollak, G. (1979). Disproportionate frequency representation in the inferior colliculus of doppler-compensating greater horseshoe bats: Evidence for an acoustic fovea, *Journal of Comparative Physiology A: Neuroethology, Sensory, Neural, and Behavioral Physiology* 132: 47–54.

Shi, R. Z. & Horiuchi, T. K. (2007). A neuromorphic vlsi model of bat interaural level difference processing for azimuthal echolocation, *IEEE Transactions on circuits and systems* 54(1): 74–88.

Shoval, S., Borenstein, J. & Koren, Y. (1994). Mobile robot obstacle avoidance in a computerized travel aid for the blind, *IEEE International Conference on Robotics and Automation*, pp. 8–13.

Siemers, B. M. & Schnitzler, H.-U. (2004). Echolocation signals reflect niche differentiation in five sympatric congeneric bat species, *Nature* 429(6992): 657–661.

Siewert I., Schillinger T., S. S. (2004). *Echolocation in Bats and Dolphins*, The University of Chicago Press, Chicago and London, chapter Forward masking and the consequences on echo perception in the gleaning bat, Megaderma lyra, pp. 266–268.

Simmons, J. A. (1973). The resolution of target range by echolocating bats, *The Journal of the Acoustical Society of America* 54(1): 157–173.

Steckel, J., Reijniers, J., Boen, A. & Peremans, H. (2008). Biomimetic target localisation using an emfi based array, *Proc. 17th IEEE International Symposium on the Applications of Ferroelectrics*, Vol. 3, pp. 1–2.

Strakowski, M. R., Kosmowski, B. B., Kowalik, R. & Wierzba, P. (2006). An ultrasonic obstacle detector based on phase beamforming principles, *IEEE Sensor Journal* 6(1): 179–186.

Surlykke, A. & Kalko, E. K. V. (2008). Echolocating bats cry out loud to detect their prey, *PLoS ONE* 3(4): e2036.

Vater, M., Feng, A. S. & Betz, M. (1985). An hrp-study of the frequency-place map of the horseshoe bat cochlea: Morphological correlates of the sharp tuning to a narrow frequency band, *Journal of Comparative Physiology A: Neuroethology, Sensory, Neural, and Behavioral Physiology* 157: 671–686.

Webb, P. & Wykes, C. (1996). High-resolution beam forming for ultrasonic arrays, *Robotics and Automation, IEEE* 12(1): 138–146.

Wykes, C., Nagi, F. & Webb, P. (1993). Ultrasound imaging in air, *Proc. International Conference on Acoustic Sensing and Imaging*, pp. 77–81.

# 11

# The Effect of Sonar on Human Hearing

Renzo Mora, Sara Penco and Luca Guastini
*ENT Department, University of Genoa*
*Italy*

## 1. Introduction

Human use of the Earth's oceans has steadily increased over the last century resulting in an increase in anthropogenically produced noise. This noise stems from a variety of sources including commercial shipping, oil drilling and exploration, scientific research and naval sonar. Sonar (for sound navigation and ranging) is a technique that uses sound propagation (usually underwater) to navigate, communicate or to localize: sonar may be used as a means of acoustic location (acoustic location in air was used before the introduction of radar). The term sonar is also used for the equipment used to generate and receive the sound. The range of frequencies used in sonar systems vary from infrasonic to ultrasonic. Sonar uses frequencies which are too much high-pitched (up to 120,000 cycles per second) for human ears to hear.
Although there is a growing concern among the public that human generated sounds in the marine environment could have deleterious impacts on aquatic organisms, only few studies address this concern on the effects of these sounds on the human auditory system. The effects of sound on the human auditory system have been the subject of several studies, but one question needs to be resolved yet: the effects caused by the naval sonar. For these reasons this chapter wants to show the effects of active middle frequency sonar on human.
Published data from humans under water in literature are scarce and sometimes use different terminology with regard to sound levels. For example sound pressure levels measured in air are normally reported with a reference pressure of 20μPa whereas levels measured in water are normally reported with a reference pressure of 1μPa. Therefore, in the diving environment it is recommended to use SPL (sound pressure level) threshold with reference pressure of one micropascal (1 μPa) for both water and air measurements in order to compare values from different sources.
The non-intuitive nature of decibels and the different reference values of air and water have led to a plethora of misconceptions concerning the magnitude and potential effects of noise levels in air and water. The magnitude of sound pressure levels in water is normally described by sound pressure on a dB scale relative to a reference root-meansquare (rms) pressure of 1 μPa (dB re 1 μPa).
For these misconceptions concerning the measurement of noise in the marine environment, studies on cetaceans have highlighted hearing damage and behavioural change at levels of sounds exposure lower than those that would cause physiological damage to the auditory system.
Continued emission of noise can increase the damage, due to the "habituation" to a familiar sound to which it is difficult to react more strongly. The "habituation" is known as being provoked by continued acoustical stimuli, reducing the hearing sensitivity to high-level

sounds; the hearing sensitivity may be regulated at both conductive (stapedial reflex) and sensorineural levels (adaptation). For these reasons, the introduction of new types of military sonar, such as low-frequency system, should proceed with caution; the low-frequency sounds produced by the systems will travel much farther than the mid-frequency sonar sounds currently causing concern. Studies on marine animals have demonstrated that changes in hair bundle density paralleled changes in hair cell nucleus density, indicating that entire hair cells disappeared following noise exposure; the inner ear damage is characterised by a permanent threshold elevation after an exposure to white noise ranging in intensity from 130 to 170 dB re 1 μPa for 24 h.
Although there are differences among the ears of different species, the basic processes of hearing are the same between marine and terrestrial mammals. For this reason, some of the previous considerations can be applied on humans. Although the true correlation between sonar and hearing damage is difficult to show, (absence of technical information, level of sound exposure and other environmental variables) this study wants to show the effects of sonar on human auditory system.

## 2. Introduction

Underwater acoustics is the study of the propagation of sound in water and the interaction of the mechanical waves that constitute sound with the water and its boundaries. The water may be in the ocean, a lake or a tank. The field of underwater acoustics is closely related to a number of other fields of acoustic study, including sonar, transduction, acoustic signal processing, acoustical oceanography, bioacoustics, and physical acoustics.
Underwater sound has probably been used by marine animals for millions of years. The science of underwater acoustics began in 1490, when Leonardo Da Vinci wrote: (Urick, 1993) "if you cause your ship to stop and place the head of a long tube in the water and place the outer extremity to your ear, you will hear ships at a great distance from you".
But only the 20th century, with the start of World War I, provided the impetus for the next wave of progress in underwater acoustics: anti-submarine listening systems were developed. In particular the development of sonar (Sound Navigation and Ranging) proceeded apace during the war, driven by the first large scale deployments of submarines.
In 1919, the first scientific paper on underwater acoustics was published, theoretically describing the refraction of sound waves produced by temperature and salinity gradients in the ocean. The range predictions of the paper were experimentally validated by transmission loss measurements (Lichte, 1919).
The next two decades saw the development of several applications of underwater acoustics: by the 1930s sonar systems were being used for passive listening systems and for active echo-ranging systems; these systems were used to good effect during World War II by both submarines and anti-submarine vessels. After World War II, the development of sonar systems was driven largely by the Cold War, resulting in advances in the theoretical and practical understanding of underwater acoustics, aided by computer-based techniques.
In the last decades, there is growing concern among the public that human generated sounds in the marine environment could have deleterious impacts on aquatic organisms: in the last years, much of interest in the effects of the human-generated sound has been focused on marine mammals (Parsons et al., 2008; Salami et al., 2010).
However, only few studies address this concern on the effects of these sounds on the human auditory system (Salami et al., 2010). The effects caused of sound on the human auditory

system have been the subject of several studies, but one question needs to be resolved yet; the effects by the naval sonar.
Although the true correlation between sonar and hearing damage is difficult to decide (absence of technical information; level of sound exposure; and other environmental variables), this chapter wants to show the effects of sonar on human hearing.

## 3. Fundamental concepts of sound propagation underwater

All sound, whether produced by a cowbell or a complicated electronic device, behaves in much the same manner. Sound originates as a wave motion by a vibrating source and requires for its transmission an elastic medium such as air or water. For example, consider a piston suspended in one of these mediums. As the piston is forced to move forward and backward, the medium is compressed on the forward stroke and decompressed or rarefied on the return stroke. Thus, a wave motion or series of compressions and rarefactions is caused to move from the source out through the medium. In the fluid medium the molecular motion is back and forth, parallel to the direction of the piston's movement. Because the fluid is compressible, this motion results in a series of detectable pressure changes. This series of compressions and rarefactions, such as is produced by the piston, constitutes a compressional wave train. Another way of explaining the phenomenon of acoustic wave propagation is to consider the medium of transmission as a loosely packed collection of mass elements connected by springy bumpers. A disturbance of the elements at some point (e.g., piston motion) moves along in the fluid by the successive extension and compression of the springs as the elements swing back and forth, each communicating its motion to its neighbor through the connecting bumpers. In this way, the agitation of a cluster of elements is propagated through the medium even though the individual elements do no more than move about their equilibrium positions without actually migrating. The sound wave propagates parallel to the source resulting in a longitudinal wave (Tindle, 2005).
A sound wave propagating underwater consists of alternating compressions and rarefactions of the water. These compressions and rarefactions are detected by a receiver, such as the human ear or a hydrophone, as changes in pressure. These waves may be man-made or naturally generated (Tindle, 2005).
Underwater acoustic propagation depends on many factors. The direction of sound propagation is determined by the sound speed gradients in the water. In the sea the vertical gradients are generally much larger than the horizontal ones (Sabra & Dowling, 2003). These facts, combined with a tendency for increasing sound speed with increasing depth due to the increasing pressure in the deep sea reverses the sound speed gradient in the thermocline creating an efficient waveguide at the depth corresponding to the minimum sound speed (Sabra & Dowling, 2003; Frosch, 1964; Snellen et al., 2001). At equatorial and temperate latitudes in the ocean the surface temperature is high enough to reverse the pressure effect, such that a sound speed minimum occurs at depth of a few hundred metres (Frosch, 1964; Snellen et al., 2001).
The presence of this minimum creates a special channel known as Deep Sound Channel, previously known as the SOFAR (sound fixing and ranging) channel, permitting guided propagation of underwater sound for thousands of kilometres without interaction with the sea surface or the seabed (Jian et al., 2009).
Another phenomenon in the deep sea is the formation of sound focussing areas known as Convergence Zones (Shvachko, 2008): in this case sound is refracted downward from a near-

surface source and then back up again. The horizontal distance from the source at which this occurs depends on the positive and negative sound speed gradients. A surface duct can also occur in both deep and moderately shallow water when there is upward refraction, for example due to cold surface temperatures.

In general, as sound propagates underwater there is a reduction in the sound intensity over increasing ranges, though in some circumstances a gain can be obtained due to focussing propagation loss (sometimes referred to as transmission loss) is a quantitative measure of the reduction in sound intensity between two points, normally the sound source and a distant receiver (Studenichnik, 2003).

The non-intuitive nature of decibels and the different reference values of air and water have led to a plethora of misconceptions concerning the magnitude and potential effects of noise levels in air and water (Chapman & Ellis, 1998). For this reason a convenient system is needed in order to measure and discuss acoustic parameters underwater.

Pressure is defined as a force per unit area. Although many people are familiar with the British units of pounds per square inch (psi), it has long been the convention in acoustics to use metric units, namely newtons per square meter ($N/m^2$), or dynes per square centimeter ($dynes/cm^2$). Of the two metric units, the $dynes/cm^2$ has been the most commonly used. It has an alternate name, microbar (bar), and is equivalent to approximately 1/1,000,000 of a standard atmosphere. For underwater sounds, a reference pressure of 1 bar was established from which all others were measured. The corresponding reference pressure for airborne sounds was 0.0002 bar, because this was the approximate intensity of a 1,000-Hz tone that was barely audible to human ears. The previously less commonly used $N/m^2$ also has an alternate name, a Pascal (Pa), and the reference standard derived from this was the micropascal (Pa), which is equivalent to $10^{-6}N/m^2$.

The decibel system was selected by acousticians for a number of logical reasons (Hood et al., 1991):

- it is a logarithmic system, which is convenient for dealing with large changes in quantities;
- it simplifies computations since multiplication and division are reduced to addition or subtraction, respectively;
- human senses have an approximate logarithmic response to stimuli such as light, sound, and heat: for example, the human ear perceives about the same change in loudness between 1 and 10 units of pressure as it perceives between 10 and 100 units of pressure;
- In the area of underwater acoustics, the primary interest is in ratios of power levels and signal levels rather than absolute numerical values.

In the decibel system, the bel is the fundamental division of a logarithmic scale for expressing the ratio of two amounts of power. The number of bels to express such a ratio is the logarithm to the base 10 of the ratio. Acousticians decided the bel was a unit too large for application in their field, and subsequently adopted the decibel (1/10 bel) as their basic logarithmic unit. The conversion factors in table 8-1 can in themselves be cumbersome to use, but when expressed in dB, only addition or subtraction is required. When converting from a pressure referenced to 1 bar to one referenced to 1 Pa, simply add 100dB. When converting from 0.0002 bar to 1 Pa, simply add 26dB. If converting from 1 Pa to the others, merely subtract the appropriate values.

With such a profusion of reference standards and measurement systems, there were ample opportunities for misunderstandings as an operator or planner consulted different sources of acoustic information. For these reason, is commonly used a measurement in terms of pressure instead of intensity: the magnitude of sound pressure levels in water is normally described by sound pressure on a dB scale relative to a reference root-meansquare (rms) pressure of 1μPa (dB re 1 μPa) (Madesen, 2005).

About the propagation of sound, approximate values for fresh water and seawater, respectively, at atmospheric pressure are 1450 and 1500 m/s for the sound speed, and 1000 and 1030 kg/m$^3$ for the density (Wolfson & Tomsovic, 2001; Leroy, 2008)

The ocean is not a homogeneous medium, and the speed of sound varies from point to point in the ocean. This variation in sound speed is one of the most important characteristics affecting the transmission of sound. The three main environmental factors affecting the speed of sound in the ocean are salinity, pressure, and temperature: the speed of sound in water increases with increasing pressure, temperature and salinity (Wolfson & Tomsovic, 2001; Leroy, 2008).

In particular (Wolfson & Tomsovic, 2001; Leroy, 2008):

- a change in salinity of one part per thousand will result in a change in sound speed of approximately 1.3 meters per second;
- pressure in most circumstances is more important than salinity, but in the sea its change is constant and thus predictable; it also causes a change in bulk modulus and density, and the result is an increase in sound speed of 0.017 m/sec for every meter of depth increase;
- temperature, the foremost factor affecting sound speed, usually decreases with depth, and this leads to an accompanying decrease in sound speed at the rate of approximately 3 m/sec per degree Celsius; below a depth of about 1,000 m, however, temperature is fairly constant, and the predominant factor affecting sound speed becomes pressure;

## 4. Sonar

Human use of the Earth's oceans has steadily increased over the last century resulting in an increase in anthropogenically produced noise. This noise stems from a variety of sources including commercial shipping, oil drilling and exploration, scientific research and naval sonar.

Sonar is an acronym for Sound Navigation and Ranging. There are two broad types of sonar (passive and active) in use (Salami et al., 2010):

- passive sonar is a listening device that can determine the presence, characteristics and direction of marine noise sources: these sources may include biological noise (animal communication) and human sounds (eg ship or submarine noise); passive sonar equipment is essentially an acoustic receiver which emits no sound and therefore has no potential to disturb marine life.
- active sonar is a technique that uses sound to determine relative positions of submerged objects (including submarines, fish, mines and wrecks of ships and aircraft) and the sea floor, by emitting a sound signal and listening for the echoes from the objects; many different types of active sonar are used throughout the world's oceans by private, commercial and military vessels; these systems mirror the purpose of sonars used by some marine animals (active sonar devices locate objects by the reflection of sound-waves and remain an important means of underwater detection and navigation).

The main types of active sonar are commercial, civilian and military sonars. The sonars used by military forces are (Gong et al., 2010; Kane et al., 2010):

- Low-frequency (LF): low frequency sonars have been defined as those that emit sound below 1000 Hz. These sonars are designed to provide theatre level protection, such as for an Aircraft Carrier Task Group out to many miles (up to 200 miles) from the ships. This is possible because of the extended propagation possible at low frequencies. Outputs are similar to medium frequency sonars (described below) but the sound travels further because of the significantly enhanced seawater propagation.
- Medium frequency (MF): medium frequency active sonars emit sounds at frequencies between (1000 and 10,000 Hz); these sonars represent a sliding scale of compromise between possible detection range and size of the transmission array; at the lower end of the frequency range (1000- 3000 Hz) the systems are capable of extended detection ranges using high output power, but the size of the transducer limits applications to large warships. These systems are designed to provide area protection for a small Task Group out to a few tens of miles.
- High frequency (HF): high frequency active sonars operate between approximately 30,000 and 500,000 Hz (30 kHz and 500 kHz); these systems allow increasingly greater resolution as the frequency increases but at the expense of range. The highest frequencies are only effective over short distances because of the rapid attenuation of high frequency sounds in seawater.

Typically high power military active sonars are operated infrequently during voyages and the sounds are not emitted continuously but as short bursts ('pings') during operation (Gong et al., 2010; Kane et al., 2010).

Commercial and civilian sonars are generally designed to detect the sea floor (echo sounders), map the sea floor and search for sunken objects (sidescan sonars) and to locate fish (fish finders). Sonars of at least one of these types are fitted to nearly all vessels. Even some small boats have fish finding and echo sounders. The characteristics of these sonars are broadly similar to the high frequency military sonars described above (Gong et al., 2010; Kane et al., 2010).

## 5. Underwater hearing

The lowest audible SPL for a human diver with normal hearing is about 67 dB re 1 μPa, with greatest sensitivity occurring at frequencies around 1 kHz (Fothergill et al., 2001).

Dolphins and other toothed whales are renowned for their acute hearing sensitivity, especially in the frequency range 5 to 50 kHz (Mooney et al., 2009). Several species have hearing thresholds between 30 and 50 dB re 1 μPa in this frequency range. For example the hearing threshold of the killer whale occurs at an rms acoustic pressure of 0.02 mPa (and frequency 15 KHz), corresponding to an SPL threshold of 26 dB re 1 μPa (Simon et al., 2005). By comparison the most sensitive fish is the soldier fish, whose threshold is 0.32 mPa (50 dB re 1 μPa) at 1.3 kHz, whereas the lobster has a hearing threshold of 1.3 Pa at 70 Hz (122 dB re 1 μPa) (Patek & Oakley 2003).

It's evident as high levels of underwater sound create a potential hazard to marine and amphibious animals as well as to human divers (Steevens et al., 1999). Recently, for these reasons, guidelines for exposure of human divers and marine mammals to underwater sound are reported by different organizations: human divers exposed to SPL above 154 dB re 1 μPa in the frequency range 0.6 to 2.5 kHz are reported to experience changes in their heart rate or breathing frequency, diver aversion to low frequency sound is dependent upon sound pressure level and center frequency (Fothergill et al., 2009; Steevens et al., 1999).

The potential for active sonar to impact on a species is dependent on the ability of the species to hear the sound. Species hear sounds over different frequencies ranges, and the efficiency of sound detection varies markedly with frequency. Additionally, species behavioural responses to a detected sound may vary according to the sensitivity of the species to disturbance and what activities the animals are engaged in at the time.
Determination of potential impact on a species must therefore include estimation of the ability of the species to detect the sound, and the likelihood of disturbance to critical activities such as feeding or parental protection of juveniles.

### 5.1 Effect of sonar on marine animals

In terrestrial habitats, increasing sound levels have been shown to induce various effects across taxa including behavioural changes, temporary physiological alterations and permanent anatomical damage. While it is apparent that anthropogenic noise may affect marine animals, we know relatively less about the actual causes or mechanisms of these effects.
We can usually see things that are miles away, but if you have ever snorkelled, you know that vision is limited to a few tens of meters underwater. Vision is the best way to sense distant objects in air, but sound is the best way to sense objects that are far away under the sea. Low frequency sounds can travel hundreds of miles in the right conditions. When mammals entered the ocean tens of millions of years ago, they evolved mechanisms to sense objects by listening for echoes from their own sounds, and to use sound to communicate over long distances.
Modern ships generate enough noise from their engines and propellers to have reduced the range over which whales can communicate. The low frequency noise from ships travels so well in the ocean that it has raised the noise levels ten to one hundred times compared to a century ago (Stocker, 2004).
Marine mammals are of particular concern regarding the effects of noise as they typically have sensitive underwater hearing and they use sound for important activities such as communicating, orienting and finding prey.
It has been suggested that overexposure to noise could induce permanent physiological damage and deleterious behavioural alterations. For these reasons: there has been growing concern that the noise humans have introduced into the sea might disrupt the behaviour of marine mammals (Salami et al., 2010).
Some marine animals, such as whales and dolphins, use echolocation systems similar to active sonar to locate predators and prey. It is feared that sonar transmitters could confuse these animals and cause them to lose their way, perhaps preventing them from feeding and mating. Recent articles report findings to the effect that military sonar may be inducing some whales to experience decompression sickness (and resultant beachings) (Parsons et al., 2008).
These temporally and spatially overlapping events seem to indicate that high-intensity sonar may instigate some marine mammal strandings. Recent work has suggested that sonar exposure could induce a variety of effects in marine mammals including changes in dive profile, acoustically induced bubble formation or decompression sickness (Salami et al., 2010).
High-powered sonar transmitters can kill marine animals. In the Bahamas in 2000, a trial by the US Navy of a 230 decibel transmitter in the frequency range 3 to 7 kHz resulted in the beaching of sixteen whales, seven of which were found dead. The Navy accepted blame in a

report published in the Boston Globe on 1/1/2002. Continued emission of noise can increase the damage, due to the "habituation" to a familiar sound to which it is difficult to react more strongly (Sypin, 2008). The "habituation" is known as being provoked by continued acoustical stimuli, reducing the hearing sensitivity to high-level sounds; the hearing sensitivity may be regulated at both conductive (stapedial reflex) and sensorineural levels (adaptation) (Sypin, 2008).

However, these hypotheses typically lack controlled experimental conditions to best evaluate potentially deleterious noise effects. Thus, the actual mechanisms that may be initiated by sonar exposure, which could actually result in multi-species strandings, have yet to be empirically supported.

Introduction of new types of military sonar, such as low-frequency system, should proceed with caution; the low-frequency sounds produced by the systems will travel much farther than the mid-frequency sonar sounds currently causing concern (Salami et al., 2010). However, at low powers, sonar can protect marine mammals against collisions with ships.

Different studies pointed out that the possible effects of the low-frequency sonar on marine mammals could include (Simmonds & Lopez-Jurado, 1991):

- Death from lung hemorrhage or other tissue trauma;
- Temporary or permanent hearing loss or impairment;
- Disruption of feeding, breeding, nursing, acoustic communication and sensing, or other vital behavior and, if the disruption is severe, frequent, or long lasting, possible decreases in individual survival and productivity and corresponding decreases in population size and productivity;
- Psychological and physiological stress, making animals more vulnerable to disease, parasites and predation;
- Changes in the distribution, abundance, or productivity of important marine mammal prey species and subsequent decreases in both individual marine mammal survival and productivity and in population size and productivity. These changes in prey species possibly could be caused both directly and indirectly by the low-frequency sonar transmissions: for example, transmissions conceivably could kill or impair development of the eggs and larval forms of one or more important marine mammal prey species; they might also disrupt feeding, spawning, and other vital functions or cause shifts in distribution patterns of certain important prey species and make some prey species more vulnerable to disease, parasites, and being eaten by other predators.

Although these evidences, recent studies showed the absence of side effects on marine animals: the sensory tissue of the inner ears did not show morphological damage even several days post-sound exposure; similarly, gross- and histopathology observations demonstrated no effects on nonauditory tissues (Popper et al., 2007).

The exposure to high frequency sonar (200-214 dB re 1 μPa) can determinate an hearing shifts of the marine animal: in particular recent report show as these data also imply that the animal must be very close to the source and/or exposed repeatedly in a short period time (Mooney et al., 2009):

- Assuming a usual sound attenuation rate of 6 dB per doubling of distance, the dB level used in high frequency sonar would be the received level approximately 40 m from the sonar source, a distance that can be considered 'close' with respect to naval ships;
- The animal would then have to maintain at most that distance for the approximate 2–2.5 min of operating the sonar to receive a level of exposure of near 214 dB;

- The animal could be located closer to the sonar source and receive a more intense signal.

However, the animal would still need to remain within a close range long enough to receive al level of exposure that would induce auditory threshold shifts, a potentially unlikely situation: all scenarios entail the subject being relatively close to the sonar source for a 'prolonged' duration (Mooney et al., 2009).

Exceptions may be if the sonar signals are rapidly repeated (which is unlikely due to overlap of returning echoes) or if oceanographic conditions are such that sound levels do not attenuate regularly over short distances (i.e. less than several 100 m) and thus remain intense. Perhaps such a situation could occur with multiple sonar sources over steep bathymetric conditions (Mooney et al., 2009).

These data show as repeated exposures are necessary to generate effects. It's evident as the effects of sound on marine animals could potentially include increased stress, damage to organs, the circulatory and nervous systems; long-term effects may alter feeding and reproductive patterns in a way that could affect the fish population as a whole. In the limited existing research on the effects of sound on marine animals hearing and behavior, different scientists have discovered that exposure to some very loud sounds, such as seismic air guns, can produce no effect, or result in a range of effects from temporary hearing loss to more lasting damage to the haircells of marine animal' inner ears. But it is hard to say that effects on one species indicate that another species will be affected in the same way by the same signal.

Furthermore, subtle behavioural changes are also associated with sonar exposure. Animals that prolong apnea must optimize the size and use of their oxygen stores, and must deal with the accumulation of lactic acid if they rely upon anaerobic metabolism (Popper et al. , 2007).

Pathologies related to effects of pressure are well known among human divers, but marine mammals appear to have developed adaptations to avoid most mechanical and physiological effect. The hazard of bubble formation during decompression is best known for humans breathing compressed gases, but empirical studies and theoretical considerations have shown that breath-hold divers can develop supersaturation and possible decompression-related problems when they return to the surface. Supersaturation has not been measured during normal diving behaviour of wild marine mammals but rather in specially designed experiments performed by trained subjects (Tyack, 2006).

Recent reports show the presence of gas and fat emboli in marine animals during exposure to naval sonar (Tyack, 2006).These reports suggest that exposure to sonar sounds may cause a decompression-like syndrome in deep-diving whales either by changing their normal diving behaviour or by a direct acoustic effect that triggers bubble growth (Tyack, 2006). The latter scenario would, however, only seem to happen for animals with 100–223% supersaturated tissues within tens of meters from a sonar where the received levels exceed 210 dB re 1 µPa (Tyack, 2006). Nonetheless, the geographical pattern of strandings suggests that animals are impacted at ranges significantly greater than those required for acoustically driven bubble growth, implying that the observed pathologies may follow from a behavioural response that has adverse physiological consequences (Tyack, 2006).

In order to further understand these pathophysiological mechanisms, recent experiences examined post-mortem and studied histopathologically different marine animals (Ziphius cavirostris, Mesoplodon densirostris and Mesoplodon europaeus) after exposure to midfrequency sonar activity: no inflammatory or neoplastic processes were noted, and no pathogens were identified. Macroscopically, whales had severe, diffuse congestion and haemorrhage, especially around the acoustic jaw fat, ears, brain, and kidneys. Gas bubble-associated lesions and fat embolism were observed in the vessels and parenchyma of vital organs. In vivo bubble formation associated with sonar exposure that may have been exacerbated by modified diving behaviour caused nitrogen supersaturation above a threshold value normally tolerated by the tissues (as occurs in decompression sickness). Alternatively, the effect that sonar has on tissues that have been supersaturated with nitrogen gas could be such that it lowers the threshold for the expansion of in vivo bubble precursors (gas nuclei). Exclusively or in combination, these mechanisms may enhance and maintain bubble growth or initiate embolism. Severely injured whales died or became stranded and died due to cardiovascular collapse during beaching. These injures are apparently induced by exposure to mid-frequency sonar signals and particularly affects deep, long-duration, repetitive-diving species like whales (Fernández et al., 2005).

### 5.2 Effects of sonar on human hearing

Relying on one's hearing it is extremely difficult to orientate oneself under water. Because of the high speed of sound under water, it is perceived by both ears virtually simultaneously and the orientation error may be possible. Bad orientation under water is also due to the prevalent bone conductivity. Sufficient audial orientation is possible to be acquired only after systematic training. The diving suit isolates the human ear from the surrounding water medium. That is why sound waves penetrate the helmet and the layer of air but reach the eardrum partly absorbed and scattered. In this case, sound perception through air conductivity is insignificant.

However, while diving without a helmet, which is possible in warm water, sound is perceived just like in the air. If the rubber helmet fits tightly, sound is well perceived because of bone conductivity - sound waves are transmitted through the bones of the human skull. With no helmet, a diver can hear very well, with a rubber helmet - fairly well, and with a metal one - very bad.

The development of underwater technology commonly results in a noisy working environment for commercial divers (Tindle & Deane, 2005). Also, the increasing use of active low-frequency sonar by submarines and ships raises the risk of accidental exposure to low frequency underwater sounds. While hearing conservation programs based on recognized risks from measurable sound pressure levels exist to prevent occupational hearing loss for most normal working environments, there are no equivalent guidelines for noise exposure underwater.

The Threshold Limit Values (TLVs) represent conditions under which it is believed that nearly all workers may be repeatedly exposed without adverse effect on their ability to hear and understand normal speech. For Threshold Limit - Ceiling Values (TLV-C) the concentration should not be exceeded during any part of the working day (ACGIH, 1998).

In particular the "American Conference of Governmental Industrial Hygienists (ACGIH)" has established permissible ultrasound exposure levels. These recommended limits (set at the middle frequencies of the one-third octave bands from 10 kHz to 50 kHz) are designed to prevent possible hearing loss caused by the subharmonics of the set frequencies, rather than the ultrasonic sound itself. These TLVs represent conditions under which it is believed that nearly all workers may be repeatedly exposed without adverse effect on their ability to hear and understand normal speech. Previous TLVs for frequencies in the 10 kHz to 20 kHz range, set to prevent subjective effects, are referenced in a cautionary note below. The 8-hour time-weighted average (TWA) values are an extension of the TLVs for noise, which is an 8-hour TWA of 85 dBA for sound below 10 kHz. The ceiling values may be verified by using an integrating sound level meter with slow detection and 1/3 octave bands. All instrumentation should have adequate frequency response and should meet the specifications of ANSI S1.4-1983 and International Electrotechnical Commission (IEC) 804 (ACGIH, 1998).

Measuring any source suspected of producing sound at levels exceeding the ACGIH recommended limits requires the use of a precision sound level meter, equipped with a suitable microphone of adequate frequency response, and a portable third-octave filter set. Consult with the Assistant Regional Administrator for Technical Support for guidance (ACGIH, 1998).

| TVLs for Ultrasound | | | |
|---|---|---|---|
| | One-third Octave-Band Level | | |
| | Measure in Air in dB Re:20 μ Pa; Head in Air | | Measure in Water in dB Re:20 μ Pa; Head in Water |
| Mid-Frequency of Third-Octave Band (kHz) | Ceiling Values | 8-hour TWA | Ceiling Values |
| 10 | 105[A] | 88[A] | 167 |
| 12.5 | 105[A] | 89[A] | 167 |
| 16 | 105[A] | 92[A] | 167 |
| 20 | 105[A] | 94[A] | 167 |
| 25 | 110[B] | -- | 172 |
| 31.5 | 115[B] | -- | 177 |
| 40 | 115[B] | -- | 177 |
| 50 | 115[B] | -- | 177 |
| 63 | 115[B] | -- | 177 |
| 80 | 115[B] | -- | 177 |
| 100 | 115[B] | -- | 177 |

[A]*Subjective annoyance and discomfort may occur in some individuals at levels between 75 and 105 dB for the frequencies from 10 kHz to 20 kHz especially if they are tonal in nature. Hearing protection or engineering controls may be needed to prevent subjective effects. Tonal sounds in frequencies below 10 kHz might also need to be reduced to 80 dB.*

[B]*These values assume that human coupling with water or other substrate exists. These thresholds may be raised by 30 dB when there is no possibility that the ultrasound can couple with the body by touching water or some other medium. [When the ultrasound source directly contacts the body, the values in the table do not apply. The vibration level at the mastoid bone must be used.] Acceleration Values 15 dB above the reference of 1g rms should be avoided by reduction of exposure or isolation of the body from the coupling source. (g = acceleration due to the force of gravity, 9.80665 meters/second; rms = root-mean-square).*

Table 1. (ACGIH, 1998).

Different studies highlighted as behavioural and memory disturbances, intellectual impairment, depression, and other long-term neuropsychiatric changes are well known in professional divers: these symptoms are probably caused by repeated focal ischemia due to intravascular gas bubbles and hyalinosis of the walls of small blood vessels (Reul et al., 1995). The lesions were predominantly in the subcortical white matter and basal ganglia, suggesting a vascular pathogenesis (Reul et al., 1995).

Other studies highlight as diving puts the inner ear at risk. Inner ear barotrauma and inner ear decompression can lead to permanent sensorineural hearing loss, tinnitus and vertigo (Klingmann et al., 2004).

Inner ear barotrauma is related to pressure changes in the middle and inner ear. Barotrauma refers to tissue damage that occurs when a gas-filled body space (e.g., lungs, middle ear) fails to equalize its internal pressure to accommodate changes in ambient pressure. The behaviour of gasses at depth is governed by Boyle's law: the volume of a gas varies inversely with pressure. During descent, as ambient pressure increases, the volume of gas-filled spaces decreases unless internal pressure is equalized. If the pressure is not equalized by a larger volume of gas, the space will be filled by tissue engorged with fluid and blood. This process underlies the common "squeezes" of descent that affect the middle ear, external auditory canal, mask, sinuses and teeth. Barotrauma of the inner ear during descent develops when middle ear clearing fails and the eustachian tube is blocked and locked (Klingmann et al., 2004). Under these conditions, the raised intracranial pressure brought about by forceful efforts to equalize pressure might be transmitted to the inner ear through a patent cochlear aqueduct. These pressure forces may cause rupture of Reissner's or the basilar membrane and/or labyrinthine window fistula with consequent impairment of inner ear functions (Klingmann et al., 2004). Symptoms often occur during ascent when expanding air in the middle ear is forced through a round window membrane fistula into the inner ear. The resulting gas bubble in the labyrinth expands during ascent and replaces the perilymph fluids. Barotrauma of the inner ear during ascent is a result of a blocked eustachian tube with air expanding in the middle ear forcing the tympanic membrane into the auditory canal. As a result, the oval window membrane is dislocated into the middle ear and the round window membrane is forced into the inner ear with increasing tension on both membranes (Klingmann et al., 2004). When there is an abrupt pressure equalization, either because of a tympanic membrane rupture or because the blocked eustachian tube releases the increased middle ear pressure, the oval and round window membranes snap back to their original position causing a pressure wave running through the inner ear.

Whether uneventful scuba diving in the absence of a decompression incident is a risk factor for cochlear disorders is a matter of debate. Most studies of diving associated hearing loss reveal an association with occupational noise exposure.

Different reports showed as divers exposed to high levels of underwater sound can suffer from dizziness, hearing damage, somnolence, lightheadedness inability to concentrate or other injuries to other sensitive organs, depending on the frequency and intensity of the sound. This may include neurological symptoms such as blurred vision, lightheadedness, vibratory sensations in hands, arms and legs, and tremors in upper extremities (Fothergill et al., 2009; Steevens et al., 1999).

Most reports of diving injury have concentrated on acute injuries rather than chronic disability e.g. deafness. Hence, while many divers reported aural symptoms, few attributed them to diving. It is possible that repeated hyperbaric exposure among very experienced divers may be responsible for their aural symptoms, despite the lack of an obvious acute injury for many. The cause(s) of the aural disorders described above are unknown. Different authors have reported that hearing loss in divers may be due to external ear canal obstruction, tympanic membrane perforation, middle ear disorders and sensorineural hearing damage (Taylor et al., 2006). However, aural barotrauma is the most likely cause (as it is a relatively common occurrence). It is known that the strain exerted upon the tympanic membrane (TM) and middle ear from minor barotrauma results in reversible impairment of the recoiling capacity of the TM elastic fibrils. It has been postulated that, if this barotrauma is repeated over lengthy periods, the TM changes could become irreversible (Taylor et al., 2006). Hence, hearing loss is a possible outcome (Taylor et al., 2006).

Sub-clinical brain and inner ear injury may offer an alternative explanation. Different authors found that divers had significantly more hyper-intense lesions of the sub-cortical cerebral white matter (on MRI) compared to controls (Taylor et al., 2006): these authors concluded that long term recreational diving may cause central nervous system degeneration even if diving incidents have not occurred. The exact mechanism of this degeneration remains unclear although paradoxical gas embolism, through a patent foramen ovale, has been postulated (Taylor et al., 2006).

However, the association between diving and hearing loss, in the absence of clinically apparent diving injury, may not be as clear cut. Therefore, the effect of acoustic trauma or potential harmful effects of increased pressure and partial pressures of breathing gases cannot be differentiated. In fact the following well-recognized factors can affect the inner ear in divers: inner ear decompression sickness, noise, and potentially chronic effects of the breathing gases.

A number of studies have compared the hearing threshold in professional divers (Klingmann et al., 2004):

- In 1961 in a group of 62 Royal Navy divers and submarine escape training instructors, a high-frequency hearing loss was found in most of the divers. However, these divers had been exposed to gunfire and machinery noise during their naval careers, and noise could not be excluded as the causative mechanism.
- An intriguing finding was a prevalence of 60% of hearing impairment in a group of abalone divers who had not been exposed to noise. These divers, however, had been subjected to an extraordinary compression decompression stress by a mean history of 6 years of diving with an average diving depth of 15 to 20 m during 4 hours on 100 days per year. The divers with recognizable hearing loss in that study remembered having barotrauma in the past. Therefore, residual damage after diving accidents may mask putative chronic effects of breathing air under hyperbaric conditions. In addition, hearing thresholds found in divers were compared with thresholds of controls from a different study.
- One hundred sixty-four professional Norwegian divers were subdivided into different age groups and hearing thresholds were compared with a standard population from Norway. Young divers were found to have better hearing compared with the reference group, and with increasing age this difference decreased. The authors claimed that hearing deteriorates faster in professional divers with increasing age. These results were confirmed when 116 divers were reexamined 5 years later. Noise at work and barotrauma were thought to contribute to the rapid deterioration of hearing in the professional divers.

It is postulated that the human hearing range is reduced from 130 dB in air to 55 to 60 dB in water. The reduction causes the diver to be less resistant to noise underwater because acoustic energy underwater does not resolve as fast as in air. In addition, sawing, drilling, and grinding underwater may give rise to noise levels of 90 to 105 dB, and noise from the air stream venting inside underwater helmets can reach average noise levels of 93 to 99.5 dB. Most of the studies examined professional divers who had been exposed to gunfire or other noise at work. Noise was likely the main cause of the altered pure-tone thresholds (Molvaer & Albrektsen, 1990; Molvaer & Lehmann 1985).

This interpretation is supported by the fact that puretone thresholds of divers who had been exposed to noise underwater are similar to those obtained from control subjects who had been exposed to noise on land. In a cross-sectional study, auditory function was compared in Norwegian construction divers and workshop workers. Both groups had been exposed to noise, and divers had less hearing impairment at low frequencies (0.25 and 0.5 kHz) (Skogstad et al., 1999).

Another study from Skogstad et al examined 54 occupational divers at the beginning of their diving career and 3 years later. That study subdivided the divers into groups of low exposure (100 dives in 3 years) and high exposure (100 dives in 3 years). Skogstad and coworkers did not find a statistically significant difference for both ears combined between both groups (Skogstad et al., 2000).

One should expect that divers with high exposure to diving should have poorer hearing levels because they have more contact with breathing gases under increased ambient pressure and work longer underwater and therefore spend more time in a noisy environment. However, the low exposure group might have worked in a noisy environment, too, when they were not underwater (Klingmann et al., 2004).

These findings are confirmed by the data of Benton, who examined 281 commercial divers. He investigated the audiometric records of a group of United Kingdom professional divers, all of whom had been examined by an approved medical examiner. All divers underwent a hearing test between 1989 and 1992 and had a minimum of 5 years of diving experience. The divers were divided into 7 age groups ranging from 25 to 60 years. The median hearing level thresholds were compared with the predicted values for otologically healthy individuals, the comparison revealed that the median hearing threshold values of the divers lay between the predicted median and predicted upper quartile values (Klingmann et al., 2004). Within the older group (40 years), the median and predicted median values of the divers were similar. The author postulated that these results show as the divers had no impairment of the inner ear function compared with a non diving control group.

This short revision of the literature highlights that the data on the effect of marine noise on diver are few and sometimes in contrast. However it's important to remember that, although different injuries (dizziness, hearing damage, etc) have been reported, the single most important issue related to diver safety resulting from low frequency sonar is that of disorientation due to vestibular stimulation. Whilst exposure to sonar transmissions below a level necessary to cause disorientation can give rise to temporary hearing threshold shifts, these are considered operationally acceptable for diving operations over limited periods (Salami et al., 2010).

This effect of sonar on diver is related to its duration too: studies on marine animals have demonstrated that changes in hair bundle density paralleled changes in hair cell nucleus density, indicating that entire hair cells disappeared after noise exposure; the inner ear damage is characterized by a permanent threshold elevation after an exposure to white noise ranging in intensity from 130 to 170 dB re 1 μPa for 24 h (Salami et al., 2010; Smith et al., 2006).

Although there are differences among the ears of different species, the basic processes of hearing are the same between marine and terrestrial mammals. For this reason, some of the previous considerations can be applied on humans (Salami et al., 2010; Popper & Fav, 2000).

In particular we had done a personal experience on ten male divers with normal hearing; the divers were exposed to active sonar of the Italian Navy for more than 100 exposures, each of at least 1-h duration, in the course of 6 months (Salami et al., 2010): all the subjects have been exposed to active sonar of the Italian Navy (Hull MF), at a frequency of 7.5 kHz and an intensity of 230 dB re 1 μ Pa. All the divers have had more than 100 exposures of at least 1 h, for six months, in the winter time (from October to April). The diver was exposed to the sonar at a constant depth of 3 m and at a distance from the sonar reducing progressively from 300 to 30 m. Each subject was instructed to stop the exposure in case of pain, tinnitus, vertigo, or hearing loss.

Before, at the end, and six months after the end of noise exposures, all the divers underwent the following instrumental examinations: pure-tone audiometry, Carhart test, Peyser test, thresholds of discomfort test (TDT), tympanometry, transient evoked otoacoustic emissions (TEOAE) with linear click emission, distortion product otoacoustic emissions (DPOAE), and auditory brainstem response (ABR) by MK 12-ABR (Amplifon—Italy) (Chapman & Ellis, 1998).

At the end of the exposure, the absence of TEOAE and DPOAE was observed in all the divers, the positive Peyser and TDT tests, observed in 7/10 and 10/10 divers, and the worsening of the mean air and bone audiometric thresholds, especially at the 4,000 and 8,000 frequencies, highlights the pathophysiologic features of continued and intense sound stimulation of the cochlea (Chapman &Ellis, 2008).

The injuries occur first in the first row of the outer hair cells, then in the inner hair cells, and subsequently in the second and third rows; the temporary threshold shift, at the Peyser test, observed in 9/10 of the divers, shows the presence of an auditory adaptation to the noise and underlines the risk of increasing the hearing damage: it is well established that a single exposure to a severe sound can result in direct mechanical damage to the delicate tissues of the peripheral auditory apparatus, including components of the middle ear (tympanic membrane, ossicles) and inner ear (organ of Corti); in contrast, regular exposure to less intense, but still noisy sounds, involves the insidious destruction of inner-ear components that eventually and unavoidably leads to an elevation in hearing levels.

The results of the TDT test confirm the correlation between the acoustic reflex threshold and the loudness discomfort level for people with hearing damage (Olsen, 1999).

Following a noise exposure, the hearing damage could also be due to the loss of the protective effect of the efferent fibres, perhaps mediated by the lateral olivocochlear neurons that synapse beneath the inner hair cells (Attanasio et al., 1999).

The transitory auditory injury observed in our test group may also be related to the hyperbaric work environment: oxygen toxicity is a problem in diving and can have fatal consequences in the water; past experiences made on divers, highlighted the significant presence of hearing disturbances and disorientation, and demonstrated changes of the Central Nervous System in hyperbaric conditions (Cakir et al., 2006).

Experiences done on animals (guinea pig) showed that repeated hyperbaric exposures that were considered to be safe did cause damage to the cochlear system (Zheng & Gong, 1992). These modifications are characterised by: alterations in the metabolism and in the concentration of neurotransmitters; block of intercellular oxidation processes; accumulation of carbon dioxide.

At the last control, the complete recovery observed in all the divers shows the temporary negative effects of repeated and lasting exposure to active sonar (Hull MF) and demonstrates the absence of permanent noise-induced hearing loss in divers exposed to active sonar (Salami et al., 2010).

The frequencies used in sonar are above the human hearing threshold (Gong et al., 2010; Kane et al., 2010): as because the power of ultrasonic sonar rapidly falls off with distance, a safe operating distance is 10 meters or greater. Diving may be conducted around this type of sonar provided the diver does not stay within the sonar focus beam. None of the above avoids the need for positive safety measures to be adopted when divers are working on or very close to sonar sources which are inactivated. The possibility of accidental activation must be precluded.

Since physical damage and impairment of the auditory system is caused both by high peak pressure and energy flux, safety limits for sound exposure should include both a maximum received energy fluid level and a maximum received peak–peak pressure level (impulse noise can have very high peak sound levels, but carry very little energy) (Madesen, 2005).

As different studies give only basic instructions governing hearing conservation and noise abatement, while they do not address exposure to waterborne sound, the instructions should provide field guidance for determining safe diving distances from transmitting sonar.

Sonar with an intensity level of about 230 dB re 1 μ Pa may cause on divers: slight visual-field shifts (probably due to direct stimulation of the semicircular canals), fogging of the face plate, spraying of any water within the mask, and other effects. In particular in the presence of long sonar pulses (one second or longer), depth gauges may become erratic and regulators may tend to free-flow. Different divers experienced these phenomena during controlled research report that while these effects are unpleasant, they are tolerable. Similar data are not available for un-hooded divers but visual-field shifts may occur for these divers at lower levels. If divers need to be exposed to such conditions, they must be carefully briefed and, if feasible, given short training exposures under carefully controlled conditions.

As the probability of physiological damage increases markedly with sound pressurese increase, fully protected divers must not be exposed to intensity level superior to 215 dB re 1 μPa for any reason.

A distinction is made between in-water hearing and in-gas hearing (Tompkins, 2007):

- in-water hearing occurs when the skull is directly in contact with the water, as when the head is bare or covered with a wet-suit hood.
- in-gas hearing occurs when the skull is surrounded by gas as in the MK 21 diving helmet.

In-water hearing occurs by bone conduction—sound incident anywhere on the skull is transmitted to the inner ear, bypassing the external and middle ear. In gas hearing occurs in the normal way—sound enters the external ear canal and stimulates the inner ear through the middle ear.

For these reasons, if the diver is helmeted, it's necessary to use greater distance from the sonar source.

It's also important to identify the type of diving equipment: wet-suit un-hooded, wet-suit hooded, helmeted; wet-suit hooded diver can safely get closer to a sonar source. If the type of sonar is unknown, start diving at 600–3,000 yards, depending on diving equipment (use greater distance if helmeted), and move in to limits of diver comfort.

Helmeted divers experience reduced sensitivity to sound pressure as depth increases.

The sonar presents different effect on divers, according to the intensity (low, medium, high):

- low-frequency sonar generates a dense, high-energy pulse of sound that can be harmful at higher power levels. As a variety of sensations may result from exposure to low-frequency sonar, it is necessary to inform divers when exposure is likely and to brief them regarding possible effects; specifically, that they can expect to hear and feel it. Sensations may include mild dizziness or vertigo, skin tingling, vibratory sensations in the throat and abdominal fullness. Divers should also be briefed that voice communications are likely to be affected by the underwater sound to the extent that line pulls or other forms of communication may become necessary (Crum & Mao, 1996).

- Medium and high frequency sonars: some military anti-submarine sonar-equipped ships do pulse high intensity pressure waves dangerous to a diver. It is prudent to suspend diving operations if a high-powered sonar transponder is being operated in the area. When using a diver-held pinger system, it is advisable for the diver to wear the standard 1/4 inch (0.64 cm) neoprene hood for ear protection. Experiments have shown that such a hood offers adequate protection when the ultrasonic pulses are of 4 ms duration, are repeated once per second for acoustic source levels up to 100 watts, and are at head-to-source distances as short as 4 inches (10 cm).

## 6. Conclusion

The power of the sonar systems, the noise that they produce and the distance they can travel can undoubtedly have an effect on marine life. Marine animals may experience gross damage to ears, damage to body tissue, masking of communication, interference with ability to acoustically interpret their environment and also interference with food finding. Long term effects caused by sonar are almost impossible to identify. Many whales that are fatally impacted can sink to the bottom of the ocean; therefore the true death toll cannot be estimated. There are widespread concerns about the danger of high intensity sonar to marine mammals, marine ecosystems and the health of our depleted oceans. Low frequency sonar can travel hundreds of miles through our oceans at considerable intensities.

It is currently difficult to provide an evaluation of the effectiveness and adequacy of the measures taken and planned for the protection of the marine environment against effects from underwater noise. One of the reasons for this is that there are still gaps in our understanding on the effects of underwater noise on marine life. There is evidence that certain activities can generate noise levels that have the potential to be harmful to marine mammals, fish and human, yet, the exact nature of the effects (temporary threshold shift, masking, behavioural response) are not totally understood. The poor understanding of effects means that any regulation and mitigation measures are likely to be based on precaution. This makes it urgent to gather data on the effects of underwater noise in order to apply appropriate regulation and /or mitigation measures.

Underwater noise has the potential to affect marine life in various ways and in some cases over relatively large areas and time scales. It is difficult to assess to what degree the introduction of underwater noise affects the overall quality status as there is little data to allow us to quantify noise levels across the exposition area. However, most of the intensities of anthropogenic sounds exceed by several order of magnitude the ambient sounds in the marine environment that occur naturally, such as sounds that are induced by rain, wind and waves. Underwater noise can have a range of impacts on marine life such as injury, permanent or temporary hearing loss, behavioural responses and masking of biological relevant signals. However, there are many uncertainties in assessing effects of noise due to the difficulties in observing individual level effects, let alone population level consequences of acoustic disturbance. From a conservation perspective, it is important to assess whether anthropogenic sound has a significant effect on populations (animals and humans). This is also important in assessing the impacts of noise in relation or addition to other stressors either to assess cumulative impacts and/or to focus protection efforts. All factors impacting on populations are cumulative and must be assessed together by discussing the significance of effects. There is currently no information available on the cumulative effects of the factors listed above. No agreed assessment framework for cumulative effects of diverse human activities exists yet.

Although these evidences, our results show the temporary negative effects of repeated and lasting exposure to active sonar on the divers. In particular our data demonstrate the absence of permanent noise-induced hearing loss in divers exposed to active sonar.

## 7. References


ACGIH. (1998). TLVs and BEIs *Threshold limit values for chemical substances and physical agents.* 117, ISBN:1-88-2417-23-2

Attanasio G.; Barbara M.; Buongiorno G.; Cordier A.; Mafera B.; Piccoli F.; Nostro G. & Filipo R. (1999). *Protective effect of the cochlear efferent system during noise exposure.* Ann N Y Acad Sci, 28:361–367, ISSN 00778923

Cakir BO.; Ercan I.; Civelek S.; Körpinar S.; Toklu AS.; Gedik O.; Işik G.; Sayin I. & Turgut S. (2006*). Negative effect of immediate hyperbaric oxygen therapy in acute acoustic trauma.* Otol Neurotol, 27:478–483, ISSN 15317129

Chapman DMF & Ellis DD (1998). *The elusive decibel: thoughts on sonars and marine mammals.* Can Acoust, 26:29–31, ISSN 07116659

Crum L.A. & Mao Y. (1996*). Acoustically enhanced bubble growth at low frequencies and its implications for human diver and marine mammal safety.* Journal of the Acoustical Society of America, 99:2898-2907, ISSN 15208524

Fav RR & Popper AN. (2000*). Evolution of hearing in vertebrates: the inner ears and processing.* Hear Res;149:1–10 ISSN 03785955

Fernández A.; Edwards JF.; Rodríguez F.; Espinosa de los Monteros A.; Herráez P.; Castro P.; Jaber JR.; Martín V. & Arbelo M. (2005). "*Gas and fat embolic syndrome*" *involving a mass stranding of beaked whales (family Ziphiidae) exposed to anthropogenic sonar signals.* Vet Pathol. Jul;42(4):446-57, ISSN 15442217

Fothergill DM.; Sims JR. & Curley MD. (2001). *Recreational scuba divers' aversion to low-frequency underwater sound.* Undersea Hyperb Med, 28:9-18, ISSN 10662936

Frosch RA. (1964). *Underwater Sound: Deep-Ocean Propagation: Variations of temperature and pressure have great influence on the propagation of sound in the ocean.* Science, 146:889-94, ISSN 00368075

Gong Z.; Andrews M.; Jagannathan S.; Patel R.; Jech JM.; Makris NC. & Ratilal P. (2006). *Low-frequency target strength and abundance of shoaling Atlantic herring (Clupea harengus) in the gulf of maine during the ocean acoustic waveguide remote sensing experiment.* J Acoust Soc Am., 2010;127:104-23, ISSN 15208524

Hood LJ.; Berlin CI. & Parkins CW. (1991). *Measurement of sound.* Otolaryngol Clin North Am., 24:233-51 ISSN 00306665

Jian YJ.; Zhang J.; Liu QS. & Wang YF. (2009). *Effect of Mesoscale eddies on underwater sound propagation.* Applied Acoustics, 70:432-440, ISSN 0003682X

Kane AS.; Song J, Halvorsen MB.; Miller DL.; Salierno JD.; Wysocki LE.; Zeddies D. & Popper AN. (2010). *Exposure of fish to high-intensity sonar does not induce acute pathology.* J Fish Biol, 76:1825-40, ISSN 0022-1112

Klingmann C.; Benton P.; Schellinger P. & Knauth M. (2004). *A safe treatment concept for divers with acute inner ear disorders.* Laryngoscope. Nov;114(11):2048-50, ISSN 0023852X.

Leroy CC.; Robinson SP. & Goldsmith MJ. (2008). *A new equation for the accurate calculation of sound speed in all oceans.* J Acoust Soc Am.;124:2774-82, ISSN 15208524

Lichte H. (1919) *On the influence of horizontal temperature layers in sea water on the range of underwater sound signals.* Physik. Z. 17 (385)

Madesen PT (2005). *Marine mammals and noise: problems with root mean square sound pressure levels for transients.* J Acoust Soc Am 117:3952–3957, ISSN 15208524

Molvaer OI. & Albrektsen G. (1990). *Hearing deterioration in professional divers: an epidemiologic study.* Undersea Biomed Res.;17:231246.

Molvaer OI & Lehmann EH. (1985). *Hearing acuity in professional divers.* Undersea Biomed Res.;12:333-349, ISSN 00935387

Mooney TA.; Nachtigall PE. & Vlachos S. (2009). *Sonar-induced temporary hearing loss in dolphins.* Biol Lett.;5:565-7, ISSN 17449561

Olsen SO. (1999). *The relationship between the uncomfortable loudness level and the acoustic reXex threshold for pure tones in normally-hearing and impaired listeners – a meta analysis.* Audiology, 38:61–68, ISSN 00206091

Parsons ECM.; Dolman SJ.; Wright AJ.; Rose NA. & Burns WCG (2008). *Navy sonar and cetaceans: just how much does the gun need to smoke before we act?* Mar Pollut Bull 56:1248–1257, ISSN 0025-326X

Patek SN.& Oakley TH. (2003). *Comparative tests of evolutionary trade-offs in a palinurid lobster acoustic system.* Evolution. 57:2082-100, ISSN 00143820

Popper AN.; Halvorsen MB.; Kane A.; Miller DL.; Smith ME.; Song J.; Stein P. & Wysocki LE. (2007). *The effects of high-intensity, low-frequency active sonar on rainbow trout.* J Acoust Soc Am;122:623-35, ISSN 15208524

Reul J, Weis J, Jung A, Willmes K, Thron (1995). *A. Central nervous system lesions and cervical disc herniations in amateur divers.* Lancet. Jun 3;345(8962):1403-5, ISSN 01406736

Sabra KG. & Dowling DR. (2003). *Effect of ocean currents on the performance of a time-reversing array in shallow water.* J Acoust Soc Am., 114:3125-35, ISSN 15208524

Salami A, Dellepiane M, Crippa B, Guastini L, Mora R. (2010). *The effect of active sonar for the protection of moored and anchored warships on the human hearing.* Eur Arch Otorhinolaryngol, 267:207–211, ISSN 0937-4477

Shvachko LV. (2008). *Structure of the sound field in convergence zones of the tropical pacific.* Acoustical Physics, 54:228-237, ISSN 10637710

Simmonds MP, Lopez-Jurado LF. (1991). *Whales and the Military.* Nature 351:448.

Simon M.; Wahlberg M.; Ugarte F. & Miller LA. (2005). *Acoustic characteristics of underwater tail slaps used by Norwegian and Icelandic killer whales (Orcinus orca) to debilitate herring (Clupea harengus).* J Exp Biol;208:2459-66, ISSN 0022-0949

Skogstad M.; Haldorsen T. & Arnesen AR. (2000). *Auditory function among young occupational divers: a 3-year follow-up study.* Scand Audiol., 29:245-252, ISSN 0105-0397

Skogstad M.; Haldorsen T. & Kjuus H. (1999). *Pulmonary and auditory function among experienced construction divers: a cross-sectional study.* Aviat Space Environ Med., 70:644-649, ISSN 0095-6562

Smith ME.; CoYn AB.; Miller DL. & Popper AN. (2006). *Anatomical and functional recovery of the goldfish (Carassius auratus) ear following noise exposure.* J Exp Biol, 209:4193–4202, ISSN 0022-0949

Snellen M.; Simons DG.; Siderius M.; Sellschopp J. & Nielsen PL. (2001). *An evaluation of the accuracy of shallow water matched field inversion results.* J Acoust Soc Am., 109(2):514-27, ISSN 15208524

Steevens CC.; Russell KL.; Knafelc ME.; Smith PF.; Hopkins EW. & Clark JB. (1999). *Noise-induced neurologic disturbances in divers exposed to intense water-borne sound: two case reports.* Undersea Hyperb Med, 26: 261–5, ISSN: 1066-2936

Stocker M. (2004). *Ocean noise could injure more than mammals.* Nature, 430:291, ISSN 14764687

Studenichnik NV. (2003). *Intensity and space–time characteristics of the sound field in the underwater sound channel of the black sea.* Acoustical physics, 49;207-16, ISSN 10637710

Sypin AY. (2008*). Hearing sensitivity during target presence and absence while a whale echolocates.* J Acoust Soc Am, 123:534-541, ISSN 15208524

Taylor D.; Lippmann J. & Smith D. (2006). *The absence of hearing loss in otologically asymptomatic recreational scuba divers.* Undersea Hyperb Med, 33(2):135-141, ISSN 1066-2936

Tindle CT. & Deane GB. (2005). *Shallow water sound propagation with surface waves.* J Acoust Soc Am., 117:2783-94, ISSN 15208524

Tompkins O. (2007). *Ears and air travel.* AAOHN J, 55:336, ISSN 0891-0162

Tyack PL.; Johnson M.; Soto NA.; Sturlese A. & Madsen PT. (2006). *Extreme diving of beaked whales.* J Exp Biol, 209(Pt 21):4238-53, ISSN 14779145

Urick, Robert J. (1983). *Principles of Underwater Sound, 3rd Edition.* McGraw-Hill, ISBN 0070660875, New York

Wolfson MA & Tomsovic S. (2001). *On the stability of long-range sound propagation through a structured ocean.* J Acoust Soc Am. 109:2693-703, ISSN 15208524

Zheng XY. & Gong JH. (1992). *Cochlear degeneration in guinea pigs after repeated hyperbaric exposures.* Aviat Space Environ Med, 63:360–363, ISSN 0095-6562

# 12

# Underwater Acoustic Detection and Signal Processing Near the Seabed

Henry M. Manik
*Department of Marine Science and Technology Faculty of Fisheries and Marine Sciences, Bogor Agricultural University Kampus IPB Darmaga Bogor 16680 Indonesia*

## 1. Introduction

### 1.1 Background

Underwater acoustic instruments has been an indispensable tool to study the ocean. Echo sounder is one of the acoustic instrument used to remotely classify distributions of biological organisms such as fish and plankton (MacLennan and Simmonds, 1992; Furusawa, 2000). Knowledge of species, location, and behavior observed by the sounders are important for fisheries, fisheries management, and ecological studies.

Echo sounders have also been used to characterize the sea bottom type such as rocks, sand, and mud (Stanton, 1994). The characterization of the sea bottom type are useful in applications such as fish habitat study, fishing port construction, geological studies, marine exploration, and mining.

With the end of World War II, studies on underwater acoustics for sea bottom interactions began extensively. Urick (1954) defined the measure of *backscatter strength* as the ratio of scattered energy to incident energy per unit area, per unit solid angle and this quantity was expressed in decibels. Urick used the frequencies of 10 to 60 kHz to measure the backscattering from the harbor bottom. Mackenzie (1961) published the results for lower frequency scattering in deep water and introduced empirical Lambertian backscatter coefficient which was to be cited often in subsequent studies. McKinney and Anderson (1964) investigated potential scattering mechanism based on the interface relief and sediment particle.

One of the methods to measure the bottom backscattering strength is to use the echo integration method (Aoyama *et al*, 1999). This method is used as the main tool for quantifying the abundance of marine fish or planktonic organisms. The echo integration output can be converted to biomass. For the general situation of arbitrary density, as with a school of fish, individual target echoes may not be resolvable. In this case, the echo voltage, after detection and application of suitable range compensation, is squared and summed over a defined interval. The resulting quantity is proportional to the volume or area backscattering coefficient. Division of this by the characteristic backscattering cross section for the target fish yields the numerical density. The echo integration technique has been originally used for fisheries surveys and at the present study we adopted it to measure backscattering strength (SS) of the sea bottom.

Acoustic scattering by the sea bottom has been studied in order to either predict the performance of the echosounder system (Aoyama, *et al.* 1999) or to quantitatively map the sea bottom. Characterization of sea bottom can be obtained by measuring the bottom backscattering strength (SS). From the measured SS we may be able to estimate the fish habitat.

### 1.2 Research objectives

The objectives of this researches are development and application of methods to characterize or quantify the fish and sea bottom by underwater acoustic instrument originally developed for measurements of fish scattering. More specifically, the objective is to extensively use multi frequency acoustics (38, 70, and 120 kHz) to measure the bottom backscattering strength (SS) and fish quantification in off Southern Jawa Island, Indonesia.
The SS can be measured by the bottom scattering theory developed by Aoyama *et al.* (1999). As the extension of this theory, a simple but powerful model for bottom scattering, that is a ring surface scattering (RSS) model was developed. The RSS model enables to interpret the bottom echo and to measure the SS value. The SS values, measured for the first time at the sea off Jawa Island, are related to the sounder frequency, beam width, and the bottom type and the properties of the SS with respect to the parameters are examined. The bottom material sampling were conducted to interpret the bottom echo. An associated objective of this study is to measure bottom depth to know the bottom topography.
The other objective is to identify the bottom material by the SS value and relate it to bottom fish habitat. We developed a new and effective method to display fish volume scattering and bottom SS simultaneously. To verify the bottom fish echo, the bottom trawling was conducted.

## 2. Bottom scattering models

### 2.1 Preface

A number of mathematical models have been developed in order to gain a better physical understanding of acoustic backscattering from the sea bottom. By observing echo fluctuations from sea bottom, Stanton (1985) determined quantitative information of bottom relief such as bottom roughness. He extended the Eckart (1953) acoustic scattering theory to show the relationship between the probability density function of the echo level and the rms roughness. Stanton model is plane wave model. Jackson *et al.* (1986) developed a bottom scattering model using the composite roughness and grazing angle parameters.
Aoyama et al. (1997) model was extended as ring surface scattering model (RSS model) in this Chapter. The extension of this model from Aoyama et al. model is described. The RSS model exhibits a raw backscattering strength (without averaging) at the peak of bottom echoes and easily convert the bottom scattering strength to the bottom backscattering strength (SS). Aoyama et al. used the approximation method for the equivalent beam angle, but the RSS model uses a strict instantaneous equivalent beam angle for surface scattering.

### 2.2 Normal incident bottom scattering theory

Normal incidence reflection of sound from the sea bottom is relatively easy to detect. It is used in depth sounders to give the water depth for ships and boats. There has been, however, little research a bottom backscattering strength measurement by normal incidence echo sounders.

Because the wave is propagating, a certain amount of power flow across a unit area normal to the direction of the propagation. This amount of power is called the sound intensity and shown as

$$I = \frac{P^2}{\rho c} \tag{2.1}$$

where $P$ is the pressure, $\rho$ is the water density, and $c$ is the sound speed in water.
Sound intensity of surface scattering from a small scattering area, $dS$ is shown by

$$dI = I_o r^{-4} \exp(-4\alpha r) b^2 S_S \, dS \tag{2.2}$$

where $I_o$ is the source acoustic intensity, $r$ is the range from transducer to the elemental scattering area, $dS$, $\alpha$ is the absorption coefficient, $b$ is the directivity function of the transducer, and $Ss$ is the bottom backscattering strength (Fig. 2.1). In decibel notation SS = 10 log Ss.
From Fig. 2.1, we have

$$dS = \frac{r \, d\theta}{\cos\theta} r \sin\theta = r^2 \tan\theta \, d\theta \, d\phi \tag{2.3}$$

where $\theta$ and $\phi$ are the polar and azimuthal angles, respectively.
Substitution of Eq. (2.3) into Eq. (2.2) yields

$$dI = I_o r^{-2} \exp(-4\alpha r) b^2 S_S \tan\theta \, d\theta \, d\phi \tag{2.4}$$

Intensity of the surface backscattering strength, $I$, is the total of $dI$, therefore

$$I = \int dI \tag{2.5}$$

In general, the $S_S$ becomes smaller when the incident angle, $\theta$ become higher. For the quantitative echo sounder with the narrow beam, however, the effect of $\theta$ to the scattering should be negligible. Therefore, Eq. (2.5) can be written as

$$I = I_o r^{-2} \exp(-4\alpha r) S_S \Phi \tag{2.6}$$

where

$$\Phi = \int_0^{2\pi} \int_{\theta_1}^{\theta_2} b^2 \tan\theta \, d\theta \, d\phi \tag{2.7}$$

and $\Phi$ is called the equivalent beam angle for surface scattering.
The integration limits $\theta_1$ and $\theta_2$ depend on the time or slant range, $r$, from the transducer to elemental scattering area and shown as

$$\theta_1 = \begin{cases} 0 & (R \le r < R + c\tau/2) \\ \cos^{-1}\left(\dfrac{R}{r - c\tau/2}\right) & (r \ge R + c\tau/2) \end{cases}$$

and

$$\theta_2 = \cos^{-1}\left(\frac{R}{r}\right) \tag{2.8}$$

where $R$ is the range from the transducer to the sea bottom, $c$ is the sound speed, and $\tau$ is the pulse width. For a circular piston transducer, the directivity of transducer is given by

$$b(\theta) = \left(\frac{2\, J_1\, (ka \sin \theta)}{ka \sin \theta}\right)^2 \tag{2.9}$$

where $J_1$ is the first order of the first kind Bessel function, $k = 2\pi / \lambda$, $k$ is the wave number and $\lambda$ is the wave length, and $a$ is the radius of the transducer.

### 2.3 Ring surface scattering model

We derived a new model called " ring surface scattering model " from the above theory. This model enables us to measure the bottom backscattering strength (SS) by using the echo sounder (Aoyama, *et al.*, 1999). Figure 2.3 shows a simplified block diagram for bottom backscattering measurement by underwater acoustic instrument.

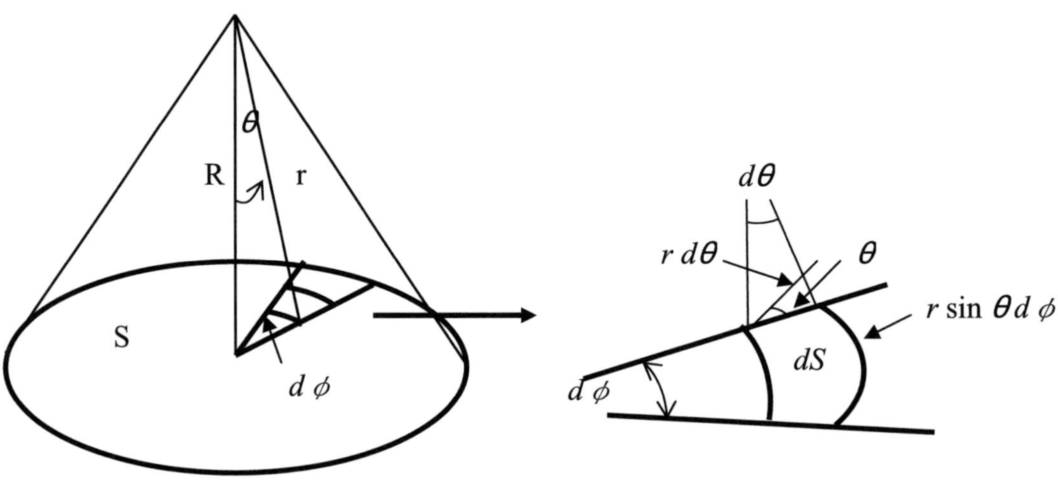


Fig. 2.1. Principle of surface scattering.

Figure 2.2 shows the geometry showing change of scattering plane with time.The surface insonified by the pulse changes from a circle (a) toa circular (b) like ring with time.
The backscattered pressure signal from the bottom received by the transducer (Fig. 2.3), $P_{RB,}$

$$P_{RB}^2 = P_o\, r^{-2}\, exp(-4a\, r)\, \Phi\, S_S \tag{2.10}$$

where $P_o$ is the source pressure level.
The bottom echo signal is amplified by the amplifier to give

$$E_{\ RB} = P_{RB}\, M\, G_R \tag{2.11}$$

where $E_{RB}$ is the echo amplitude at the preamplifier output, $M$ is the receiving sensitivity of the transducer, and $G_R$ is the preamplifier gain.
The echo amplitude, $E_{RB}$, is shown from Eqs. (2.10) and (2.11) as

$$E_{RB}^2 = K_{TR}^2\, r^{-2} \exp(-4\alpha\, r)\, \Phi\, S_S \tag{2.12}$$

where $K_{TR} = P_o\, M\, G_R$ is the transmitting receiving coefficient. In this reduction we assumed $r \simeq R$ except for $r$ and $R$ in $\Phi$ (Eq. 2.7).
The time varied gain (TVG) amplifier output of $E_{RB}$ corrected for an absorption and spreading losses, $E_{TB}$, is

$$E_{TB} = G_{TM}\, r \exp(2\alpha\, r) E_{RB} \tag{2.13}$$

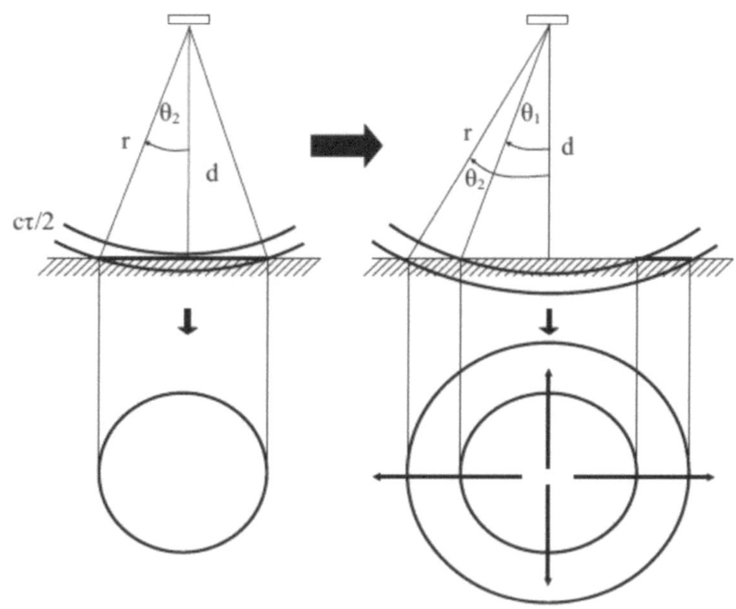


**(a) circle surface** **(b) circular rings**

Fig. 2.2. Geometry showing change of scattering plane with time. The acoustic wave hit the bottom at the bold lines.

where $G_{TM}$ is the coefficient for the 20 log $r$ TVG amplifier gain. From Eqs. (2.12) and (2.13) we obtain

$$E_{TB}{}^2 = (K_{TR} G_{TM})^2 \Phi\, S_S\,. \tag{2.14}$$

The "raw" SV value of the bottom echo, $S_{VB}$, is

$$S_{VB} = \frac{E_{TB}{}^2}{K_M{}^2} \tag{2.15}$$

where $K_M$ is the multiple echo coefficient and given by

$$K_M{}^2 = (K_{TR}G_{TM})^2 \Psi(c\tau/2) \tag{2.16}$$

where

$$\Psi = \int_0^{2\pi}\int_0^{\pi/2} b^2 \sin\theta \, d\theta \, d\phi \tag{2.17}$$

The above $\Psi$ is the equivalent beam angle of the volume scattering.
Substitution of $E_{TB}$ of Eq. (2.14) and $K_M$ of Eq. (2.16) into Eq. (2.15) yields

$$S_{VB} = \frac{Ss\,\Phi}{\Psi\,(c\tau/2)} \tag{2.18}$$

This equation is called " *ring surface scattering model* ".
The echo wave form simulation is possible by this model. The results of the wave form simulation are compared with the wave forms from the actual data obtained by the echo sounder to interpret the data. When our purpose is to simulate the wave form and echo level, we use the instantaneous equivalent beam angle $\Phi$, Eq. (2.7), as a function of range.

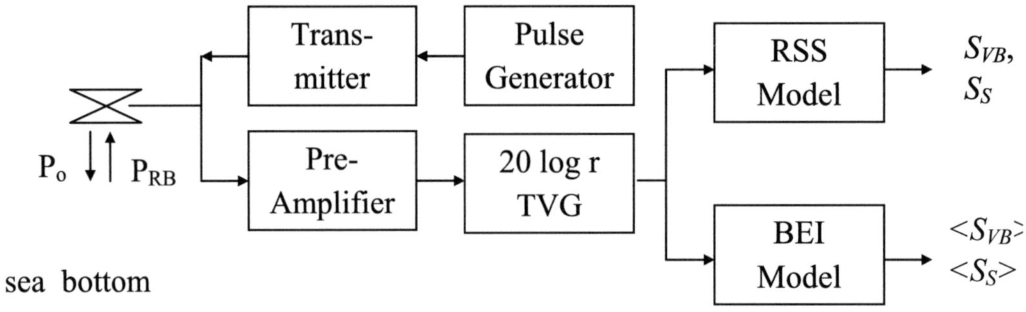


Fig. 2.3. The simplified block diagram of underwater acoustic instrument to measure bottom backscattering strength (SS). RSS is ring surface scattering and BEI is bottom echo integration.

### 2.4 Bottom echo integration

In the bottom echo integration method (Aoyama, *et al.*, 1999) we average the bottom echoes for a predefined depth layer ($r$ to $r+r_w$) including the bottom echoes and for a ping sequence to obtain the average bottom echo integration strength.
The average squared voltage of TVG outputs with respect to the ping (index $i$ ) and the range is

$$\left\langle E_{TB}{}^2 \right\rangle = \frac{1}{r_w}\int_r^{r+r_w}\left(\frac{1}{m}\sum_{i=1}^{m} E_{TBi}{}^2\right)dr \tag{2.19}$$

where $m$ is the integration period in ping. The average bottom echo integration strength ("bottom SV"), $< S_{VB} >$, is obtained as done in the ordinary echo integration method :

$$\langle S_{VB} \rangle = \frac{\langle E_{TB}{}^2 \rangle}{K_M{}^2} \tag{2.20}$$

Substitution of the Eq. (2.19) into Eq. (2.20) yield

$$\langle S_{VB} \rangle = \left( \frac{K_{TR} G_{TM}}{K_M} \right)^2 \frac{1}{r_w} \int_r^{r+r_w} \left[ \frac{1}{m} \sum_{i=1}^{m} \Phi_i \, Ss_i \right] dr \tag{2.21}$$

When $\Phi$ is independent of $i$, which is a reasonable assumption for a case of a small integration period, we have

$$\langle S_{VB} \rangle = \left( \frac{K_{TR} G_{TM}}{K_M} \right)^2 \langle Ss \rangle \frac{1}{r_w} \int_r^{r+r_w} \Phi \, dr \tag{2.22}$$

where

$$\langle Ss \rangle = \frac{1}{m} \sum_{i=1}^{m} Ss_i \tag{2.23}$$

is the average bottom scattering strength.
In the case of sharp beam, we have (Aoyama *et al.* 1997)

$$\int_r^{r+r_w} \Phi \, dr = \Phi_o \frac{c\tau}{2}$$
$$\Phi_o = \int_0^{2\pi} \int_0^{\pi/2} b^2 \, \theta \, d\theta \, d\phi \tag{2.24}$$

where $\Phi_0$ is the asymptotic value of the equivalent beam angle for the surface scattering. We find from Eq. (2.24) :

$$\Phi_o \cong \Psi \, . \tag{2.25}$$

Introducing Eq.(2.25) into Eq. (2.22), we get

$$\langle S_{VB} \rangle = \frac{K_{TR}{}^2 G_{TM}{}^2 \Phi_o c\tau / 2 \langle Ss \rangle}{(K_{TR}{}^2 G_{TM}{}^2 \Psi \, c\tau / 2)' r_w} \tag{2.26}$$

where the prime means values which should be given to the underwater acoustic instrument as parameters. Finally we have

$$\langle Ss \rangle = r_w \langle S_{VB} \rangle \, , \tag{2.27}$$

if the parameter values have no error. This is the simple relationship between the average SS and the bottom SV.

**2.5 Simultaneous display of fish and bottom scattering**

Using the RSS Model we have a relationship between the raw (or not averaged) SV value of the bottom echo, $S_{VB}$ , and the raw bottom backscattering strength, $S_S$, as

$$S_S = \frac{S_{VB}\,\Psi(c\tau/2)}{\Phi} \tag{2.28}$$

where $\Phi$ and $\Psi$ are the equivalent beam angle for surface and volume scattering, respectively, $c$ is the sound speed, and $\tau$ is the pulse width. At the peak of the bottom echo, we have

$$\Phi = \Phi_0 \cong \Psi \tag{2.29}$$

where $\Phi_o$ is the asymptotic value of the equivalent beam angle for surface scattering (Fig. 2.4). Introducing Eq. (2.29) into Eq. (2.28) gives

$$S_S = (c\,\tau/2)\,S_{VB} \tag{2.30}$$

and we can easily convert the raw SV to raw SS. An application of this formula for all $S_{VB}$ yields a convenient measure called "instantaneous" SS. Figure 2.5 show the raw SV of bottom echo ($S_{VB}$), raw SS ($S_S$), and instantaneous SS.

In decibel unit,

$$SS = 10\log(c\,\tau/2) + SV_B \tag{2.31}$$

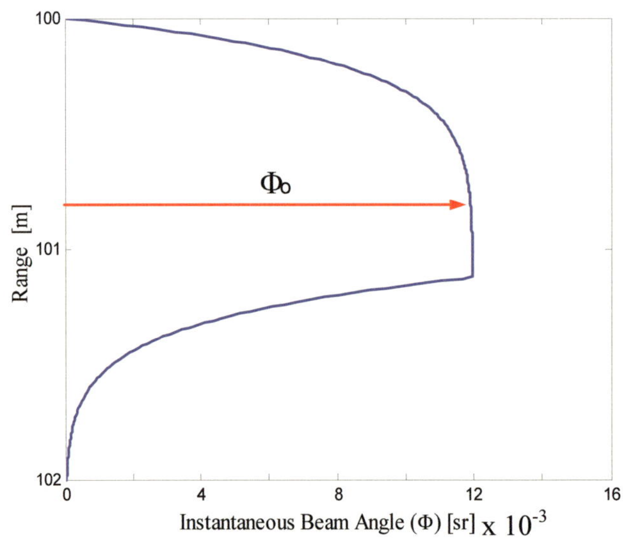


Fig. 2.4. The instantaneous equivalent beam angle $\Phi$and the asymptotic value of $\Phi_o$.

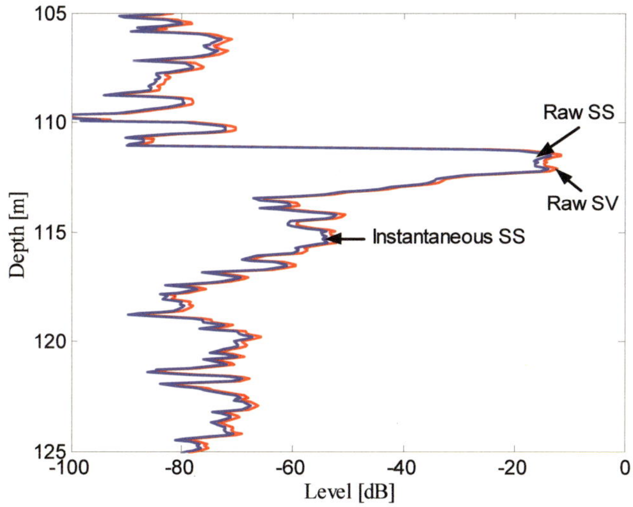


Fig. 2.5. The raw SV, raw SS, and instantaneous SS.

## 3. Survey methods

The echo sounder is employed in this survey to explore fish resources, bottom topography, and bottom backscattering strength. Additional instrument, such as Global Positioning System (GPS) and data storage are included as part of the acoustic system. The raw data of echo sounder is processed using Echoview and Matlab program. The sphere calibration was conducted to guarantee a high quality of acoustic data (Fig. 3.1).

Bottom material sampling using an anchor dredge was conducted to interpret the bottom echo. The obtained bottom material was saved for further processing in laboratory. The acoustic data collection and the bottom material sampling were conducted simultaneously. The bottom material was analyzed using the sieving and pipette methods. Then, the analysis of particle size was done using Wentworth scale.

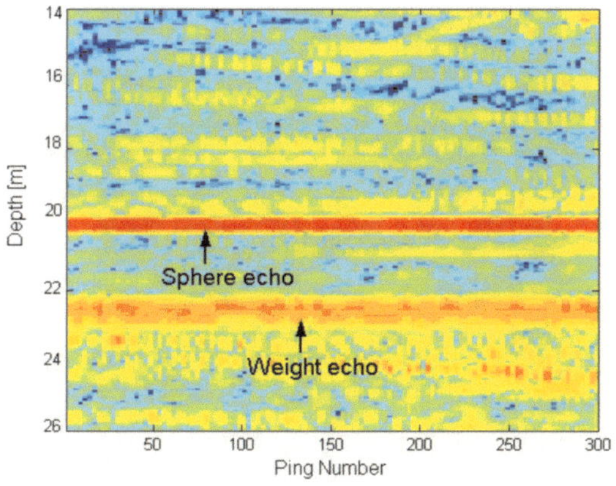


Fig. 3.1. The sphere echo during calibration.

# 4. Results and discussions

## 4.1 Bottom echo characteristics

It is shown that the underwater acoustic instrument can rather easily measures echoes by reflection of sounding pulses from the bottom. The sounder is especially useful when both fish and sea bottom should be quantified. The quantitative echogram of the It is shown that the underwater acoustic instrument can rather easily measures echoes by reflection of sounding pulses from the bottom. The sounder is especially useful when both fish and sea bottom should be quantified. The quantitative echogram of the echosounder showing raw SV as shown in Fig. 4.1 enables us to interpret the backscattered signal from any object. showing raw SV as shown in Fig. 4.1 enables us to interpret the backscattered signal from any object.

Figure 4.1 shows typical echograms at 38 kHz for sand, silt, and clay seabed. The scale shows the raw SV value ranging from -70 to -10 dB. The red belts represent the sea bottom. Because of the same range width of the echograms we can easily compare the bottom echoes and find that the thickness of the reddish belts for sand is thinner than those of silt and clay bottoms; this mainly comes from the difference of the bottom depth rather than of bottom material as discussed later.

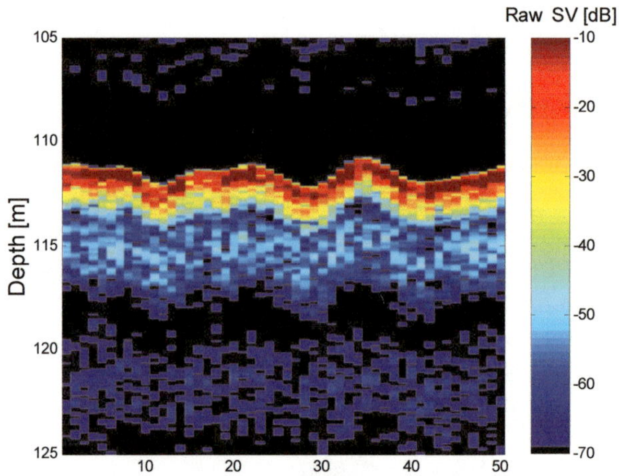


(a) Sand

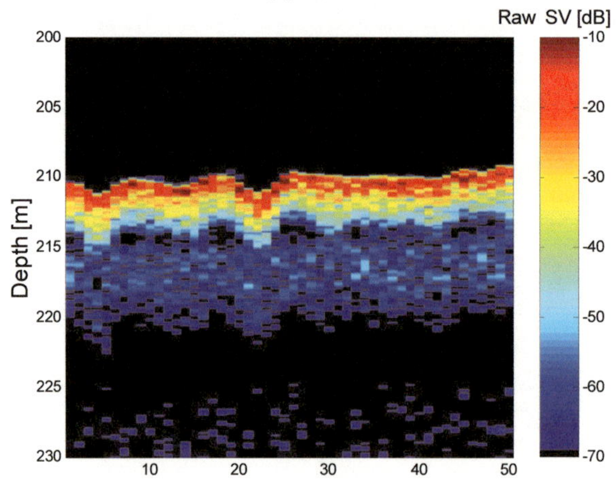


(b) Silt

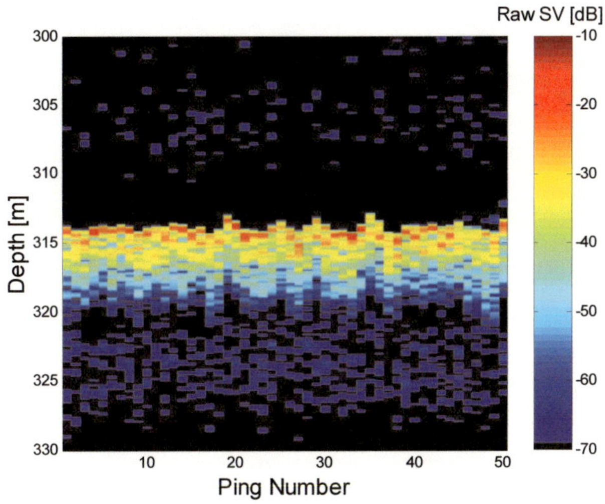


(c) Clay

Fig. 4.1. Typical echograms in raw SV for sand, silt, and clay. The range width and ping number are the same for three figures.

Figure 4.1 shows the bottom echoes in the same range scale width and easily reveals differences or similarity among the three echograms. The first red belts correspond to the contribution from the main lobe as can be seen from the simulation results in Fig. 4.2; the difference in the thicknesses comes from the depth differences but not from the material differences. We also observe two or three more weaker bluish belts beneath the main belt and these are contributions from side lobe(s) as also can be seen from Fig. 4.2.

Figure 4.2 shows several typical wave forms of the bottom echoes, with absolute raw SV scale for the three frequencies and the three materials near the points whose echograms are shown in Fig. 4.1. Figure 4.2 also shows the simulated wave forms by Eq. (2.18) using the SS values of each bottom type shown in each figure : the SS values were obtained from the peak raw SV values of bottom echo and giving $\Psi$, $\Phi$, $c$, and $\tau$ using Eq. (2.28). At the peak of bottom echo, the echo level of sand was higher than silt and clay by more than 8 dB.

As shown in Fig. 4.2, the model calculations and measurements agree rather well in the wave form. The width change of the bottom echo was caused by the increasing depth. The bottom material is reflected only in the SS value in the present model. It was supported by the simulation results that the beam spreading caused the increase of the bottom echo width with increasing depth. The present model can well predict the scattering by the side lobes ; this figure also shows that the narrow beam and short pulse configuration receives the contributions of the bottom echoes separately by the side lobes. Comparing with the more thorough model as by Chotiros (Chotiros, 1994), this model can be used easily, sacrificing a little bit rigorousness.

#### 4.2 Examination of bottom echo integration method

This section describes the examination of the bottom echo integration method. We examined the relation of measured bottom backscattering strength (SS) on integration width $r_w$ for the three bottom materials of sand, silt, and clay (Fig. 4.3).

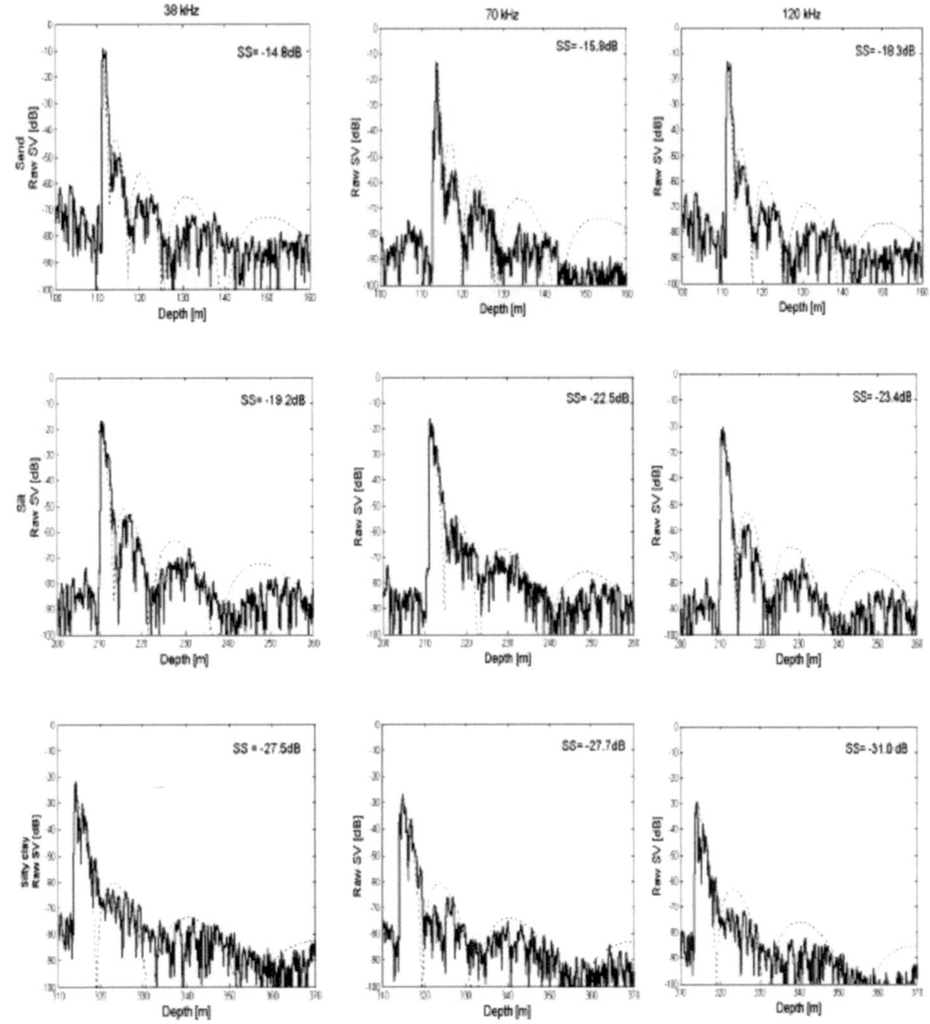


Fig. 4.2. Wave forms of sand (top), silt (middle), and clay (bottom) in raw SV by measurements (—) and simulations (-----) at 38 kHz (left), 70 kHz (middle), and 120 kHz (right).

We find from Fig. 4.3, the average SS values by the bottom echo integration are independent from the change of integration layer, $r_w$.

The concept of SA was originally developed for estimating the fish biomass by acoustic method. For the echo integration, the received signal is once converted to through the volume backscattering strength, $S_V$. The integration of $S_V$ over the range interval [$r_1$, $r_2$] is called the area backscattering strength $S_A$

$$S_A = \int_{r_1}^{r_2} S_v \, dr \tag{4.1}$$

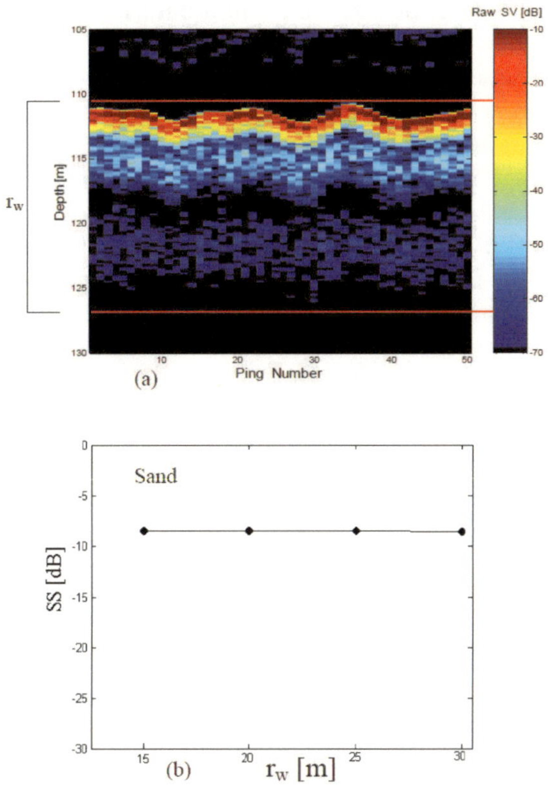


Fig. 4.3. The echogram in raw SV scale (a) and the relation of measured SS on $r_w$ values (b).

In decibel scale $SA = 10 \log S_A$.
From Eq. (4.1) we have

$$S_A = \Sigma r_w <S_V> \tag{4.2}$$

for the actual processing in which average SV, that is $<S_V>$, are computed for each integration cell.
By Eq. (4.2) and Eq. (2.27 ) yields for one layer

$$S_A = r_w <S_V> = <S_S> \tag{4.3}$$

We find that the average SS value $<SS>$, is equivalent to the area backscattering strength of the bottom echo (SA). Therefore, we can easily get the value by the normal function of the echo integration by selecting a belt like integration layer including the bottom echo.

### 4.3 Comparison of ring surface scattering model and the bottom echo integration

Comparison of the SS values by the ring surface scattering (RSS) model as shown in Fig. 4.2 and bottom echo integration methods by Eq. (2.27) are shown in Table 4.1 to confirm a good agreement. The SS values were obtained from the peak raw SV values of bottom echo and giving $\Psi$, $\Phi$, $c$, and $\tau$ using Eq. (2.28). The average SS by the bottom echo integration (BEI) model used the integration period of 0.1 nmi.

### 4.4 Frequency characteristics of bottom scattering backstrength

We examine the frequency characteristics of the bottom scattering backstrength by underwater acoustic instrument frequencies of 38, 70, and 120 kHz. The multi-frequency approach will increase the accuracy of the target (sea bottom) estimation due to different response of the target to the frequencies used.

Figure 4.4 shows the average SS for 100 number of data and standard deviation of the SS as a function of frequencies for each bottom type. The standard deviation is obtained by taking the root mean square (RMS) deviation of the average SS. There are several same data of mean diameter particle. The SS decreases with increasing frequency (Fig. 4.4). The results show that the SS for 38 kHz are 2 to 3 dB greater than for 70 and 120 kHz. This is consistent with the conclusions of Stanic *et al.,* (1988) who showed that the SS at normal incidence decreased with increasing frequency by 1.5 dB.

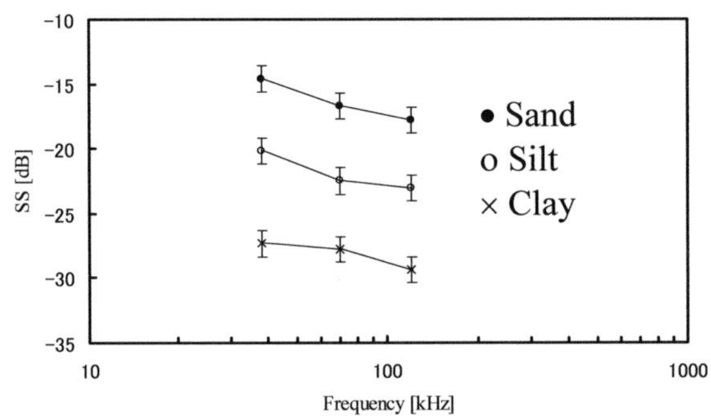


Fig. 4.4. Frequency dependence of measured bottom backscattering strength for sand, silt, and clay with ± 1 standard deviation indicated.

| Bottom material | SS [dB] | | | | | | | | |
|---|---|---|---|---|---|---|---|---|---|
| | 38 kHz | | | 70 kHz | | | 120 kHz | | |
| | BEI | RSS | Δ | BEI | RSS | Δ | BEI | RSS | Δ |
| Sand | -12.6 | -12.8 | 0.2 | -15.7 | -15.8 | 0.1 | -18.8 | -19.3 | 0.5 |
| Silt | -21.2 | -20.2 | 1.0 | -22.6 | - 22.6 | 0 | -24.0 | -24.4 | 0.4 |
| Clay | -28.3 | -28.5 | 0.2 | -28.8 | - 28.7 | 0.1 | -29.4 | - 31.4 | 1.0 |

Table 4.1. The average SS values obtained by bottom echo integration method (BEI) and ring surface scattering (RSS) method and the difference (Δ=BEI-RSS). The number of data for each bottom type is 100.

### 4.5 Bottom material characteristics

Ten bottom material samplings were conducted during the survey. The bottom materials of sand, silt, and clay were determined observing physical characteristics of the samples and mean diameter values. Particle size analysis were conducted by Wentworth scale. The mean diameter of the bottom materials from each 100 data are 223 to 301 μm for sand, 58 μm for silt, 36 to 43 μm for sand-silt-clay, and 9 to 10 μm for clay.

To examine the correlation with bottom properties, the measured average (integrated) SS values were plotted against the mean diameter of the bottom material (Fig. 4.5). The regression line shown is given by

$$SS\,[\mathrm{dB}] = 9.0 \log_{10} (d\,[\mu\,\mathrm{m}]) - 45.8 \tag{4.4}$$

with a high correlation coefficient of 0.98, where $d$ is mean diameter of particle size in $\mu$ m. The data were merged for frequencies in the regression because of the small frequency dependence, but the frequencies are discriminated in Fig. 4.5. The larger the grain size, the stronger the bottom backscattering strength.

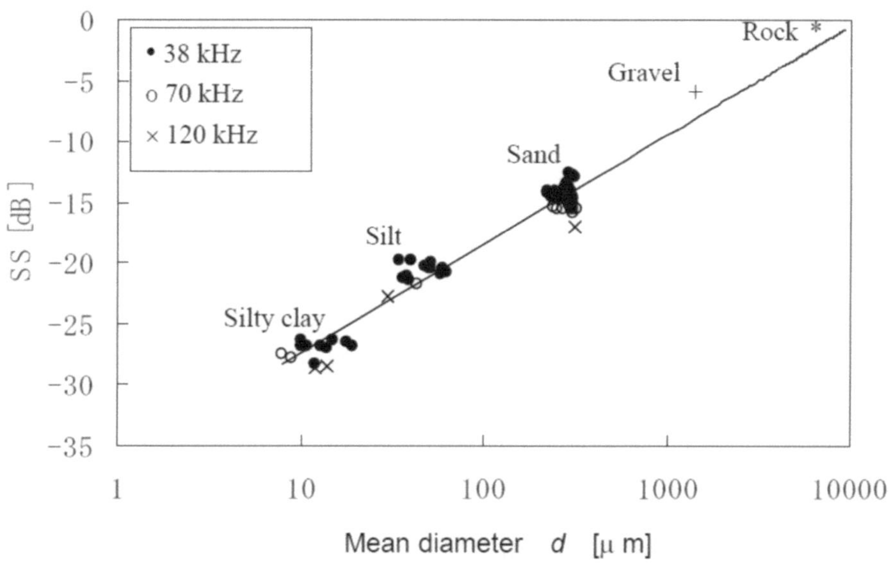


Fig. 4.5. Bottom particle diameter dependence of bottom backscattering strength. The regression line and correlation coefficient, $r_c$, are indicated.

An increase in mean diameter is accompanied by a higher backscattering strength (Fig.4.5). The reason for this is related to the bulk density of the sediments. The sand (grain sizes from 62 to 2000 μm, from the smooth sand to coarsest sand) has a bulk density between 1.9 and 2.1 g/cm³, silt and clay (grain sizes < 62 μm) have much lower values of bulk density, usually varying between 1.2 and 1.6 g/cm³. The bulk density of the sediment is determined principally from porosity. The porosity values of silt and clay are higher than sand; this is due to the fact that the silt and clay bind water more than the sand. Therefore, there are more water-filled voids per given volume of silt and clay than sand. Bulk density determines the acoustic impedance of sediment ; the higher the sediment density, the higher the impedance, and the greater the scattering backstrength.

As shown in Figs. 4.4, and 4.5 the average SS of sand is higher than silt and clay by more than 5 dB. To some extent, it was possible to relate SS to mean diameter, suggesting the possibility of bottom type classification. We examined our regression to estimate the bottom backscattering strength (SS) for rock and gravel. For this purpose, the regression line is extended without disturbing the original data. By our regression, the estimated SS is -1.3 dB for rock with diameter 8 mm. Urick (Urick,1954) used frequency of 55 kHz and found the SS of rock is -0.8 dB. The estimated SS is -6.7 dB for gravel with diameter 2 mm. The measured SS for gravel is -7.8 dB (Applied Physics Laboratory, UW, 1994).

Although the present data lack particle sizes from 70 $\mu$ m to 220 $\mu$ m, further investigation is necessary to clarify the particle size dependence of SS, Fig. 4.5 or Eq. (4.4) gives a guide to relate the SS and the bottom material.

## 5. Bottom survey off java island

### 5.1 Sea bottom depth measurement

The echo sounder measured sea bottom depths and the result of the quantitative echogram in instantaneous SS scale is shown in Fig. 5.1 as examples.

The maps show that most of the floors have irregular topography and that the surveyed areas were steeply sloping bottom. According to the depth, the survey area is the continental shelf and the other survey areas are the continental slopes. These are due to that the bottom morphology of this areas are mostly consists of troughs.

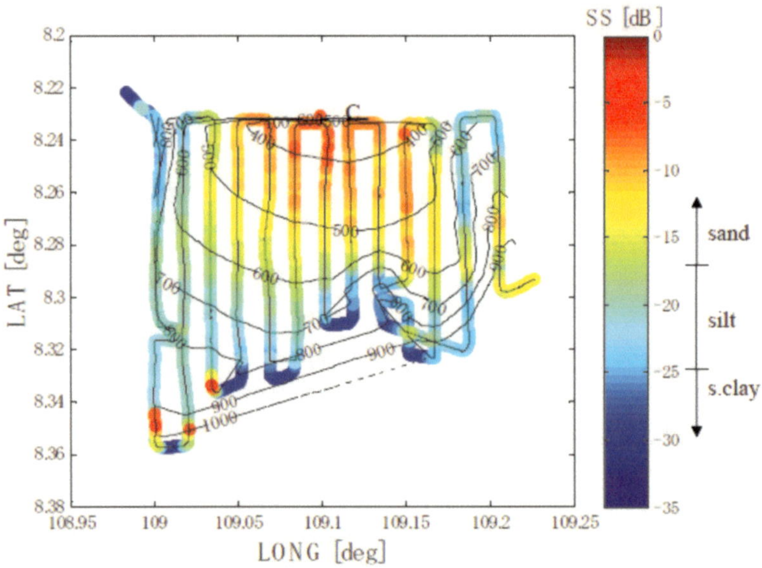


Fig. 5.1. The acoustic track, the bathymetry map (a) and SS distribution (b).

### 5.2 Bottom backscattering strength by bottom echo integration

The measurement of bottom backscattering strength (SS) using the bottom echo integration method along the acoustic track were conducted. The bottom echo integration (BEI) method measures the average bottom SS by averaging the bottom echoes for a predefined depth layer including the bottom echoes and ping sequence. We found that this process is theoretically same to the ordinary process of the echo integration to get the area backscattering strength (SA) and that the average bottom SS can be easily obtained by a commercial software such as Echoview. The SS map results for the integration period of 0.1

nmi of bottom echo are shown in Fig. 5.1. together with depth contour. The general correlation between SS and depth contour for the present survey was the increasing depth followed by decreasing SS. Figure 5.2 show the bottom trawling, SS distribution, and detected fish school (•) of the survey area.

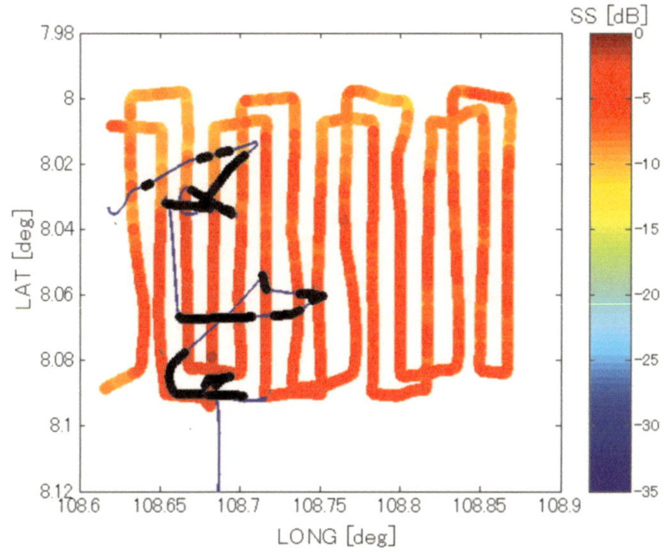


Fig. 5.2. The bottom trawling (—), SS distribution, and detected fish school (•) of survey area

### 5.3 Identifying bottom material by the SS value and relate it to fish habitat

This section shows an investigation of the relationship between sea bottom material and fish close to sea bottom information obtained by using echo sounder. From the bottom surface to upper depth is SV value and to the lower depth is instantaneous SS.

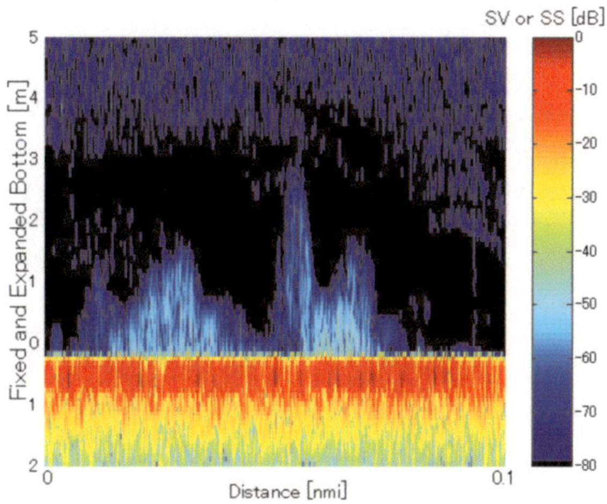


Fig. 5.3. Measured scattering strength of fish (SV) close to the sea bottom and the SS sea bottom.

Figure 5.3 shows simultaneously the scattering from fish school (SV) close to the sea bottom and the SS values by RSS model (Eq. (2.30)) as the fixed and expanded bottom at trawl survey area. The example of echograms is shown in Fig. 5.3. The increasing of fish density is followed by the increasing receiving voltage of acoustic instrument (Fig. 5.4).
The estimated bottom material is conducted by the SS value.In our results,, the SS value higher than -14.0 dB we categorized as a sand. The SS lied between – 20.0 to -15.0 dB is the silt and the SS less than -22.0 dB is as clay.

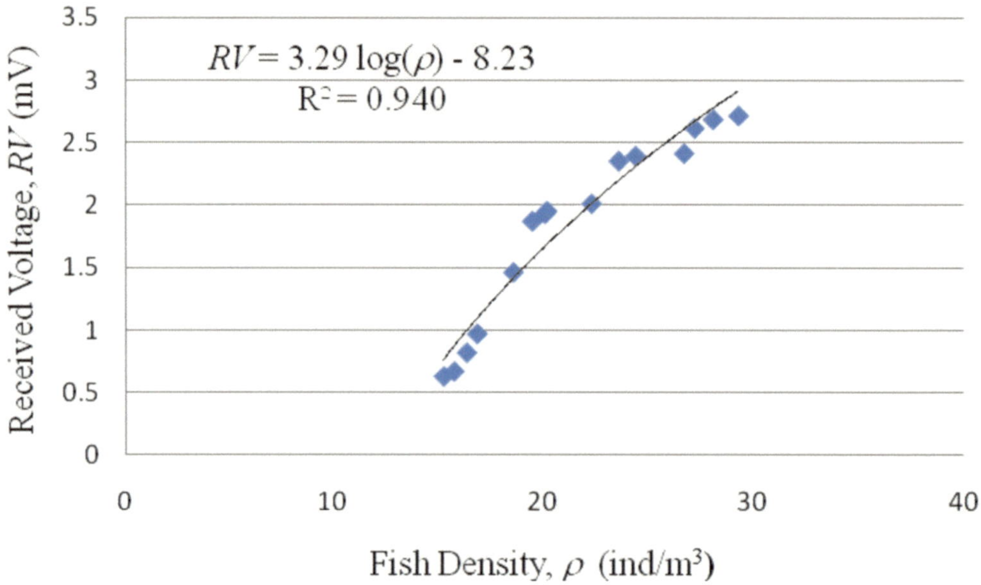


Fig. 5.4. The relationship between received acoustic voltage and fish density.

**5.4 Consideration**

The bottom material is estimated by the SS value. The measured SS is about -12.0 to -5.0 dB when the bottom fish schools were existences. By this value, the bottom material is estimated as sand. The measured SS is about -28.0 to -16.0 dB when the bottom fish schools were absent. The bottom material is estimated as silt and clay, respectively. From this results, the bottom fish schools tend to be associated with the main areas of sand. This interpretation is also supported by the trawl data showing the fish caught in this sand area. The scattering from fish (SV) and bottom (SS) are simultaneously display in echogram. Among the total 98 fish schools examined 95 schools were on the sand bottom. The SV values of fish close to the sea bottom are ranging from -70.0 to -35.0 dB.
The distribution of fish reflect variability in the surrounding environment, and those area normally occupied by a species constitute its habitat. Generally, individual species are not ubiquitous, suggesting that habitat is restrictive.
Bottom type has a major influence on the fish distribution patterns (Gibson and Robb, 1992;). Bottom type preference related to the distribution of suitable prey and the ability of fish to effectively bury themselves, which means they are protected from predation and achieve a strategic advantage in feeding (Gibson and Robb, 1992).

There may also be important differences in the availability and quality of bottom fish associated with different bottom (Gray, 1974). He reported that fish are abundant in sand and decreases in abundance in the silt and clay. The highest densities of fish in sand, since sand are stable from water currents. The other fish is abundant in silt and clay in which they feed. Our case fish is abundant in sand, may be for their living and burying to protect from predation, but further investigation is necessary.

## 6. Conclusions

The quantification technique of echoes using underwater acoustic instrument can rather easily measure echoes by reflection of sounding pulses from fish and sea bottom.
The conclusions from the results of the experiments, assessment, and analysis performed in this study are described below :

1. Underwater acoustic instrument is a reliable tool to measure bottom backscattering strength and useful when to observe both fish and sea bottom ;
2. The ring surface scattering (RSS) model expressed by the instantaneous equivalent beam angle can well predict echo shapes of the actual bottom echo ;
3. Higher bottom backscattering strength is followed by the higher mean diameter of particle as shown by Eq. (4.4) ;
4. The bottom backscattering strength a little bit decreases with increasing frequency ;
5. The SS is independent with the integration width, $r_w$ ;
6. The area backscattering strength (SA) is equivalent to the bottom backscattering strength (SS);
7. The RSS model is able to measure the SS and the result is nearly equal with the bottom echo integration model ;
8. The bottom material sampling in the study area show that bottom type is sand, silt, sand-silt-clay, and clay.
9. The bottom fish were existence with the sand bottom material ;
10. The bottom echo integration (BEI) model enables us to study bottom material in the synoptic area while the RSS model for bottom material in detail beneath the specified fish.

## 7. Acknowledgment

The research was supported by Directorate General of Higher Education Ministry of National Education Indonesia and Japan Society for the Promotion of Science (JSPS).

## 8. References

Aoyama, C., E. Hamada, M. Furusawa, K. Saito. (1997). Calibration of Quantitative Echo Sounders by Using Echoes from Water Tank Surface, *Nippon Suisan Gakkaishi.*; 63:570-577. (*in Japanese*).

Aoyama, C, E. Hamada, M. Furusawa. (1999). Total Performance Check of Quantitative Echo Sounders by Using Echoes from Sea Bottom. *Nippon Suisan Gakkaishi*, 65(1), 78-85 (*in Japanese*).

Applied Physics Laboratory, University of Washington. (1994). High Frequency Ocean Environmental Acoustic Models Handbook.

Chotiros, N.P. (1994). Reflection and reverberation in normal incidence echo sounding. *J. Acoust. Soc. Am.*; 96: 2921-2929.

Eckart, C. (1953). The scattering of sound from the sea surface. : *J. Acoust. Soc. Am.*; 25: 566-570.

Furusawa, M. (2000). New technologies for quantitative echo sounders. Meeting of Scientific Committee for Ocean Research (SCOR).

Gibson, R.N., Robb, L. (1992). The relationship between body size, sediment grain size and burying ability of juvenile plaice. Fish Biol. 40, 771-778.

Gray, J.S. (1974). Animal-sediment relationships. Oceanogr. Mar. Biol. Annu. Rev. 12:222-261.

Jackson, D.R, S.P. Winebrenner, and A. Ishimaru. (1986). Application of composite roughness model to high frequency bottom scattering. *J. Acoust. Soc. Am.*; 79: 1410-1422.

MacKenzie, K.V. (1961). Long range shallow water bottom reverberation. *J. Acoust. Soc. Am.* 34. 62.

MacLennan, D.N and E.J. Simmonds. (1992). *Fisheries Acoustics*, Chapman & Hall, ISBN-13: 978-0-632-05994-2. London.

McKinney, C.M and C.D. Anderson. (1964). Measurements of backscattering of sound from ocean bottom. *J. Acoust. Soc. Am.* 36. 1.

Stanic, S., Briggs, K.B., Fleischer, P., Sawyer, W.B. and Ray, R.I. (1989). High Frequency Acoustic Backscattering from a Coarse Shell Ocean Bottom. *J. Acoust. Soc. Am.* 85: 125-136.

Stanton,T.K : Sound scattering by marine objects. (1994). Lecture Notes. Meeting of Marine Acoustic Society of Japan. 21(4).

Stanton, T.K. : Sea surface scattering - Echo PDF. (1985). *J. Acoust. Soc. Am.* 77: 1367-1369.

Urick, R.J. (1954). The backscattering of sound from a harbor bottom. *J. Acoust. Soc. Am.* 26. 231.

# 13

# Accuracy of Acoustic Methods in Fish Stock Assessment Surveys

S. Georgakarakos[1], V. Trygonis[1] and J. Haralabous[2]
[1]*Fisheries Acoustics Laboratory, Department of Marine Sciences, University of the Aegean*
[2]*Institute of Marine Biological Resources, Hellenic Centre for Marine Research*
*Greece*

## 1. Introduction

Acoustic methods have been widely used in fisheries research for pelagic fish biomass estimation, lately including very sophisticated techniques, such us multi-frequency, wide band, multibeam, vertical and horizontal echosounding. Moreover, in the new era of ecosystem-based management, developments in acoustic technology could extend our knowledge from the stock to the ecosystem (Bertrand, 2003) and enhance our understanding of the ecosystem structure (Koslow, 2009). Until now, biologists utilised acoustic technology mainly for fish biomass estimation. Normally, acoustics are superior to other methods for pelagic fish stock assessment (Simmonds, 2003); acoustic surveys are therefore often used to tune the VPA or other classical biomass estimation methods. The main advantage of fisheries acoustics is its ability to estimate most measurements errors and provide a level of the total accuracy of the abundance, which only crudely can be asserted by the other methods. This error analysis and de-biasing approach is not easy to design and less easier to implement in a real situation. However, it is worth to investigate each error factor affecting the measurement, estimate its nature (random or systematic) or its magnitude and try to minimise its impact if possible. Finally, a procedure known as intrinsic error analysis takes into account all errors sources and estimates the total error, hence revealing the quality of the final results. Detail studies of error analysis are recently published for krill (Demer, 2004) and Norwegian spring-spawning herring assessment (Løland et al., 2007).
The chapter will review some of the most important sources of errors and their impact on acoustic biomass estimation, with emphasis on the assessment of pelagic species and the development of methods aiming at relevant de-biasing approaches. Simmonds & MacLennan (2005) reviewed this problem and provided some indicators of how much error might be expected in a typical acoustic survey, with optimum sampling design and proper instrument preparation. The expected error magnitudes are reproduced in Table 1 slightly modified. The errors are divided in two groups, those generated due to the instrumentation uncertainty and others caused by the living resource complexity of behaviour. Absolute biomass estimations expressed in weight per unit area have a higher uncertainty compared to the estimates of relative indices, namely acoustic integration values per unit area. According to Table 1, large systematic errors such as these generated by bubble attenuation, hydrographic conditions and vessel avoidance can underestimate the biomass up to 10% of its original value. However, if the same research vessel is used under similar speed, weather

conditions and instrumentation, particularly the same frequency and beam angle, it can be assumed that the relative biomass estimates are less affected by these error sources. In any case, the decision for using relative or absolute estimates should carefully be made, taking into account both the survey conditions, as well as the requirements of the stock assessment and its management.

| **A. Errors affecting both relative and absolute biomass estimation** | | | |
|---|---|---|---|
| **SOURCE OF ERROR** | **Random** | **Bias** | **Comment** |
| Instrument platform | | | |
| Instrument calibration | ± 2 to 5% | ± 2 to 5% | Worse at higher frequencies |
| Vessel motion | | 0 to 25% | Narrow beams are more biased |
| Bubble attenuation | | 0 to 90% | Keel mounted and deep towed systems are less sensitive |
| Hydrographic conditions | ± 2 to 5% | 0 to 25% | High frequencies and deep targets are more biased |
| Fish behaviour | | | |
| Target Strength | ± 5 to 25% | | Uncertainty in fish size and fish orientation |
| Species identification | ± 0 to 50% | | Depends on species and level of species mix |
| Random sampling | ± 5 to 20% | | Depends on spatial distribution and school size distribution |
| Migration | | 0 to 30% | Depends on timing between survey and fish movement |
| Vertical movements | ± 0 to 50% | | Depends on TS change due to pressure variations |
| Vessel avoidance | | 0 to 50% | Stronger in shallow areas and noisy vessels |
| **B. Additional errors affecting only the absolute biomass estimation** | | | |
| **Source of error** | | **Bias** | **Comment** |
| Instrument calibration | | ± 3 to 10% | Worse at higher frequency and narrow beam |
| Hydrographic conditions | | ±2 to 25% | High frequencies and deep targets are more biased |
| Target Strength | | 0 to 50% | Best for well investigated swimbladdered species. |

Table 1. Expected errors in well designed and properly calibrated acoustic surveys (modified from Simmonds & MacLennan, 2005)

The sampling error caused by the acoustic measurements being stochastic samples of the true mean density is affected by the survey design, the spatial structure of fish and the encountered intrinsic variability among samples. It is worth noticing, that depending on bathymetry, acoustic sampling volume and ping interval, the insonified water volume is often 10 000 – 20 000 smaller than the volume of interest in a typical survey, and therefore a high uncertainty should be expected due to this small sampling portion.

In this chapter, we focus on the three most important sources of errors which affect the assessment accuracy, namely: target strength (TS) estimation, species identification - biomass allocation per species and fish avoidance reactions to survey vessels.

## 2. Target strength estimation

The unbiased target strength (TS) estimation is a prerequisite for any approach of fish stock abundance monitoring. It is a key parameter that transforms backscattering into biomass. The TS of an individual fish depends on the sound frequency, as well as on some fish characteristics such as size, anatomy, physiology and swimming behaviour. It is suggested that environmental factors cause different morphological adaptations within species (Fässler et al., 2008) and consequentially backscattering properties. For that reason it is necessary to elaborate on multivariate relationships between TS and its possible regressors, concurrently to the biomass estimation surveys. Certain detailed investigations have been recently published on herring (Ona, 2003), anchovy (Zhao et al., 2008) and squid (Soule et al., 2010).

Modern echosounders perform a direct compensation of the directivity pattern effect to determine beam angles and size of single targets. The method is based on some "single target detector" algorithms, which filter out multiple echoes from the total echo population. The performance of these algorithms is acceptable only in very low fish density conditions (Ehrenberg & Torkelson, 1996; Simmonds & MacLennan, 2005), contrary to the biological sampling efficiency, which is higher by well defined fish layers. The resulting bias in target strength estimation due to failure to reject multiple echoes can be very high, especially in the case of small pelagic fish species, where the signal to noise ratio is lower and the expected packing densities prevent the successful operation of the single target detectors (Soule et al., 1997). Fish biomass is mixed with plankton, particularly in the tropics, and thresholding is used for reducing plankton echoes. Unfortunately, thresholding biases the target strength distribution, because a high proportion of small targets are compensated. For relative large fishes (L ~50 cm) this bias is negligible, however for small fish, such as the small pelagic species (10 cm <L<30 cm) the bias can be increased to a TS overestimation of 50% (Simmonds & MacLennan, 2005).

The *in situ* TS measurements of wild fish, despite their practical difficulties, are superior to other methods where fish is immobilised or confined within a case, since swimming behaviour affects considerably the fish body tilt angle and the backscattering echo. The goal of *in situ* measurements is to estimate an average swimming pattern, as this can be expressed in backscattering. Either implemented in software (Ona & Hansen, 1991; Schell et al., 2004) or hardware (Hedgepeth et al., 2002), single fish tracking algorithms isolate single fish echoes and provide TS point estimates as well as further information concerning the swimming pattern (Furusawa & Amakusu, 2010).

### 2.1 Target strength experiments

The acoustic data were recorded in Thermaikos Gulf during the "FISHCAL" research project, aiming at the TS estimation of the three most commercially important species, namely sardine (*Sardina pilchardus*), anchovy (*Engraulis encrasicolus*) and horse mackerel (*Trachurus trachurus*). A modified French pelagic trawl was employed for species identification and fish length measurements. The trawl was monitored by a FURUNO Net Data Recorder to increase catch performance.

The hydroacoustic measurements were carried out by use of a calibrated SIMRAD EK-500 echo-sounder (Foote & MacLennan, 1984; Bodholt et al., 1989), with mounted circular 38 and 120 kHz split beam transducers. The pulse duration was 1.0 and 0.5 ms respectively for the two frequencies. Data were collected using a -60 dB noise threshold, from standard EK-500 telegrams. Since the internal single-echo-detection (SED) algorithms are unable to reject all multiple targets within the sample (Sawada et al., 1993; Soule et al., 1995), some post processing is required, implementing tracking filters (Ona & Hansen, 1991; Handegard et al., 2005). The detection volume for tracking was calculated by applying the maximum angle (± 5 degrees) of the split beam angle detectors. However, in order to reduce the impact of the noise threshold, only beam corrections less than 1 dB were included in the final TS estimation statistics. Before tracking, all data were scrutinised and multiple targets were identified by plotting the TS [dB] measurements against the integration values $s_A$ [$m^2$ $n.mi^{-2}$].
Echo tracking algorithms were developed (initially in standard C and later in R), which scrutinise the raw data and extract the echoes with the highest possibility to belong to the same fish. User defined threshold criteria are: instantaneous and average depth angle, horizontal angle, deviations of mean depth angle and mean horizontal angle. Finally, the following parameters are calculated:

- Fish identification code and number of echoes belonging to the same fish body
- Compensated and non-compensated TS values of each single echo
- Compensated less than 1 dB and their mean acoustic cross section $\langle\sigma\rangle$
- Variability of the sampled σ, within the same target
- Angles of the fish track in relation to the transducer position

The algorithm is applied off-line on the 3-dimensional information provided by the split beam concerning each encountered echo. For each echo, the fish position relative to the ship is represented in a 3 dimensional coordinate system by the vector P as:

$$R\cos(\phi)\sin(\theta),\ R\sin(\phi)\sin(\theta),\ R\cos(\theta) \tag{1}$$

where R is the range between transducer and fish, (φ) and (θ) define the target direction relative to the acoustic axis and the vessel speed vector respectively.
The mean Target Strength <TS> is usually expressed in a linear relation to the fish length:

$$\langle TS\rangle = a * Log_{10}(L_{cm}) + TS_0 \tag{2}$$

where $TS_0$ is the intercept of the regression and represents the TS of a hypothetical fish of length L=1 cm. The scope in the following experimental design was the identification and quantification of the parameters that affect the unbiased estimation of mean TS and finally the estimation of parameters defining the linear regression in (2).
Target echoes plotted according to their beam angles (a and b) show a uniformly spatial distribution (Fig. 1a), whilst the filtered by tracking echoes reveal a longitudinal distribution, due to the ship's movement (Fig. 1b).

### 2.2 Fish tracking results

Following the tracking procedure, echoes could be identified as belonging to the same fish, once, twice or more times inside the beam. By isolating targets which occurred more than 3 times inside the lobe, it is possible to study the variation of back-scattering energy per individual fish. Mostly, increasing the sample (n) of insonifications per fish, the statistics

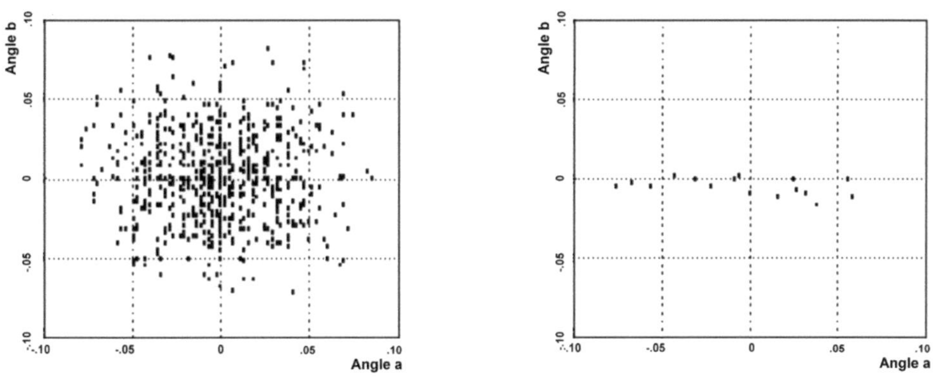


Fig. 1. Filtering echoes using tracking algorithms. 1a, left: The same dataset before filtering and 1b, right: After filtering by tracking, where each point on the angle scatterplot reveals that fish was encountered 6 times inside the beam.

remain relative stable (Fig. 2). Both mean and maximum values converge to their maximum values, where the mean difference between maximum and mean TS per fish stays stable around the 3 dB. Certain factors are responsible for the gradual change of TS values. There is a preference in the sampling procedure for bigger targets, due to their higher signal to noise ratio and increased sampling volume, especially in deeper waters. Furthermore, it is known that the fish has a specific angle which is responsible for the maximum back scattering energy and that this value is an upper limitation for the recorded TS maxima.

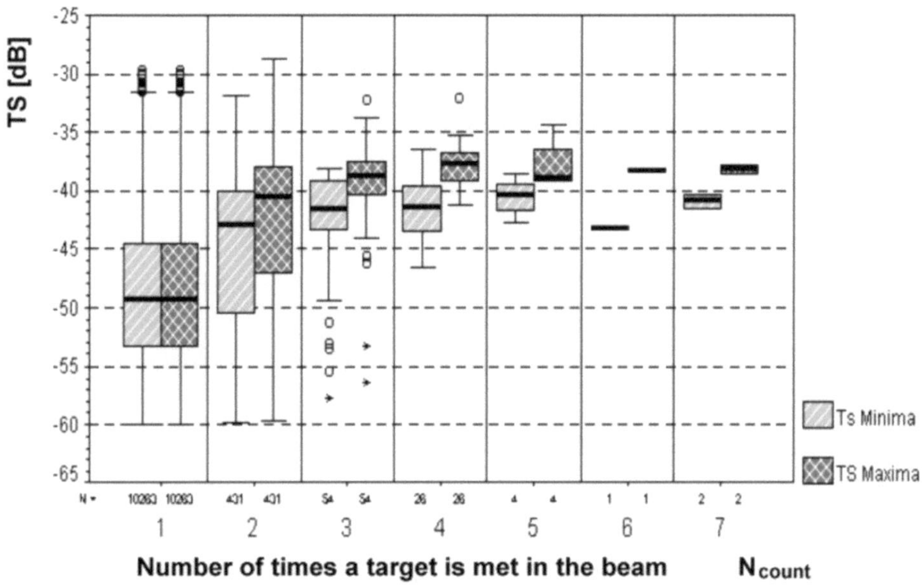


Fig. 2. Echoes encountered more times inside the beam belong to fish with higher backscattering cross sections, measured as average or as maximum observed value.

By increasing the echo samples per fish (n>3) the possibility of including multiple echoes is minimised. Unfortunately, as mentioned above, the lower voltage echoes that belong to smaller targets have fewer chances to be detected in successive pings and therefore are underestimated by tracking. This is a limitation of the tracking procedure and we next describe a de-biasing technique based on a simulation approach.

**Simulated fish tracking**

In order to study the tracking-induced bias a simulated procedure was performed, using targets with TS obeying a normal distribution (mean= -39.0 dB and sd = 10.0 dB).
After fish tracking the mode of the observed echoes, corrected for the beam effect (on-axis) is increased by 2 dB (Fig. 3), whilst the calculated TS mean:

$$\hat{TS} = \log_{10}\langle \sigma_{bs} \rangle \tag{3}$$

is overestimated by 1.5 dB. The standard deviation in the logarithmic form was reduced to 8 dB, namely adjusted to the same average dispersion level observed in the *in-situ* data.

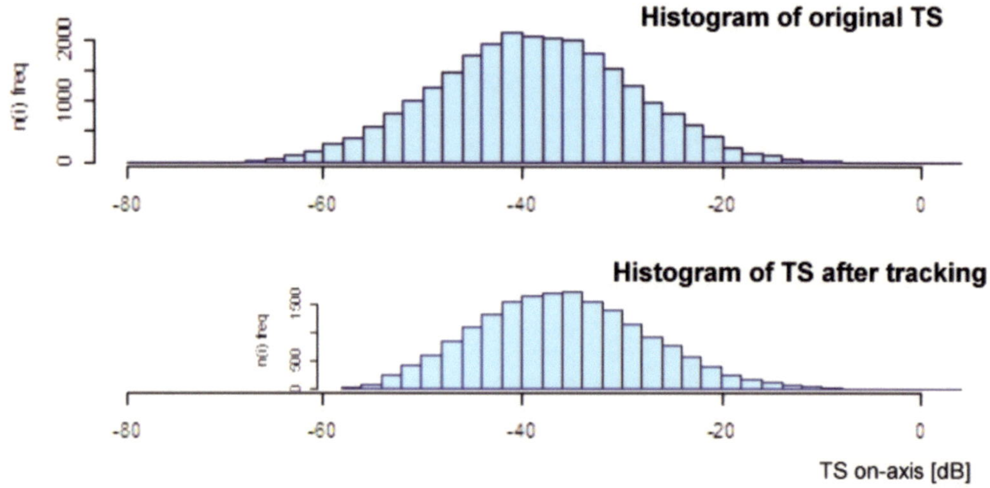


Fig. 3. Histograms of TS data: Original values before tracking (above) and "observed" values after fish tracking and beam angle correction (below).

Specifically, small pelagic species backscatter echoes with low signal-to-noise ratio (SNR). Experimental (Reynisson & Sigurdsson, 1996) as well as simulation studies (Ehrenberg & Torkelson, 1996; Kieser et al., 2005) predict that a decrease in SNR tends to increase the bias in TS estimates. Additionally to the mentioned loss of small fish near the beam border, the additive noise increases the average TS level and affects the single fish detectors of the split or dual beam techniques. The expected additional bias is about 0.2 to 0.5 dB, for a standard $7^0$ beam-width transducer, with 38 kHz pulses on targets of -30 dB size (on-axis values), insonified at 150-200 m depth (Kieser et al., 2005). For smaller targets this bias can be higher. Therefore, all TS estimations based on tracking should be finally corrected for the total error by reducing their TS values by 1.0 - 2.0 dB.
It is recommended to quantify the working limits of the fish density in order to avoid overlapping echoes (Ona, 1999; Ona & Barange, 1999). The probability to encounter multiple

echoes inside the acoustic beam is expected to be 5%, assuming a fish Poisson distribution with mean number of targets about 0.10 individuals. This relationship between integrated values $s_A$ and TS has been calculated for targets in 50 and 100 m depths (Fig. 4).
It can be seen that for a given average observed TS different maximum $s_A$ limits are allowed, according to the accepted probability to have multiple echoes (2% and 5%).

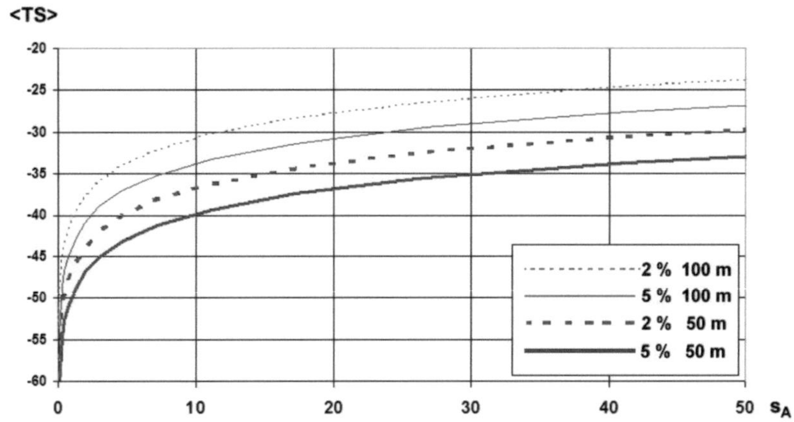


Fig. 4. Observed average TS [dB] vs. maximum allowable nautical area back scattering coefficient $s_A$ [$m^2$ $nmi^{-2}$]. Solid lines represent 5% and dashed lines 2% probability to have multiple echoes.

Further analysis including both frequencies of the isolated post-tracking single fish echoes brings out a significant relationship between the two frequencies (Fig. 5).

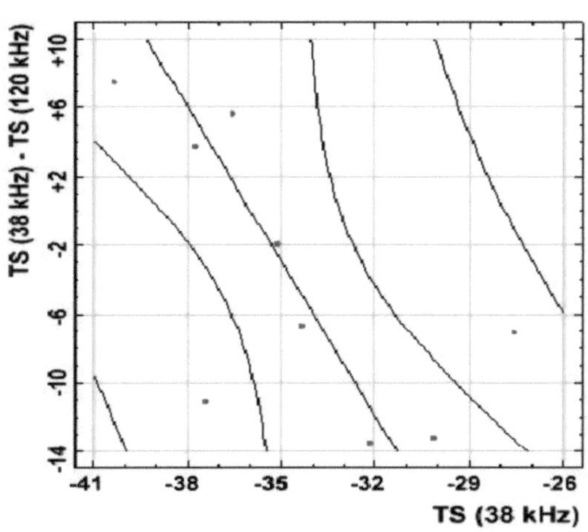


Fig. 5. Comparison of fish TS measured by two frequencies 38 and 120 kHz (detail in text).

The TS difference between 38 and 120 kHz expressed in dB is related to the echo level itself, which indicates different resonance and fish beam characteristics for different fish sizes (correlation coefficient r=0.67, p<5% ). Each point on the scatterplot represents data sets with echo tracks encountered simultaneously by 38 and 120 kHz and regarding each time the same fish. The estimated correlation coefficients among different data sets varied between r=0.38 and r=0.7.

### 2.3 TS estimations relative to the fish body length

According to common practice, especially valid for the pelagic fish species, we forced the slope of the regression between the average TS and the body length L, in equation (2), to be 20 (Simmonds & MacLennan, 2005). The resulted intercepts $TS_0$ have been summarised by categories and are displayed in the box plots of the figure 6.

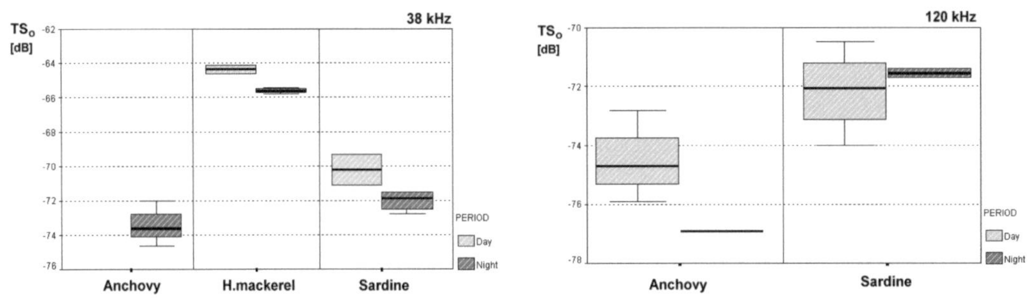


Fig. 6. Box plots of tracked echoes during *in-situ* measurements of free-swimming fish by 38 kHz (left) and 120 kHz (right) split beam transducers.

The box plots of tracked echoes (3-5 times per fish) display values not yet corrected for the bias mentioned in 2.1.1. According to these results significant TS differences exist among species, periods of the day and insonification frequency. Another important outcome of the study is that daytime acoustic data display systematically higher values compare to night time. In most cases the amount of data groups was sufficient to prove the significance of this hypothesis at the 5% level. Unfortunately anchovy was not always present, because the collected data very often showed a higher fish density and therefore were extremely biased.

### 2.4 Target strength conclusions

As is well known, the backscattering energy from the swimming fish is a dynamic quantity, highly dependent on fish behaviour, particularly the swim tilt angle (Huse & Ona, 1996). The different behaviour is also responsible for the measured systematic differences in estimated mean TS values between day and night. The validity of this finding was proved for all three target species. This phenomenon was well known from cage experiments (Simmonds & MacLennan, 2005), but for the *in-situ* measurements it is relative new (Axenrot et al., 2004) and should be taken into account during the fish stock assessment estimation.

Detailed TS estimations of sardine, anchovy and horse mackerel echoes are very rare and biomass estimations especially in Mediterranean Sea are based on a general equation established for Clupeoids (Simmonds & MacLennan, 2005) or on closely related species from other regions (Bonanno et al., 2005). Given that biomass estimation is carried out mainly applying 38 kHz, the final $TS_0$ estimations of different species are summarised in Table 2 for this frequency only.

| **Intercepts ($b_{20}$) of target strength estimations relative to fish body length** | | | |
|---|---|---|---|
| **SPECIES** | **$b_{20}$** | **kHz** | **REFERENCES** |
| Anchovy (*Engraulis encrasicolus*) | - 71.2 | 38 | ICES (1983) |
| Anchovy *(Engraulis capensis)* | - 76.1 | 38 | Barange et al. (1996) |
| Anchovy *(Engraulis japonicus)* | - 71.6 | 38 | Zhao et al. (2008) |
| Anchovy (*Engraulis encrasicolus*) | - 74.8 | 38 | This study (night) |
| Sardine (*Sardina pilchardus*) | - 72.6 | 38 | Ben Abdallah et al. (2000) |
| Sardine (*Sardina pilchardus*) | - 71.2 | 38 | ICES (1983) |
| Sardine (*Sardina pilchardus*) | - 70.4 | 38 | Patti et al. (2000) |
| Sardine (*Sardina pilchardus*) | - 74.0 | 38 | Hannachi et al. (2005) |
| Sardine (*Sardina pilchardus*) | - 72.5 | 38 | This study (day) |
| Sardine (*Sardina pilchardus*) | - 74.0 | 38 | This study (night) |
| H. mackerel (*Trachurus trachurus capensis*) | - 66.8 | 38 | Barange et al. (1996) |
| H. mackerel (*Trachurus trachurus*) | - 66.5 | 38 | This study (day) |
| H. mackerel (*Trachurus trachurus*) | - 67.0 | 38 | This study (night) |

Table 2. Comparisons of the estimated target strength values with available references for European anchovy, European pilchard and Atlantic horse mackerel or closely related species.

The $b_{20}$ intercepts for anchovy were estimated only from night time data, since day time data were biased from high fish density. The estimated value in this study (-74.8) is lower than the ICES recommendation (-71.2), however it is very likely that the day intercepts are higher, like those observed in sardine echoes. The median of night-TS of sardine is similar to the recently measured average value by Hannachi et al. (2005), whilst the day estimation is in generally lower than the rest of the published values. Regarding horse mackerel, the results are consistent with the published data of the closely related sub-species (*Trachurus trachurus capensis*).

As demonstrated by the historical cage experiments of Edwards & Armstrong (1983), the TS of a fish with swimbladder is depth dependent, following Boyle's law. Consequently, comparisons between echoes from different depths should be carefully interpreted. Unfortunately, TS is affected by the cross-section of the fish swimbladder, which does not follow its volumetric change. Furthermore, it is not clear if the swimbladder compression alone is responsible for TS reduction or additional the fish swimming behaviour changes with depth. Recently, detailed studies on the compression mechanism of the swimbladder of herring (Ona, 2003) and anchovy (Zhao, 2008; Sawada et al., 2009) demonstrated the importance and complexity of the TS variability by increasing depth.

## 3. Species identification

Traditionally, in biomass estimation surveys, acoustic backscattering is allocated to different fish species by taking into account the species composition of trawl samples. Obviously, this procedure is time consuming, subjective and not testable for bias.

The basic idea for a more objective approach derived in the early 90s, with the development of specific software for automatic extraction of school parameters and by further testing the hypothesis that school features can be used for school species identification (Scalabrin et al., 1992, 1996; Reid & Simmonds 1993; Georgakarakos & Paterakis 1993). This approach triggered the development of sophisticated expert systems for species identification, where the artificial neural networks (ANNs) provided a promising solution (Ramani & Patrick, 1992; Dunne, 2007). The first approach of fish school species identification applying ANNs was implemented in early 90s (Haralabous & Georgakarakos 1993, 1996). There were feed-forward neural networks (also called Multi-Layered Perceptrons, MLPs), using as species predictors morphometric, energetic and bathymetric school descriptors.

Meanwhile, modern systems were developed, utilising faster school scrutinisation algorithms (Reid et al., 2000, Fernandes et al., 2002), multi-frequency (Korneliussen & Ona, 2002, Korneliussen et al., 2009), multi-dimensional school descriptors (Paramo et al., 2007; Trygonis et al., 2009) and advanced artificial neural networks (Cabreira et al., 2009). In the following we describe an advanced species classification system based, among other tools, on Bayesian Neural Network techniques.

### 3.1 Data collection and pre-processing

Data were acquired in two different geographic areas, namely in Thermaikos Gulf (Northern Aegean Sea) and in North Sea (Shetland Island and Norway). Emphasis was given on cross-validating school echograms with trawl samples. Schools' images matched one by one the respective trawl samples were characterised with a species identification attribute (species id).

Greek data were obtained during three surveys (October 1996a and b, May 1997) carried out in Thermaikos Gulf, utilising a SIMRAD EK-500 echosounder with two mounted split-beam transducers (38 and 120 kHz). The surveys were mainly devoted to the selection of acoustic information related to the schools of the target species: sardine, anchovy and horse mackerel.

The second data set was obtained from two different surveys in the North Sea (Shetland Islands, July 1996) and Norway (November 1997), operating the same instrumentation and software. The school species identity has been supplied by the Marine Laboratory of Aberdeen and the Institute of Marine Research Bergen by means of trawling. Target species were herring (*Clupea harengus*), sprat (*Sprattus sprattus*), blue whiting (*Micromesistius poutassou*), Norwegian pout and 0-group pout (*Trisopterus esmarkii*).

#### 3.1.1 Software development

The developed in-house software implemented in C the following three software modules: an echo data pre-processing system, as an interface to the standard Bergen Integration System (Korneliussen, 2004), a SCHOOL algorithm for school isolation and parameterisation and certain ANNs implementations, mostly based on the Bayesian approach. The interface module allows data extraction from the echograms, using windows that cover certain school traces of interest. The interface outputs the echogram volume backscattering values together with ping and depth information to the second software module (SCHOOL), which is responsible for school isolation, parameterisation and quantification of school descriptors. Finally, these school parameters compose the input data for the Artificial Neural Networks during training or application processes. The SCHOOL module is a set of algorithms written in standard C, that are responsible for detection of school aggregations inside a given echogram or window in an echogram.

| Abbreviation | | School Descriptor | Abbreviation | | School Descriptor |
|---|---|---|---|---|---|
| **Time & Space** | | | ***Bathymetric*** | | |
| 1 | YEAR | Year | 21 | DMEAN | Mean school depth |
| 2 | DAY | Day of the year | 22 | DMIN | Minimum school depth |
| 3 | HOUR | Time | 23 | DMAX | Maximum school depth |
| 4 | LAT | Latitude | 24 | BOTT | Bottom depth |
| 5 | LON | Longitude | 25 | AMIN | Minimum altitude |
| **Morphological** | | | 26 | AMAX | Maximum altitude |
| 6 | N | Samples above threshold (N) | ***Energetic*** | | |
| 7 | H | Height (H) | 27 | EMEAN | Mean school energy (m = E / N) |
| 8 | L | Length (L) | 28 | EMAX | Max. energy sample |
| 9 | PERI | Perimeter (P) | 29 | EMIN | Min. energy sample |
| 10 | AREA | Cross-sectional area (A) | 30 | ESD | SD of energy sample |
| 11 | ELON | Elongation (L/H) | 31 | EVAR | Variance of energy sample (V) |
| 12 | CIRC | Circularity ($P^2/4\pi A$) | 32 | ECV | CV of energy sample |
| 13 | RECT | Rectangularity (LH/A) | 33 | E | Total school energy (E) |
| 14 | RMEAN | Mean radius | 34 | CROWD | Mean crowding of school ($m_c$= m +(V/m -1)(1+V/N$m^2$) |
| 15 | RMIN | Min. radius | | | |
| 16 | RMAX | Max. radius | 35 | PATCH | Patchiness of school ($m_c$ / m) |
| 17 | RVAR | Variance of radius | 36 | I | Index of dispersion (NV/ m) |
| 18 | RSD | SD of radius | 37 | K | Clumping coefficient ($m^2$ / (V-m)) |
| 19 | RCV | CV of radius | | | |
| 20 | FRAD | Fractal (2 *ln*(P/4) / *ln*A) | | | |

Table 3. Data output of the SCHOOL module

The main routine of the algorithm "scrutinises" the values of the pixels and filters out all pixels with values lower than a basic threshold (default -60 dB). The filtered pixels are then tested for vertical and horizontal continuity, according to the minimum allowed vertical and horizontal gap between neighbourly pixels inside an isolated single school. Finally, a second threshold is applied filtering out low-density aggregations (such as plankton) with very low mean integration level. A procedure similar to the SCHOOL algorithm is used for bottom detection in combination to the nominal bottom depth provided by the EK-500.

### 3.1.2 The applied neural network simulators

Certain ANNs simulators have been tested in order to assess their performance. Most of the non-Bayesian ANN implementations were based on the commercial 32-bit neural network simulator "NeuroShell 2" (Ward Systems Group©). All Bayesian ANNs presented here are developed under the UNIX environment using the SFBM simulator (Neal, 2006).

**General design of Neural Networks**

The SCHOOL module generates a set of school descriptors (Table 3), which are used as a whole or as a subset in the neural network module. Three examples of descriptor patterns of sardine schools are displayed in Figure 7.

The input layer is always a subset (13-20) of the descriptors extracted from the "known" fish schools. The amount of the available "known" schools limits the number of the concurrently in use descriptors. Bayesian Neural networks are less sensitive to this limitation. Non-Bayesian ANNs have been developed using a complex hidden layer, containing 3 sublayers. Different activation functions were applied to the 3 sublayers in order to detect different features in a pattern processed. In addition to the classical sigmoid function, a Gaussian function was used on the second sublayer in order to detect features in the mid-range of the data. In the third sublayer, a Gaussian complement was used for detecting features from the upper and lower extremes of the data set. In the output layer, each unit corresponds to a certain species, taking the value of 1 if the school belonged to this category and 0 otherwise. To measure the classification efficiency of the ANNs we compare the actual output of the network to the correct output over a number of testing trials. The most widely used method to obtain this test set is to reserve a separate representative subset of the available examples. We experimented with testing subsets ranging from 5% to 30% of the available data.

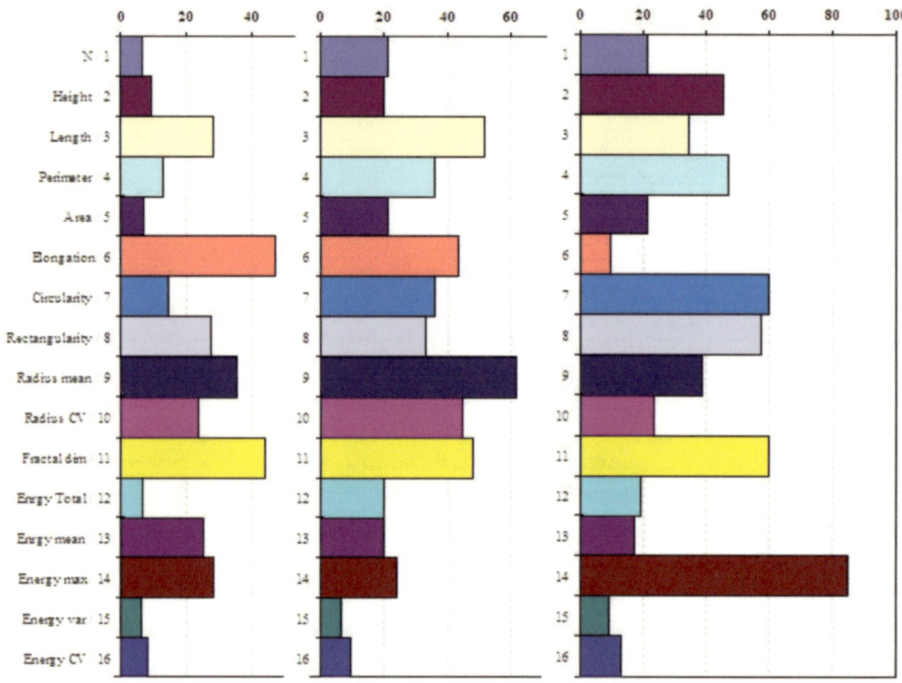


Fig. 7. Three example-patterns from sardine schools, in Thermaikos Gulf: each pattern is the vector of input values (16 descriptors) extracted using the SCHOOL module. Values are rescaled into a common range of 0-100.

For the evaluation of the accuracy of a trained ANN we calculated certain statistics, the most interesting of which is the $R^2$: the coefficient of multiple determination. $R^2$ was not the ultimate measure of whether a net produced good results, especially for classification nets such as those used. For example, if the ANN generated output values of 0.5, 0.6 and 0.4 in the three outputs, the $R^2$ value would not be very high, but the classification would be correct if the second output was the answer. In addition, a confusion matrix was established counting the number of correct classifications.

### 3.2 Results

The target species were predicted applying ANNs and Discriminant Function Analysis (DFA) methods. The results of this comparative study are summarised in Tables 4 (Greek data) and 5 (Shetland data).

| COUNTS | | | DFA predictions | | | NN predictions | | |
|---|---|---|---|---|---|---|---|---|
| | Species | TRUE | AN | SA | HM | AN | SA | HM |
| Train | Anchovy | 313 | 284 | 23 | 6 | 313 | 0 | 0 |
| | Sardine | 159 | 41 | 109 | 9 | 0 | 159 | 0 |
| | H. mackerel | 44 | 9 | 17 | 18 | 0 | 0 | 44 |
| | Total | 516 | 334 | 149 | 33 | 313 | 159 | 44 |
| | **Total correctly** | | **411 (79.65%)** | | | **516 (100%)** | | |
| Test | Anchovy | 170 | 144 | 19 | 7 | 165 | 4 | 1 |
| | Sardine | 93 | 22 | 66 | 5 | 3 | 89 | 1 |
| | H. mackerel | 25 | 11 | 8 | 6 | 2 | 0 | 23 |
| | Total | 288 | 177 | 93 | 18 | 170 | 93 | 25 |
| | **Total correctly** | | **216 (75.00%)** | | | **277 (96.18%)** | | |

Table 4. Classification results of 804 known-species "Greek" schools using both DFA and NN predictions. A total of 516 schools are used for training (~ 64%) and 288 for testing (~ 36%). Abbreviations used: AN=anchovy, SA=sardine, HM=horse mackerel.

| COUNTS | | DFA predictions | | | | | NN predictions | | | | |
|---|---|---|---|---|---|---|---|---|---|---|---|
| **Species** | **TRUE** | **0-P** | **NP** | **BW** | **HR** | **SP** | **0-P** | **NP** | **BW** | **HR** | **SP** |
| **Train** | | | | | | | | | | | |
| 0-P | 31 | 22 | 2 | 0 | 7 | 0 | 31 | 0 | 0 | 0 | 0 |
| NP | 83 | 1 | 65 | 3 | 12 | 2 | 0 | 83 | 0 | 0 | 0 |
| BW | 14 | 0 | 4 | 8 | 2 | 0 | 0 | 0 | 14 | 0 | 0 |
| HR | 111 | 6 | 10 | 0 | 95 | 0 | 0 | 0 | 0 | 111 | 0 |
| SP | 3 | 0 | 2 | 0 | 1 | 0 | 0 | 0 | 0 | 0 | 3 |
| Total | 242 | | | | | | | | | | |
| **Total correctly** | | **190 (78.51%)** | | | | | **242 (100.00%)** | | | | |
| **Test** | | | | | | | | | | | |
| 0-P | 4 | 3 | 0 | 0 | 1 | 0 | 3 | 0 | 0 | 1 | 0 |
| NP | 12 | 0 | 11 | 1 | 0 | 0 | 0 | 12 | 0 | 0 | 0 |
| BW | 3 | 0 | 0 | 2 | 1 | 0 | 0 | 0 | 3 | 0 | 0 |
| HR | 19 | 1 | 0 | 0 | 18 | 0 | 0 | 0 | 0 | 19 | 0 |
| SP | 0 | 0 | 0 | 0 | 0 | 0 | 0 | 0 | 0 | 0 | 0 |
| Total | 38 | | | | | | | | | | |
| **Total correctly** | | **34 (89.47%)** | | | | | **37 (97.37%)** | | | | |

Table 5. Classification results of 280 known-species "Shetland Islands" schools using both DFA and NN predictions. A total of 242 schools are used for training and 38 for testing. Abbreviations used: 0-P= 0-group pout, NP= Norwegian pout, BW= blue whiting, HR= herring, SP=sprat.

In the Greek data set, the neural network was very adaptive to the data during the training procedure, providing a higher classification rate (100%) than the DFA (79.65%). The classification of horse mackerel schools by DFA was the worst (40.91%). Testing the two models (ANN vs. DFA), the ANN achieved scores between 92% (horse mackerel) and 97.06 % (anchovy), whilst the DFA varied between 24% (horse mackerel) and 84.71% (anchovy).

Both models provided lower prediction scores for the classification of horse mackerel. Note however that the total amount of schools of this species is relative low (69 schools).

In the Shetland data set, some species had very few examples to satisfy both training and testing requirements, as for example no data for testing sprat schools were available. During training, the neural network achieved very high classification (100%) in all species while the DFA varied from 85.59% (herring) to 57.14% (blue whiting). The score provided for sprat was not taken into account due to its limited presence in the data set. Testing the two models, the ANN achieved 100% scores for three species but only 75% for the 0-group pout, whilst the DFA varied between 94.74% (herring) and 66.67% (blue whiting). Both models provided lower scores (75%) for the 0-group pout, which could be explained by the limited amount of available data.

### 3.2.1 Prediction probability

The application of Bayesian ANNs also provided each prediction with the probability of the inference. This allowed a decision with different confidence limits on the probability threshold of accepting or not the network prediction. An example of box plots displaying the distribution of the school-species prediction probabilities is shown in Fig. 8. The median of the provided predictions varied from 0.975 (horse mackerel) to 1 (anchovy). In all species, the distribution was skewed to the high values (close to 1).

The evaluation of the discrimination power of each descriptor was achieved by means of two coefficients, namely the "contribution factors" and the distribution of the "sigma" values (Fig. 9). The first was specific for each species and represented the weights used by the ANN after a successful application. The weights were based on the neural implementation run during the training procedure as generated from a standard ANN.

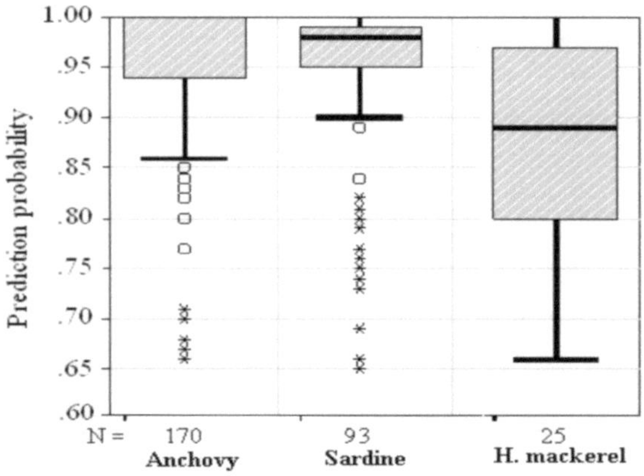


Fig. 8. Prediction probabilities given by the Bayesian NN for the "Greek" schools

The second coefficient was based on the automatic relevance determination (ARD) method and is complementary to the first one. The hybrid Markov Chain Monte Carlo (MCMC) simulation algorithm produced a sample of several network implementations, which showed the distribution of the variance of the synaptic weights (MacKay, 1992a, 1992b; Neal, 2006). It should be mentioned that, for ease of different trials comparison, the school descriptors were logarithmically transformed and standardised and that the relative importance of input descriptors increased by higher sigma values. The results showed that the two data sets contained some common and highly discriminating descriptors: mean bottom depth, mean school depth, mean and variance of school energy. In particular, the Crowding and Index of dispersion in the case of the Greek data, as well as other depth descriptors for the Shetland data could be of importance. For both data sets, the bathymetric descriptors seemed to be more important than the energetic ones while the latter were more important than the morphometric descriptors.

### 3.2.2 The generalisation problem

Bayesian networks had an increased performance in recognising schools of the same species, but encountered in different survey areas. This is important, taking into account that schools have been insonified in different depths and their morphological and energetic descriptors were distorted, due to the increased beam effect in deeper waters. This generalisation performance of the Bayesian networks has been tested during the joint project ACOUSTICS (Ona et al., 1998), applying school data collected in two different environments, namely training the network with schools from Shetland Islands (UK) and predicting schools in Norway.

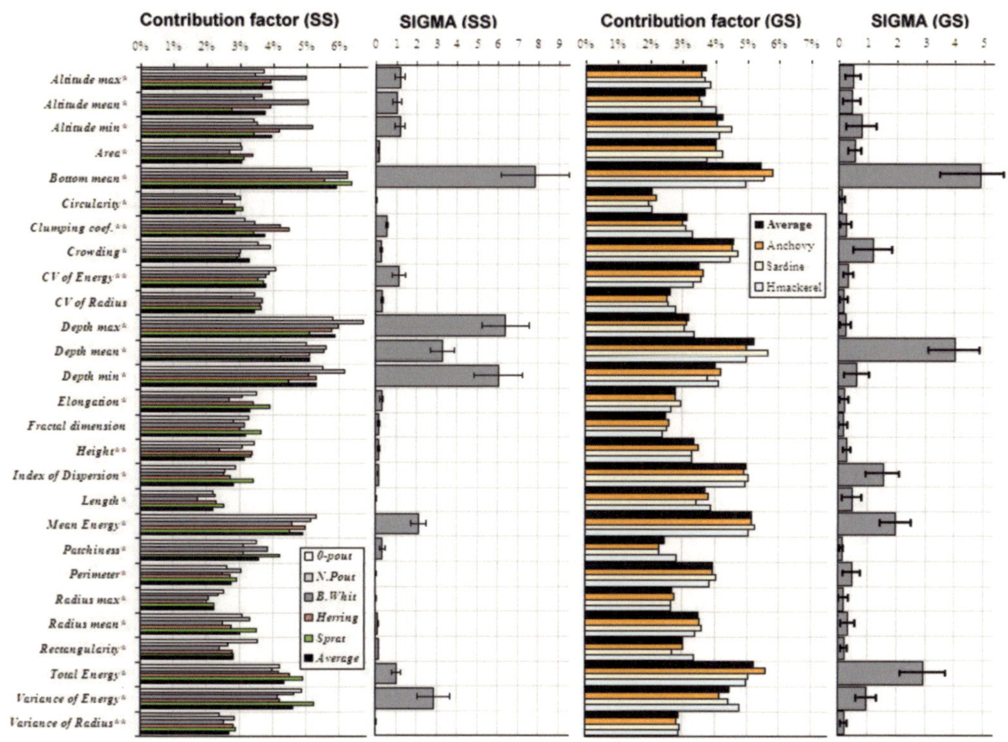


Fig. 9. Contribution factors of school descriptors extracted by the standard ANNs and sigma values extracted by the Bayesian ANNs, for the Greek (GS) and the Shetland Islands (SS) schools. The higher the contribution factor and the sigma value the higher the relevance of the descriptor. The error bars on sigma values represent the standard deviation of a distribution from 100 hybrid Markov-chain Monte-Carlo samples (* descriptor Ln transformed, ** Ln+1 transformed).

As example, the test output of a Bayesian Neural Network is summarised in Table 6a. The used network was trained and tested with data collected from the same area at Shetland in 1996. The overall correct classifications on the test set were above 97%. The performance was perfect (100%) for the classification of herring and Norwegian pout, i.e. the two species that had enough representatives in the training set (45.8% and 34.3% of cases respectively).

| **Shetland** | | | **Predicted species** | | | | | |
|---|---|---|---|---|---|---|---|---|
| **True species** | | | **0 Pout** | **N. Pout** | **B. whiting** | **Herring** | **Total correct predicted** | |
| 0 Pout | 4 | 10.5% | 3 | 0 | 0 | 1 | 3 | 75% |
| N. Pout | 12 | 31.6% | 0 | 12 | 0 | 0 | 12 | 100% |
| B. whiting | 3 | 7.9% | 0 | 0 | 3 | 0 | 3 | 100% |
| Herring | 19 | 50.0% | 0 | 0 | 0 | 19 | 19 | 100% |
| **TOTAL** | **38** | **100.0%** | **3** | **12** | **3** | **20** | **37** | **97.4%** |

Table 6a. Summary results of the species prediction of 38 test schools (14%) out of 280 total schools from the Shetland 1996 survey, using a Bayesian Neural Network

In a second experimental step, this already trained and validated network with data collected from Shetland Islands was applied on the "different" school data set collected in Norway. Therefore, this test functioned as measure of the "generalisation power" of the Bayesian Network, namely to predict the unknown Norwegian herring schools. The resulted predictions of this network on the Norway data are given in Table 6b.

| **Norway** | | | **Predicted species** | | | | |
|---|---|---|---|---|---|---|---|
| **True species** | | | **N. Pout** | **B. whiting** | **Herring** | **Total correct predicted** | |
| B. whiting | 36 | 12.29% | 1 | 13 | 22 | 13 | 36.1% |
| Herring | 257 | 87.71% | 20 | 10 | 227 | 227 | 88.3% |
| **TOTAL** | **293** | **100.0%** | **21** | **23** | **249** | **240** | **81.9%** |

Table 6b. Summary results of the species predictions of 293 Norway schools using the Bayesian Neural Network that had been trained with 5 species of Shetland schools.

When a non-Bayesian ANN was used, the percent of correct predicted herring schools was about 60%. The high prediction rate (88.3%) of the Bayesian Neural Network concerning herring schools of another survey area (Norway) emphasises its good generalisation properties. One important difference between the collected school data from the two areas was the average school depth. Taking Norwegian schools only from depths similar to those from Shetlands, the prediction score of herring has been improved from 88.3% to 97.8%, indicating that deeper encountered schools can be hard recognised by the neural networks. These results suggested that if we want to mix data from schools insonified in different depth ranges, any attempt to reduce the beam effect on school parameters is a promising way to overcome this problem (Diner, 2001; Georgakarakos, 2005).

### 3.2.3 Species identification conclusions

The presented methodology demonstrated how school species identification is feasible by extracting specific descriptors from echograms, generated from single-beam and narrow-band transducers. The identification efficiency was related to the ability of non-linear

modelling; a fact that explains the higher predictive ability of the ANNs compared to the classic discrimination procedures (DFA).
The proper application of both methods required a dataset containing the details of schools of well-defined species composition, obtained by means of aimed trawling in the target area. The problem of selecting the appropriate size of the training and the testing data sets was more crucial for the classical (DFA) and the standard ANN analysis. The Bayesian neural networks appeared to be less sensitive, which theoretical considerations lead us to expect. However, predictions based on a small number of cases (for instance horse mackerel), independent of the applied methodology, carry the risk of emphasising non-homogeneous cases relative to the rest of the statistical population. This is expressed by the larger deviation between the medians of the probability distributions. Bayesian networks due to their probabilistic approach tend to "generalise" the sampled features of schools working on wider distributions than that empirically observed. This capability can also be used to confront the problem of applying a neural network to a different spatio-temporal domain.

## 4. Fish avoidance reaction to survey vessels

Acoustic fish stock assessment methods are based on the assumption that survey vessels do not affect the behaviour and distribution of fish, and that de-biasing techniques can be alternatively applied to correct the respective errors. This is emphasised and reviewed in the earlier literature by Olsen (1990), followed by a series of studies (Aglen, 1994; Misund, 1997; Mitson & Knudsen, 2003). After the investigations of Olsen (1971) however, several researchers have reported the avoidance reaction of fish due to vessel or fishing-gear induced noise (Ona & Godø, 1990; Gerlotto & Fréon, 1992), and noted its complex dependency on species, feeding behaviour, migration, visual and auditory stimuli, hydrographic conditions, or water surface and bottom boundary interactions (Fréon et al., 1993; Levenez et al., 1990; Misund, 1990).
This behavioural pattern is known in fisheries acoustics as "fish avoidance effect", and involves an initial vertical fish movement towards the sea bottom, followed by a lateral displacement perpendicular to the approaching vessel's course. Thorough experiments with split-beam sounder buoys (Handegard et al., 2003) and direct comparisons between conventional and low-noise vessels (Ona et al., 2007) revealed several aspects of this behaviour. The process alters fish target strength via compression of the swimbladder (Boyle's law) and changes in tilt angle, reduces density in the acoustic beam compared to its undisturbed value, and further biases the acoustic observation through disposition, shape, and density distortion of fish schools near the surveying vessel. It is generally accepted that vertical echosounders are inherently prone to avoidance reaction bias due to their insonification geometry, and that vessel-induced noise is an important underlying factor, particularly at lower frequency noise bands (Vabø et al., 2002; Skaret et al., 2005). Following the ICES recommendations (Mitson, 1995), modern research vessels exhibit substantial reductions in noise levels compared to previous constructions (Mitson & Knudsen, 2003). Nonetheless, the question whether fish do or "do not avoid survey vessels" (Fernandes et al., 2000) can not yet be decisively addressed, nor its magnitude be quantitatively estimated under all possible sampling conditions.
In this context, multibeam sonars are powerful tools for studying fish school reactions to research vessels, fishing gears or predators, due to their large sampling volume and ability for long-range concurrent insonification of multiple schools. Diner & Masse (1987) used

sonar systems to record complex reactions of clupeoid schools relative to a research vessel, and Hafsteinsson & Misund (1995) reported species-specific differences: almost 20% of encountered North Sea herring schools exhibited some degree of reaction to the approaching vessel, but this behaviour was not observed in Barents Sea migrating capelin. Further scientific modification of commercial sonars provided quantitative measurements of fish school internal structure (Gerlotto et al., 2000), typology (Paramo et al., 2007) and spatial behaviour (Soria et al., 1996).

### 4.1 Multibeam experiments

The acoustic data were recorded during the AVITIS-98 survey in Greece (Thermaikos Gulf, April-May 1998) by utilising a 450 kHz RESON SEABAT 6012 multibeam sonar (60 beams over a $90^0$ swathe). The transducer was mounted on a pole on the starboard side of the research vessel with the first beam of the sonar vertical and the 60th beam horizontal by $0^0$ roll ship movements.

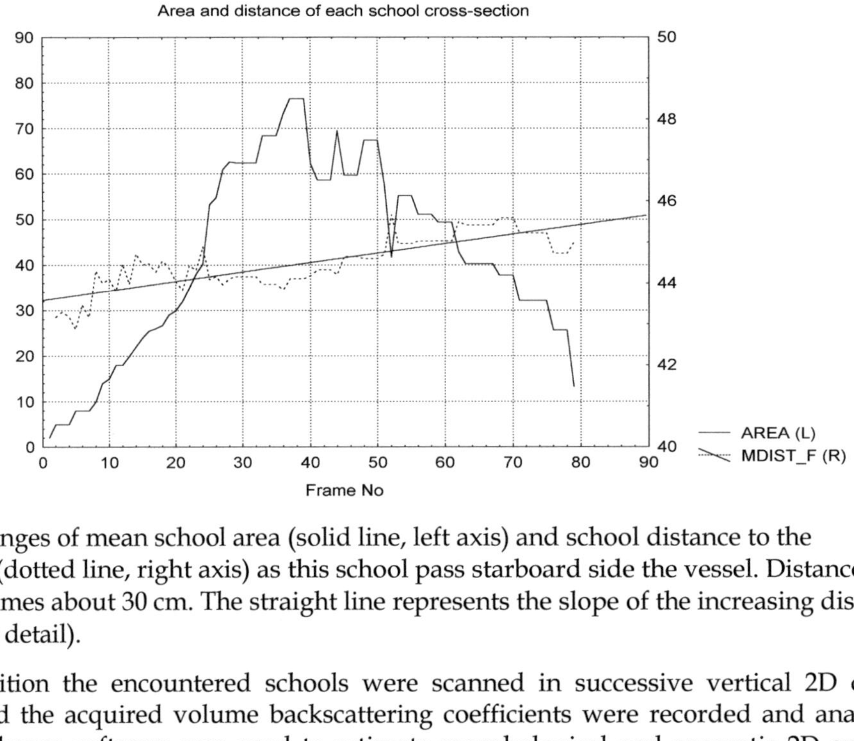


Fig. 10. Changes of mean school area (solid line, left axis) and school distance to the transducer (dotted line, right axis) as this school pass starboard side the vessel. Distance between frames about 30 cm. The straight line represents the slope of the increasing distance (see text for detail).

In this position the encountered schools were scanned in successive vertical 2D cross-sections and the acquired volume backscattering coefficients were recorded and analysed off-line. In-house software was used to estimate morphological and energetic 2D and 3D school descriptors. Most of the observed schools showed the earlier first edge (Fig.10 first frame) closer to the ship than their later and final edge (frame 80). This increasing distance indicated a mean school movement of 0.2 $ms^{-1}$ perpendicular to the vessel. A similar increasing in depth is also observed. Using joint echosounder and sonar observations, Soria et al. (2003) formulated a behavioural model for sardine and anchovy Mediterranean schools, and argued that school length variations in echosounder data are initially related to the avoidance reaction in front of the research vessel, while environmental factors like the thermocline or halocline depth may also have a subsequent effect.

### 4.2 Fish avoidance conclusions

Recent technological developments such as high resolution sonar hardware with calibrated output (Trenkel et al., 2008) or operational software systems that can quantitatively process large amounts of multibeam raw data (Balabanian et al., 2007; Trygonis et al. 2009) promise further insights into the fish avoidance effect, and the provision of *in situ* measurements of pelagic fish aggregative dynamics. Such a case is illustrated in Fig. 11, where the pelagic aggregation associated to a drifting Fish Aggregating Device (dFAD) was observed *in situ* with a vessel-mounted long range multibeam SP90 sonar; the research vessel drifted alongside the dFAD at a relatively constant distance of 400 to 500 m. Acoustic data were processed using the *Multibeam Sonar Tracer system* (Trygonis et al., 2009), which can detect, measure, and track pelagic fish schools and the dFAD echotraces, subsequently transform fish school echotrace coordinates to the Cartesian system centred on the dFAD position per ping, and reveal the spatiotemporal distribution of the aggregation in its natural, undisturbed state.

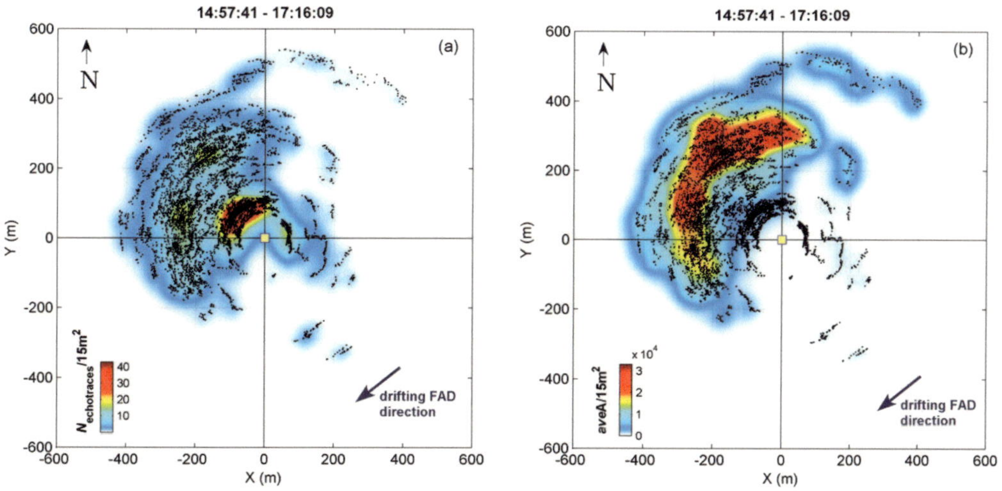


Fig. 11. Distribution of large pelagic echotraces around a dFAD in the Indian ocean, pooling all tracked acoustic observations within a sampling period of approx. two hours. Colour in Fig.11a denotes the number of tracked echotraces per 15m×15m bin ($N_{echotraces}/15m^2$), and Fig. 11b is coloured by the respective average area (aveA/$15m^2$); the drifting FAD resides at the centre of the Cartesian system.

## 5. References

Aglen, A. (1994). Sources of error in acoustics estimation of fish abundance. In: *Marine Fish Behaviour in Capture and Abundance Estimation*, A. Fernø, & S. Olsen (Eds.), 107-133, Fishing News Books, Oxford

Axenrot, T.; Didrikas, T.; Danielsson, C. & Hansson, S. (2004). Diel patterns in pelagic fish behaviour and distribution observed from a stationary, bottom-mounted, and upward-facing transducer. *ICES Journal of Marine Science*, 61: 1100-1104.

Balabanian, J.-P., Viola, I., Ona, E., Patel, R., & Gröller, M.E. (2007). Sonar Explorer: A new tool for visualization of fish schools from 3D sonar data. *Proceedings of Data Visualization - EuroVis 2007*, ISBN: 9783905673456, Norköpping, Sweden, May 23-25 2007.

Barange, M.; Hampton, I. & Soule, M. (1996). Empirical determination of in situ target strengths of three loosely aggregated pelagic fish species. – *ICES Journal of Marine Science,* 53: 225–232.

Ben Abdallah, L.; Ben Salem, S.; Gaamour, A. & El Abed, A. (2000). Estimation de la biomasse et distribution des petits pélagiques dans les eaux tunisiennes. *Atelier national organisé par le gouvernement tunisien et JICA sur le développement durable de pêcherie en Tunisie*, Mahdia 23–25 octobre 2000. AVFA, Tunis, JICA, Tokyo.

Bertrand, A.; Josse, E.; Bach P. & Dagorn, L. (2003). Acoustics for ecosystem research: lessons and perspectives from a scientific programme focusing on tuna-environment relationships. *Aquatic Living Resources* 16, pp. 197–203

Bodholt, H.; Nes, H. & Solli, H. (1989). A new echosounder system. *Proceedings of the Institute of Acoustics,* 11: 123–130.

Bonanno, A.; B., Patti; S. Goncharov; G. Basilone; A. Cuttitta; G. Buscaino; J. García Lafuente; A. García; V. Palumbo; M. Cancemi & S. Mazzola. (2005). Fluctuation of sardine and anchovy abundance in the Strait of Sicily investigated by acoustic surveys. *MedSudMed Technical Documents*, 5: 39–47.

Cabreira, A. G.; Tripode, M. & Madirolas, A. (2009). Artificial neural networks for fish-species identification. – *ICES Journal of Marine Science*, 66: 1119–1129.

Demer, D. (2004). An estimate of error for the CCAMLR 2000 survey estimate of krill biomass. *Deep Sea Res. Part II Top. Stud. Oceanogr.* 51: 1237-1251.

Diner, N. 2001. Correction on school geometry and density: approach based on acoustic image simulation. Aquatic Living Resources, 14: 211–222.

Diner, N., & Masse, J. (1987). Fish school behaviour during echo survey observed by acoustic devices. *Proceedings of the International Symposium on Fisheries Acoustics*, Seattle, USA, June 22-26, 1987

Dunne, R. A. (2007). *A Statistical Approach to Neural Networks for Pattern Recognition*, Wiley Series in Computational Statistics, ISBN: 9780471741084

Edwards J. I. & F. Armstrong (1983). Measurement of the target strength of live herring and mackerel. *FA0 Fish. Rep.* 300, 69-77.

Ehrenberg J. E. & Torkelson T. C. (1996). Application of dual-beam and split-beam target tracking in fisheries acoustics. *ICES Journal of Marine Science*, 53: 329-334.

Fässler, S. M. M.; Gorska, N.; Ona, E. & Fernandes, P. G. (2008) Differences in swimbladder volume between Baltic and Norwegian spring-spawning herring: consequences for mean target strength. *Fisheries Research*, 92: 314–321.

Fernandes P.G., Brierley, A.S., Simmonds, E.J., Millard N.W., McPhail S.D., Armstrong F., Stevenson P., & Squires, M. (2000). Fish do not avoid survey vessels. *Nature,* Vol.404, pp. 35-36

Fernandes, P.G.; Gerlotto, G.; Holliday, D.V.; Nakken, O. & Simmonds, E.J., (2002). Acoustic applications in fisheries science: the ICES contribution. *ICES Mar. Sci. Symp.* 215, 483–492.

Foote, K. G., & MacLennan, D. N. (1984). Comparison of copper and tungsten carbide spheres. *Journal of the Acoustical Society of America*, 75: 612–616.

Fréon, P., Soria, M., Mullon, C., & Gerlotto, F. (1993). Diurnal variation in fish density estimate during acoustic surveys in relation to spatial distribution and avoidance reaction. *Aquatic Living Resources*, Vol.6, No.3, pp. 221-234

Furusawa, M., & Amakusu, K. (2010). The analysis of echotrace obtained by a split-beam echosounder to observe the tilt-angle dependence of fish target strength in situ. *ICES Journal of Marine Science*, 67: 215–230.

Georgakarakos, S. (2005). Investigation of the beam effect on fish school parameters. A simulation approach. *In Proceedings of the 1st International Conference on Underwater Acoustic Measurements: Technologies and Results, 28 June – 1 July 2005.* J.S. Papadakis & L. Bjørnø (Ed), FORTH/IACM, Heraklion, Greece. 6 pp. ISBN 960-88702-08

Georgakarakos, S. & Paterakis, G. (1993). "SCHOOL": A software for fish-school identification. *ICES CM* 1993/B:8.

Gerlotto, F., & Fréon, P. (1992). Some elements on vertical avoidance of fish schools to a vessel during acoustic surveys. *Fisheries Research*, Vol.14, No.4 , pp. 251-259

Gerlotto, F., Georgakarakos, S., & Eriksen, P.K. (2000). The application of multibeam sonar technology for quantitative estimates of fish density in shallow water acoustic surveys. *Aquatic Living Resources*, Vol.13, No.5, pp. 385-393

Hafsteinsson, M.T., & Misund, O.A. (1995). Recording the migration behaviour of fish schools by multi-beam sonar during conventional acoustic surveys. *ICES Journal of Marine Science*, Vol.52, No.6, pp. 915-924

Handegard, N.O., Michalsen, K., & Tjøstheim, D. (2003). Avoidance behaviour in cod (Gadus morhua) to a bottom-trawling vessel. *Aquatic Living Resources*, Vol.16, No.3, pp. 265-270

Handegard, N. O.; Patel, R. & Hjellvik, V. (2005). Tracking individual fish from a moving platform using a split-beam transducer. *Journal of the Acoustical Society of America*, 118: 2210–2223.

Hannachi, M.S.; Ben Abdallah, L. & Marrakchi, O. (2005). Acoustic identification of small pelagic fish species: target strength analysis and school descriptor classification, *MedSudMed Technical Documents* 5: 90–99.

Haralabous, J. & Georgakarakos, S. (1993). Fish-school species identification using a neural network. *ICES CM* 1993/B:9.

Haralabous, J. & Georgakarakos, S. (1996). Artificial neural networks as a tool for species identification of fish schools. *ICES Journal of Marine Science*, 53: 173-180.

Hedgepeth, J.B.; G.E. Johnson; J.R. Skalski & J. Burczynski. (2002). Active fish sonar (AFTS) for assessing fish behavior. *Acta Acoustic United with Acustica* 88:739-742.

Huse, I., & Ona, E. (1996). Tilt angle distribution and swimming speed of overwintering Norwegian spring spawning herring. *ICES Journal of Marine Science*, 53: 863-873.

ICES. (1983). Report of the 1983 Planning Group on ICES-Coordinated Herring and Sprat Acoustic Surveys. *ICES Document CM* 1983/H: 12.

Kieser, R.; Reynisson, P. & Mulligan, T. J. (2005). Definition of signal-to-noise ratio and its critical role in split-beam measurements. *ICES Journal of Marine Science*, 62:123-30.

Korneliussen, R. J. (2004). The Bergen echo integrator post-processing system, with focus on recent improvements. *Fisheries Research*, 68: 159–169.

Korneliussen, R. J., Heggelund, Y., Eliassen, I. K. & Johansen, G. O. (2009). Acoustic species identification of schooling fish. *ICES Journal of Marine Science*, 66: 1111–1118.

Korneliussen, R. J., & Ona, E. (2002). An operational system for processing and visualizing multi-frequency acoustic data. *ICES Journal of Marine Science*, 159: 293–313.

Koslow, J. A. (2009). The role of acoustics in ecosystem-based fishery management. *ICES Journal of Marine Science,* 66: 966–973.

Levenez, J.-J., Gerlotto, F. & Petit, D. (1990). Reactions of tropical coastal pelagic species to artificial lighting and implications for the assessment of abundance by echo integration. Rapports et Procés-Verbaux des Réunions du Conseil International pour l'Exploration de la Mer, Vol.189, pp. 128–314

Løland, A.; Aldrin, M.; Ona, E.; Hjellvik, V. & Holst, J. C. (2007). Estimating and decomposing total uncertainty for survey-based abundance estimates of Norwegian spring-spawning herring. *ICES Journal of Marine Science*, 64: 1302–1312.

MacKay, D.J.C. (1992a). A practical Bayesian framework for backpropagation networks. *Neural Computation*, 4, 448-472.

MacKay, D.J.C. (1992b). The evidence framework applied to classification networks. *Neural Computation,* 4, 720-736

Misund, O.A., 1990. Sonar observation of schooling herring: school dimensions, swimming behaviour and avoidance of vessel and purse seine. Rapports et Procés-Verbaux des Réunions du Conseil International pour l'Exploration de la Mer, Vol. 189, pp. 135-146.

Misund, O.A. (1997). Underwater acoustics in marine fisheries and fisheries research. Reviews in Fish Biology and Fisheries, Vol.7, No.1, pp. 1-34

Mitson R. B. (1995). Underwater noise of research vessels: review and recommendations. *ICES Cooperative Research Report*, Vol. 209, 61 pp.

Mitson, R.B., & Knudsen, H.P. (2003). Causes and effects of underwater noise on fish abundance estimation. *Aquatic Living Resources*, Vol.16, No.3, pp. 255-263

Neal, R. M. (2006) "Classification with Bayesian Neural Networks", in *Evaluating Predictive Uncertainty* J. Quinonero-Candela; I. Dagan; B. Magnini & F. D'Alche-Buc (editors), Visual Object Classification and Textual Entailment, Springer.

Olsen, K. (1971). Influence of vessel noise on the behaviour of herring. In: *Modern Fishing Gear of the World*, H. Kristjonsson (Ed.), 291-293, Fishing News Books, London

Olsen, K. (1990). Fish behaviour and acoustic sampling. *Rapports et Procés-Verbaux des Réunions du Conseil International pour l'Exploration de la Mer*, Vol.189, pp. 147-158

Ona, E. (1999) Methodology for target strength measurements. *ICES Coop. Res. Rep.*, 235.

Ona, E. (2003). An expanded target-strength relation for herring. *ICES Journal of Marine Science*, 60: 493–499.

Ona, E. & Barange, M. (1999) Single target recognition. In: *Methodology for Target Strength Measurements (with Special Reference to In Situ Techniques for Fish and Mikronekton,* Ona E., editor.). Vol. 235. *ICES Cooperative Research Report*; 1999. p. 28-43.

Ona, E., Georgakarakos, S. & Simmonds, E. J. (1998). Acoustic properties of fish and their exploitation in classification schemes for fish stock assessment. In (ed.), *Third european marine science and technology conference MAST conference*, Lisbon, 23 27 May 1998: Project synopses Vol. 5: Fisheries and Aquaculture AIR: 1990 94 Selected projects from the research programme for Agriculture and Agro Industry including Fisheries, 241-244.

Ona, E. & Godø, O.R. (1990). Fish reaction to trawling noise: the significance for trawl sampling. *Rapports et Procés-Verbaux des Réunions du Conseil International pour l'Exploration de la Mer*,Vol.189, pp. 159-166

Ona, E., Godø, O.R., Handegard, N.O., Hjellvik, V., Patel, R., & Pedersen, G. (2007). Silent vessels are not quiet. *Journal of the Acoustical Society of America*, Vol.121, No.4, pp. 145-150

Ona, E. & Hansen, D. (1991). Software for Target Tracking with Split-Beam Echo Sounders: User's Manual. *Institute of Marine Research*, Bergen, Norway.

Paramo, J., Bertrand, S., Villalobos, H. & Gerlotto, F. (2007). A three-dimensional approach to school typology using vertical scanning multibeam sonar. *Fisheries Research*, 84: 171–179.

Patti, B.; Mazzola, S.; Calise, L.; Bonanno, A.; Buscaino, G. & Cosimi, G. (2000) Echo-survey estimates and distribution of small pelagic fish concentrations in the Strait of Sicily during June 1998. *GFCM/ SAC Working Group on Small Pelagics*, Fuengirola, Spain, 1–3 March 2000. 8 pp.

Ramani, N. & Patrick, P. (1992). Fish detection and classification using neural networks—some laboratory results. *IEEE Journal of Oceanic Engineering*, 17: 364–368.

Reid, D. G.; Scalabrin, C.; Petitgas, P.; Massé, J.; Aukland, R.; Carrera, P. & Georgakarakos, S. (2000). Standard protocols for the analysis of school based data from echo sounder surveys. *Fisheries Research*, 47: 125–136.

Reid, D. G. & Simmonds, E.J. (1993) Image analysis techniques for the study of school structure from acoustic survey data. *Can. J. Fish. Aqu. Sci.* 50, 1264-72.

Reynisson, P. & Sigurdsson, Th. (1996). Diurnal variation in acoustic intensity and target-strength measurements of oceanic redfish (Sebastes mentella) in the Irminger Sea. *ICES Demersal Fish Committee. ICES C.M.* 1996/G:25, 15 pp.

Sawada, K.; Furusawa, M. & Williamson, N. J. (1993). Conditions for precise measurements of fish target strength in situ. *Journal of the Marine Acoustical Society of Japan* (E), 20: 15–79.

Sawada, K.; Takahashi, H.; Abe, K.; Ichii, T.; Watanabe, K. & Takao, Y. (2009). Target-strength, length, and tilt-angle measurements of Pacific saury (*Cololabis saira*) and Japanese anchovy (*Engraulis japonicus*) using an acoustic-optical system. *ICES Journal of Marine Science*, 66: 1212–1218.

Scalabrin, C.; Diner N.; Weill A.; Hillion A. & Mouchot M-C. (1996). Narrowband acoustic identification of monospecific fish shoals. *ICES Journal of Marine Science*, 53:181-188.

Scalabrin, C.; Weill A. & Diner N. (1992). The structure of multidimensional data from acoustic detection of fish schools. In *Weydert M. (Ed.) European Conference on Underwater Acoustics*. Elsevier Applied Science, London, pp. 141-146.

Schell, C.; Linder, S. P. & Zeidler, J. R. (2004). Tracking Highly Maneuverable Targets With Unknown Behavior. *Proceedings of the IEEE*, 92, NO. 3.

Simmonds E. J. (2003) Weighting of acoustic- and trawl-survey indices for the assessment of North Sea herring. *ICES Journal of Marine Science* 60: 463-471.

Simmonds, E. J. & MacLennan, D. N. (2005). *Fisheries Acoustics: Theory and Practice*, 2nd edn. Blackwell Publishing, Oxford. 456 pp.

Skaret, G., Axelsen, B.E., Nøttestad, L., Fernö, A., & Johannessen, A. (2005). The behaviour of spawning herring in relation to a survey vessel. *ICES Journal of Marine Science*, Vol.62, No.6, pp. 1183-1192

Soria, M., Bahri, T., & Gerlotto, F. (2003). Effect of external factors (environment and survey vessel) on fish school characteristics observed by echosounder and multibeam sonar in the Mediterranean Sea. *Aquatic Living Resources*, Vol.16, No.3, pp. 145-157

Soria, M., Fréon, P., & Gerlotto, F. (1996). Analysis of vessel influence on spatial behaviour of fish schools using a multi-beam sonar and consequences for biomass estimates by echo-sounder. *ICES Journal of Marine Science*, Vol.53, No.2, pp. 453-458

Soule, M.; Barange, M. & Hampton, I. (1995). Evidence of bias in estimates of target strength obtained with a split beam echosounder. *ICES Journal of Marine Science*, 52: 139–144.

Soule, M.; Barange, M.; Solli, H. & Hampton, I. (1997). Performance of a new phase algorithm for discriminating between single and overlapping echoes in a split-beam echosounder. *ICES Journal of Marine Science*, 54: 934–938.

Soule, M. A., Hampton, I. & Lipiński, M. R. (2010). Estimating the target strength of live, free-swimming chokka squid *Loligo reynaudii* at 38 and 120 kHz. *ICES Journal of Marine Science*, 67: 1381–1391.

Trenkel, V.M., Mazauric, V., & Berger, L. (2008). The new fisheries multibeam echosounder ME70: description and expected contribution to fisheries research. *ICES Journal of Marine Science*, Vol.65, No.4, pp. 645-655

Trygonis, V.; Georgakarakos, S. & Simmonds, E. J. (2009). An operational system for automatic school identification on multibeam sonar echoes. *ICES Journal of Marine Science*, 66: 935–949.

Vabø, R., Olsen, K., & Huse, I. (2002). The effect of vessel avoidance of wintering Norwegian spring spawing herring. *Fisheries Research*, Vol.58, No.1, pp. 59-77

Zhao, X.; Wang, Y. & Dai, F. (2008). Depth-dependent target strength of anchovy (*Engraulis japonicus*) measured in situ. *ICES Journal of Marine Science*, 65: 882–888.

14

# Side-Scan Sonar as a Tool for Seafloor Imagery: Examples from the Mediterranean Continental Margin

Alessandra Savini
*Dept. of Geological Sciences and Geotechnologies, Milano-Bicocca University*
*Italy*

## 1. Introduction

The majority of the ocean world's features are being discovered, identified, characterized and imaged by their interactions with sound (Medwin and Blue, 2005). Often a specifically designed sound source is used to learn about the sea and its boundaries (Medwin and Clay, 1998), as in the case of the use of acoustical geophysical devices that provide information about the seafloor and the strata below, transforming the things that we cannot see into numerical data and pictures which give us a model that is able to visually represent the seabed and to outline its physical proprieties and processes (Morang et al., 1997). Such a model, based on the behaviour of the sound pulses and their interaction with water, sediment, rocks and the whole marine environment in which they are emitted, reveals diverse seascapes from shallow water to deep sea and allows us to determine the nature and characteristics of the seabed and additionally, to promote a wide range of practical applications. The research in the marine environment (from oceanography to marine geology, benthic ecology, marine archaeology etc.) now depend heavily on such tools, that provide, as the first step in obtaining the information, the geological framework that can further promote focussed scientific investigations and also better drive seabed utilization for social and economical purposes.

The indirect techniques of investigation used for exploration and study of the submerged environment are different and change according to particular research purposes (Jones, 1999). Generally, acoustical geophysical devices are sound sources that work by transmitting and receiving the sound waves that are reflected, diffracted or scattered off the bottom. They differentiate themselves from each other by the different frequencies employed and thereby the different information that are able to provide about the seabed's proprieties (table 1, Morang et al., 1997). Echo-sounders, side-scan sonar and sub-bottom profilers are the three principal acoustic devices, used to map the seabed in sea exploration programmes, collecting geophysical data that in turn produce morphological and sedimentary models which became instrumental in determining and understanding the geomorphology and the present-day (Holocene) sedimentary processes of the investigated areas. Echo-sounders with a single-beam (*Single-Beam Echo-Sounder* – SBES) are used for hydrographic (bathymetric) surveys. The *Multi-Beam Echo-Sounders* (MBES) represent an improvement upon the traditional SBES and allow the reconstruction of high resolution 3D views of submarine structures and of topography. The side-scan sonar provides an image (*sonograph*) from which an understanding of the nature of materials on the seafloor can be drawn (Fish and Carr, 1990; Blondel and Murton, 1997; Blondel, 2009). The *sub-bottom profilers* are used for examining the high resolution sismo-stratigraphy of the seabed for better interpreting sedimentary processes acting upon the seafloor (Damuth, 1980; Flood 1980).

Imaging of the seabed started in the 1940s, when the first side-scan sonar (SSS) sonograph (hard copy output of sonar data) was recorded (Fish and Carr, 1990). The first sonographs had low resolution and were only able to detect and to image large physical targets on the seafloor, such as big shipwrecks. Then, the rapid developments in acoustic electronics of the 1970s and 1980s led to the production of high-resolution images of the seabed. The SSS system is considered the most relevant acoustic device able to produce images of the seafloor: pursuant to its emission of a pulse of acoustic energy, it amplifies and records the intensity of backscattering from the seafloor, generating the sonograph, that may be considered analogous to a continuous aerial photograph (see Fish and Carr, 1990; Blondel and Murton, 1997 and Blondel, 2009 for details in generating a sonograph).

We can distinguish between high frequency SSS systems, that work roughly between 100 and 1000 kHz (Table 1), and low frequency systems that operate with lower frequencies, down to less than 10 kHz (see Table 1 from Blondel 2009 for the technical specifications of the most common low frequency SSSs).

High frequency SSSs generally work between 25 and 600m of operated range and sonograph resolution is range dependent. The choice of the operated range depends on the purposes of the survey and on the environment conditions. A number of sonographs, recorded by high frequency SSS systems during different national and international research programmes, have been recovered and examined in order to produce a representative collection of sonographs collected within the Mediterranean setting. The dataset presented in this chapter was ground-truthed with sediment samples and/or video data acquired by means of *Remotely Operated Vehicles* (ROVs). This chapter does not pretend to be an exhaustive list of everything from the Mediterranean seascape, although the sonographs here presented will attempt to outline the most significant features that characterise the Mediterranean continental shelves. A brief description of the side-scan sonar performance and its technical specifications is given in the following paragraph, to help in the understanding of the organization of the data presented, although it is recommended to refer to authoritative books (such as Fish and Carr, 1990; Blondel and Murton, 1997 and Blondel, 2009) for more detailed and precise explanations on the technical aspects related to the acquisition and processing of sonar imagery.

## 2. The side-scan sonar as a tool to image the seafloor

SSS has been defined as an acoustic imaging device used to provide wide-area, high-resolution pictures of the seabed (Kenny et al., 2003). This technique was developed by Professor Harold Edgerton and others in the 1960s and is based on the *Anti-Submarine Detection Investigation Committee* (ASDIC) system built during World War II to detect submarines (Fish and Carr, 1990; Jones, 1999). Its first applications mainly pertained to the

| Brand | Name of product | Actual frequency kHz | Max range m | Max depth rating | Max length of cable | Horizontal beam width | Vertical beam width | Max horizontal resolution | Max operating speed |
|---|---|---|---|---|---|---|---|---|---|
| C-MAX Ltd. | CM2 | 325/780 | 150 | 2000m | 3000m | 0.2° | Full below horizontal | 4 cm | 1-12 kn |
| DSME E& R Ltd. | SonarBeam S-150 series | Single frequency: 100, 400, 900, 1250 Dual frequency: 100/400; 400/900; 400/1250 | 500 | 300m | 350m | 100kHz:1.2° 400kHz:0.3° 900kHz:0.3° 1250kHz: 0.3° | 100kHz:40° 400kHz:40° 900kHz:40° 1250kHz: 30° | Range 100m – 3cm | 8 kn |
| EdgeTech | 4125 | Choice of 400/900 or 600/1600 | 150 | 200m | 600m | 0.2° | 50° | 0.5 cm | 10 kn |
| EdgeTech | 4200 | Choice of 100/400 or 300/600 or 300/900 | 500 | 2000m (SS); 500 (Allum.) | 10000m | 0.2° | 50° | 1 cm | 10 kn |
| GeoAcoustic A Kongsberg Company | 2094 Digital | 114/410 | 600 (114kHz); 150 (410kHz) | 1000m (2000m optional) | 6000m | 1° (114kHz) 0,3° (410kHz) | 50° (114kHz) 40° (410kHz) | 4 cm | 1-12 kn |
| Imagenex Technology Corp. | SportScan | 330 or 330/800 | 120 | 30m | 60m | 0,7° to 1,8° | 30° to 60° | Range scale÷250 (both side displayed) Range scale÷500 (one side displayed) | 5 |
| Imagenex Technology Corp. | YellowFin | 260/330/ 770 nominal | 200 | 300m | 600m | 0,5° to 2,2° | 30° to 75° | Range scale÷1000 | 5 |
| IXEA | SHADOWS mapping sonar | 100/300 | 300 | 300m | 1200m | 6° | 60° | 15*15cm square pixel constant across and along track | |
| JW Fisher Mfg. | SSS – 100K/600K | 100 and 600 | 600 and 75 | 150m | 300m | 1° and 1° | 40° and 40° | 6cm and 3cm | 3 |
| L-3 Klein Associated, Inc. | System 3000 digital side-scan sonar | 100/500 | 600 (100kHz); 150 (500kHz) | 1500m | 300m coaxial; 5000m armoured | 0,7° (100kHz) – 0,21° (500kHz) | 40° | 2,5cm | - |
| L-3 Klein Associated, Inc. | System 3900 dual-frequency side-scan sonar | 445/900 | 150 (445kHz); 50 (900kHz) | 200m | 250m Ltwt. coaxial cable | 0,21° (900kHz) – 0,21° (445kHz) | 40° | 7,5cm | - |
| L-3 Klein Associated, Inc. | System 4000 multibeam side-scan sonar | 455 | 125 | 200m | 250m Ltwt. coaxial cable; 900m armoured | Dynamically focused, 40° Vertical, tilted 20° down | 40° | Along track 20cm @75m, 30cm@125m (across track) determined by selected pulse length (from 3.75cm to 15cm) | - |
| L-3 Klein Associated, Inc. | Klein Hydrochart 5000 | 455 | 250 | - | - | | 0.4° | 10cm at 38m, 20cm at 75m, increasing to 36cm at 150m and 61cm at 250m | - |
| L-3 Klein Associated, Inc. | System 5000 V2 Multibeam Side-scan Sonar | 455 | 250 | 500m | 250m Ltwt. coaxial cable; 900m armoured | 10cm along track or 20cm across track - focused | Nominal 40° | 10cm at 38m, 20cm at 75m, increasing to 36cm at 150m and 61cm at 250m | 2 to 10 kn |

| | | | | | | | | | |
|---|---|---|---|---|---|---|---|---|---|
| Systems Engineering & Assessment Ltd. | SWATHplus | 117/234/ 468 | 350 (117kHz); 200 (234kHz); 90m (468kHz) | 1000m | - | 1,7° (117kHz); 1,1° (234kHz); 1,1° (468kHz) | 240° | 5cm (117kHz); 2cm (234kHz); 1cm (468kHz) | 6kn |
| Tritech International | StarFish 450F | 450 | 1m to 100m | 50m | 20m 50m available | 1,7° | 60° | - | - |
| Tritech International | StarFish 452F | 450 | 1m to 100m | 50m | 20m 50m available | 0,8° | 60° | - | - |
| Tritech International | StarFish 990F | 1000 | 1m to 35m | 50m | 20m 50m available | 0,3° | 60° | - | - |
| Teledyne Benthos | C3D-LPM Side-scan Imaging with Bathymetry | 200kHz standard 100kHz optional | 25-300m (200kHz) 25-600m (100kHz) | Surface vessel mounted | 20m | 1° | 100° | Side-scan across track 4,5cm; Bathymetry across track 5,5cm | 10kn |
| Teledyne Benthos | C3D-TOW Side-scan Imaging with Bathymetry | 200kHz standard 100kHz optional | 25-300m (200kHz) 25-600m (100kHz) | 3000 | 200m Kevlar; 10000m double armoured cable | 1° | 100° | Side-scan across track 4,5cm; Bathymetry across track 5,5cm | 10kn |

Table 1. Examples of some of the most common models of high frequency SSS systems and their different technical specifications (data are from a survey of products carried out by *Hydro International Journal* and available at http://www.hydro-international.com/productsurvey).

search for objects on the seabed, subsequent works have led to far-reaching advances in the use of sonar in marine geophysics and geology (see Belderson et al., 1972; McQuillin and Ardus 1977; Blondel and Murton, 1997; Jones 1999 and Blondel, 2009 to get an historical perspective on the evolution of side-scan sonar systems applied to the exploration of the seafloor).

Due to its capacity to provide seabed images, SSS has become an instrumental tool in seafloor mapping, with a number of applications, such as:

- Investigation of seafloor morphology and sediment characteristics (i.e. occurrence of reliefs, depressions, sedimentary structures etc.)
- Map the distribution of marine sediments and even peculiar biocenosis like seagrass meadows, coral banks etc.
- Detect specific targets on the seafloor such as shipwrecks, mines, sunken objects, etc.
- Identify the proper placement of infrastructures (oil drilling, platforms, cables etc.)
- Monitoring the seafloor for environmental applications, such as coastal and deep environment management.

The system typically consists of an underwater transducer connected via a cable to a shipboard recording device. The emitting lobe of sonar energy (narrow in azimuth from the transducer) has a beam geometry that insonifies a wide swath of the seabed, particularly when operated at low frequencies (<100 kHz). The returning echoes from the seafloor are received by the transducers over a short period of time (from a few milliseconds to 1 s), amplified on a time-varied gain curve and then transmitted to the recording unit. Most technological advances relate to the control of the phase and amplitude of the emitting sonar signal, and in the precise control of the time-varied gain applied to the return signals. In the case of a non-digital transducer, the analogue signal is converted into a digital format, the proper position for each signal in the final record (pixel by pixel) is calculated and then these echoes are printed on electro-sensitive or thermal paper, one scan or line at a time. Modern high (dual) frequency digital SSS devices offer high-resolution images of the seabed on which objects in the order of few centimetres may be detected at a variable range, from a few tens of metres up to 600 m either side of the tow fish (in this latter case the total swath width is 1200 m and horizontal resolution can reach less then 100cm).

The sonographs present as grey level (or optionally as colour scale) the local backscattering coefficients. The different backscattering intensities received by the sonar depend on a number of physical (the frequency of the system - which significantly varies from 100 up to 1000 kHz - and other technical configurations), geometric (angle at which the acoustic wave reaches the seafloor) and geological parameters (porosity and roughness of the seafloor that define the acoustic proprieties of the surface sediments in accordance to their physical and chemical nature). A copious literature (see, among others, Jackson and Briggs, 1992; Boyle and Chatiros, 1995; Blondel et al., 1998; Goff et al., 2000; Holland et al., 2000) has been developed to define the dependence of the acoustic backscattering on the different factors mentioned before, providing *backscattering models*. Nevertheless, no consensus exists on which model is the better one for side-scan sonar backscattering and research is still very active (Blondel and Murton, 1997; Blondel, 2009). Besides, different seabed types can be seen on the sonographs as different textures of the various parts of the image. In an effort to make side-scan sonar interpretation a function of quantitative analysis, a lot of authors (among others see Blondel et al., 1998; Huvenne et al., 2002; Cochrane and Lafferty, 2002) have reported the use of some image texture processing algorithms (i.e. co-occurrence matrix) to discriminate the regions of different seabed types.

Since different seabed types can be seen on the sonographs as different textures on the various parts of the image; the first step of sonograph interpretation consists of the discrimination of the different textures that compose the sonographs that in turn produce the identification of different *acoustic facies*. This process can be carried out through supervised or unsupervised automatic classification (of image textures) operated by proper software or by human discrimination of different textures, which is often the case, although qualitative interpretations of acoustic imagery are increasingly supported by quantitative analyses of backscatter and by the study of the role that the seafloor plays in its variations (Loiacono et al., 2009). The association of ground-truth information (sediment samples and/or video recording and still images) to the *acoustic facies* that have been picked out with both the analysis, provide then aid classification of the seabed.

### 2.1 The side-scan sonar performance

Precision and accuracy of sonographs depend on a number of factors and they are determined according to SSS performance. Defining the SSS performance is thus necessary to understand what SSS can image from the seafloor environments and how. For instance, the resolution is an important feature in sonographs, because it provides the measure of the detail that is possible to detect on the sonograph. The resolution of SSS can be meant in a different way:

- The *system resolution*: governed by the shape of the acoustic beam and the length of the transmitted pulse (Jones, 1999). So it depends on the three-dimensional distribution of the acoustic energy of the system that affects the size of the *footprint*. If the side-scan sonar has low frequency sources (10-30 kHz), the sound pulse will be transmitted and received at long range (covering a large area in a short time), but this configuration provides lower resolution in the sonographs. If the needs of the survey are for accurate, fine image details, then it is better to use a higher frequency sonar (100 – 500 kHz), even if the short wavelengths generated cannot be transmitted long distance and thereby limit the usable range. For this reason these kinds of systems are employed mainly on continental shelves. Flemming et al. (1976) well described the characteristic *transverse resolution* and *range resolution* of the side-scan sonar systems; they can be meant as across-track footprint and along-track footprint respectively (Blondel and Murton, 1997). Obviously each transducer capability can offer different resolution, depending on the technical specification of each sonar (Table 1) and also the speed of the transducer in the water influences the along-track resolution. Lastly, for each system, the across-track resolution strictly depends on the range setting employed (nowadays, for new models, the along-track resolution can be independent of range).

- The *sonograph resolution*: governed by the pixel size. Indeed the successive across-track profiles are 'plotted' into sonographs by computation of square picture elements (pixels).

The resolution is not the only parameter that affects the system performance and data quality. The underwater environment (current, density, salinity etc.) can considerably impact the data, and even more importantly, so too can the operator management of the system. Survey vessel course, transducers speed, tow fish altitude above the seabed and the range setting will determine the quality of the sonar data, according to the aim of the survey.

## 3. Data and methods

The data presented here were collected at different sites located on the continental shelf of the Mediterranean sea. Different range settings are presented, illustrating the different resolution capacity of the employed high frequency (100 – 500 kHz) sonar systems. The sonographs are arranged in three paragraphs, according to their range, for a total of three sections (wide-, medium-, small-range). An additional paragraph discusses some peculiar situations that can occur during a side-scan sonar survey, as noise comes from the underwater environment (e.g. signal generated from other ship propellers and/or from interference with other acoustic sources, artefacts created by objects or fishes in the water column). In each sonograph, where it is not specified, high levels of backscatter are shown as dark tones. Each section is introduced with an overview on the different types and spatial scale of seafloor features that can be investigated, according to the employed range setting.

It is worth noting that the range setting also determines the operational conditions of the survey, because the tow fish must be located at a precise distance above the bottom, approximately equivalent to between 10% and 20% of the range. So, the geomorphological setting and the survey environment do not always allow a free choice of the range setting. With complex topography and in a deep environment it is not possible to use narrow range that requires to have the tow fish at a short distance from the bottom (this situation indeed does not guarantee the safety condition of towed equipment), to operate in a deep environment, high frequency SSS must be integrated within AUV or specific ROV. In shallow water we cannot use a wide range, because it is recommended to have more than 10% of the range setting as a distance above the bottom. Therefore, the range setting not only influences the aim of the survey (because of the associated sonograph resolution), but also the depth range in which the survey can be performed, according to the environment condition (Table 2).

| Range setting (m) | Optimum two fish distance from the bottom | Resolution (m x pixel) | Geomorphological setting | Main aims of the survey |
|---|---|---|---|---|
| From few metres to 150 | 5-50m | <1 | Continental shelf in shallow water, between -10 and -30 m up to -50 m if the topographic features are known and without reliefs. | Research of specific target, such as small shipwrecks, Monitoring of underwater infrastructures, such as pipelines or other small structures. |
| 150-300 | 15-30m | Ca. 1 | From 30 m of water depth down to the shelfbreak. | Meso-scale maps of the seafloor to represent a number of environment features (geological, sedimentological, geomorphological, habitat maps, etc.). |
| 300-600 | 30-60 m | >1 | From -40 m up to the inner slope (600 m). | Meso-scale maps of the seafloor. |

Table 2. Summary of the three main classes of high frequency SSS range settings (from top to bottom: small-, medium and wide-) with indication of the associated sonograph resolution, optimum geomorphological setting and aim of surveys (from Savini, 2004).

### 3.1 High frequency SSS imagery through wide-range setting

SSS surveys through wide-range setting aim to cover wide areas of the seafloor on the outer continental shelf and the upper slope. Range from 300 up to 600 m are considered, so an operating frequency not higher than 100kHz should be employed with the sonograph's resolution reaching roughly 1 m per pixel. Wide-range settings, through high frequency SSS systems, are adopted when medium to small-scale maps of the seafloor must be produced and water depth is more than a minimum of 40 m (because the distance of the tow fish from the bottom should not be less than the 10% of the operating range). The outer continental shelf and the upper continental slope are candidate areas to be investigated by wide-range setting, where the roughness of the seafloor can vary a lot, outlining homogeneous seafloors, with a gently sloping shelf-break and muddy bottom, or various types of articulated seabeds with reliefs and/or depressions of different sizes, rocky outcrops and abrupt changes in slope gradient. If the seafloor morphology is very rough (especially in an active geodynamic context), huge shadows and repeated echoes can characterize the associated sonographs. Geological attributes that can be relevant to map with wide-range setting can be:

- Canyon heads and scarps of different origin and sizes (Fig. 1).
- Rocky outcrops where typical small scale faults and other structural features can be detected.

- Boulders and/or masses of different sizes, often associated with failure deposits on the upper slope.

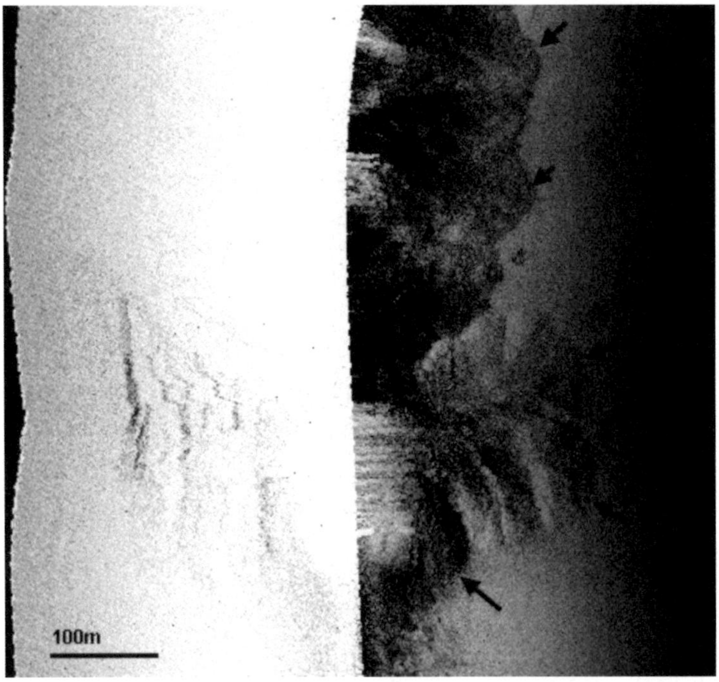


Fig. 1. Edge of canyons on the shelf break (Thyrrenian Sea). The left side of the sonographs has very low levels of backscatter due to the high steepness of the bottom. The sonographs follows the shelf break, the continental shelf is toward the right.

- Confined seafloor morphologies, such as gullies and channels (i.e. gateways and channels - Fig. 2).

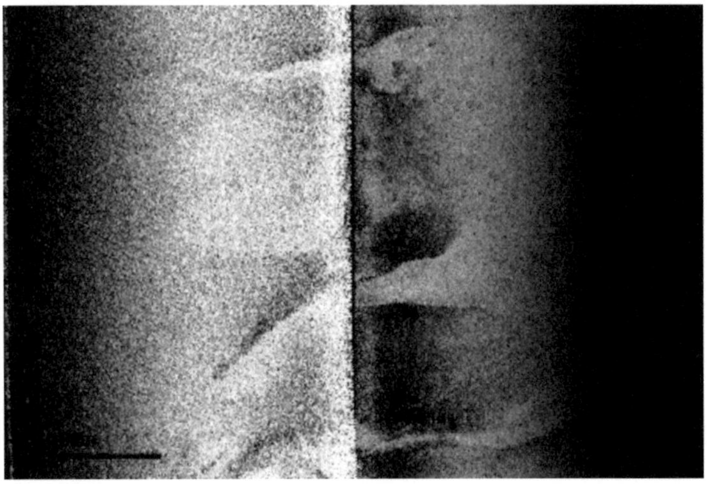

Fig. 2. Confined erosional bedforms on the shelf break (Thyrrenian Sea). The continental shelf is toward the right.

- Erosive seafloor features related to bottom current e.g. scours, furrows etc. (Fig. 3).

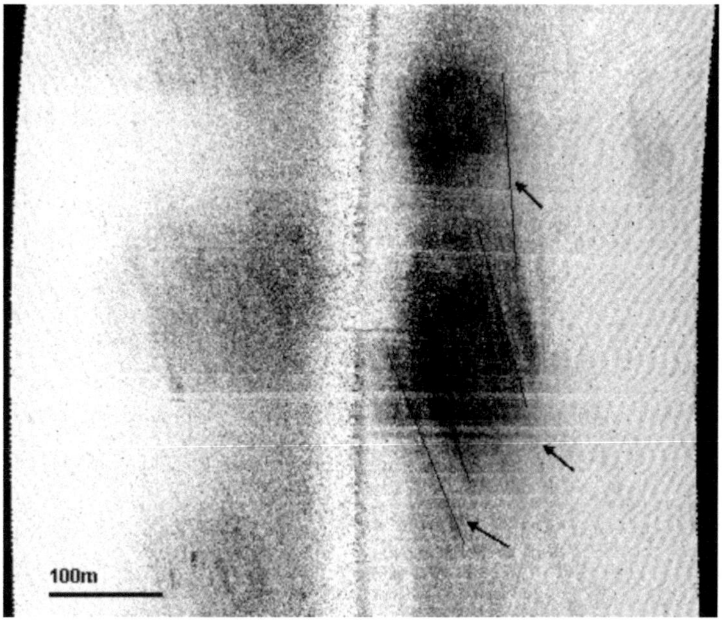


Fig. 3. Furrows (highlighted by straight black lines) and patches of coarse grained sediment (rounded areas with high levels of quite homogeneous backscatter). Thyrrenian Sea, continental shelf, 40m of water depth.

- Positive reliefs of biogenic origin (carbonate mounds, carbonate build-ups etc. Fig. 4).

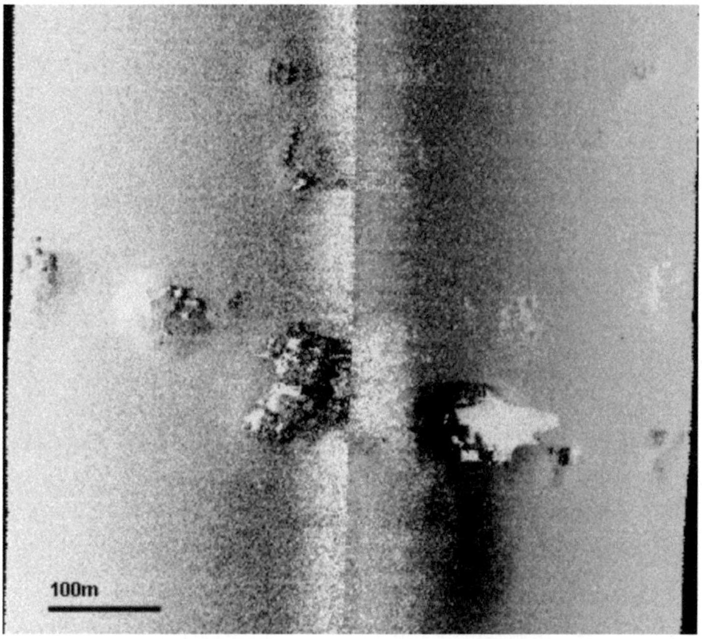


Fig. 4. Positive reliefs on the seafloor (Thyrrenian Sea – 80m of water depth).

- Mud volcanoes and pockmarks, methane-derived carbonate heterogeneous crusts and slabs, carbonate build-ups associated with seepage (Fig. 5).

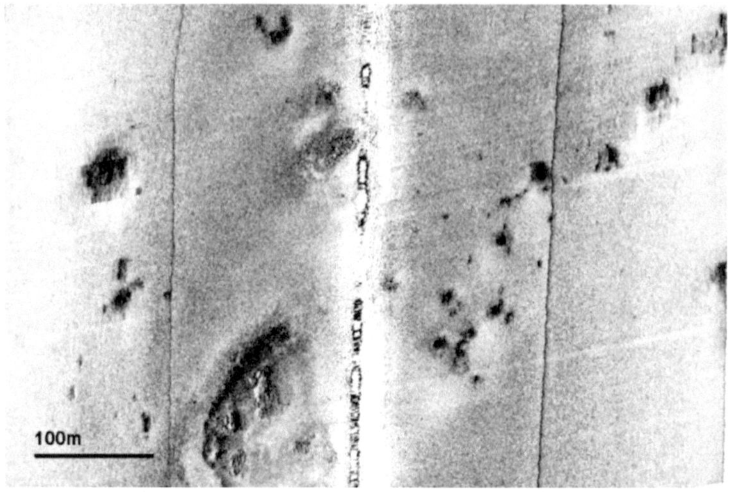


Fig. 5. Small mud volcanoes and seepage-related seafloor features (Sicily channel - 160m of water depth).

- Sedimentary structures: from small to large dunes in their different morphologies (barchan dunes, lunate dunes etc. Fig. 6), sediment waves of metric scales, gravel/sand bars, irregular gravel/sand patches (Fig. 3), sand/gravel ribbons, obstacles and scours (comet marks) etc.

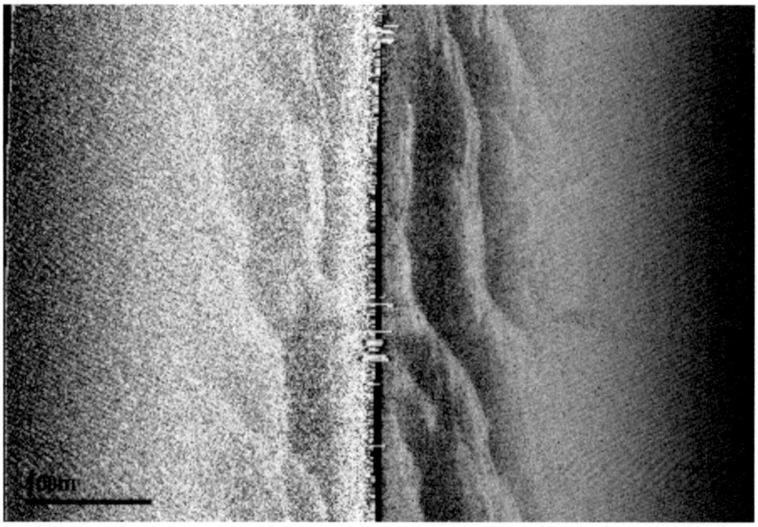

Fig. 6. Depositional bedforms on the outer continental shelf (Thyrrenian Sea - 80m of water depth).

- Failure-related seafloor features, such as blocks of detached sediment, pressure ridges, tension cracks (Fig. 7) etc.

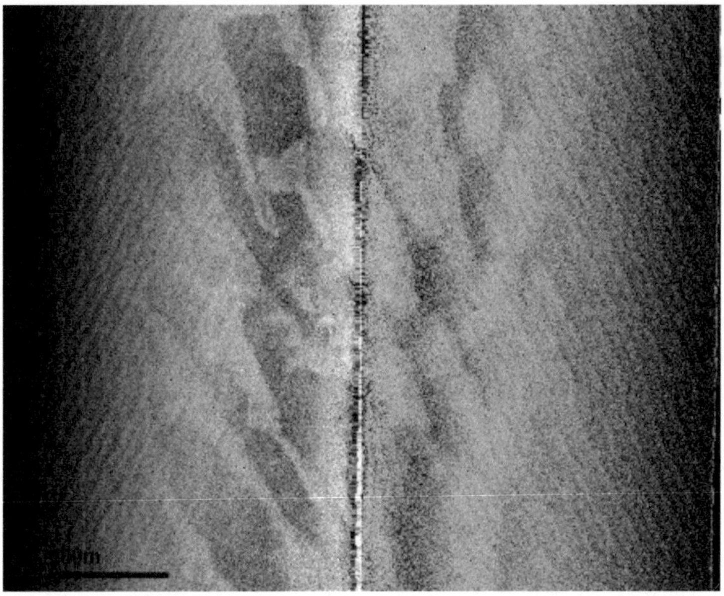

Fig. 7. Depositional bedforms on the outer continental shelf (Thyrrenian Sea - 60m of water depth).

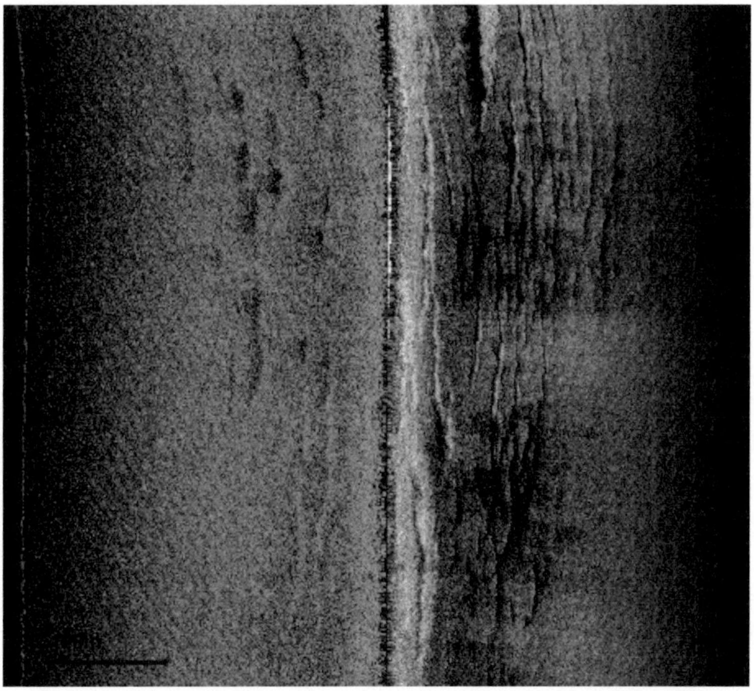

Fig. 8. Fractures on the seafloor (Thyrrenian Sea - 180m of water depth). The sonographs run parallel to the shelf break, the orientation of the fractures suggests a possible inception of mass-wasting, probably due to seabed creeping.

### 3.2 High frequency SSS imagery through medium-range setting

SSS surveys through a medium-range setting aim to cover relatively wide areas of the seafloor on the continental shelf. Range between 150 and 300 m are considered. The sonograph's resolution can be less than 1 m per pixel. SSS sonographs of medium-range setting produce medium-scale maps of the seafloor to image the local geological/geomorphological attributes of the investigated areas. Indeed the medium-range setting is relevant for mapping seafloor morphologies and sediment characteristics of the whole continental shelf, from roughly 20 m of water depth down to the shelf break. It is worth considering that seafloor mapping at meso-scales is important because many of the solutions to environmental problems (from geo-hazards to offshore resources management) require changes in management strategies at this 'seascape scale'. In addition, backscattering data along with bathymetry at meso-scale provide a basis for integrating seabed samples and biological information from which habitat may be interpreted (Kenny et al., 2003). Standards and protocols for habitat mapping are starting to emerge at international level (see MESH EU-project - http://www.searchmesh.net) and they strongly consider that backscattering information from multibeam data or SSS sonographs provide the crucial spatial and structural context for more detailed investigation which can reveal the benthic ecology. Therefore, marine habitat maps are usually based on a combination of wide-area data sets (e.g. side-scan sonar data), able to illustrate the abiotic environment, and point- or line-based information (e.g. photo and video data, seabed samples) providing 'ground-truthing' of the substrate and biological information.

On the Mediterranean continental shelves, the phanerogam biocenosis are well detectable on medium-range SSS sonographs, along with most of their physiographic attributes, that can be relevant to assess their health state (Figs. 9, 10, 11 and 12).

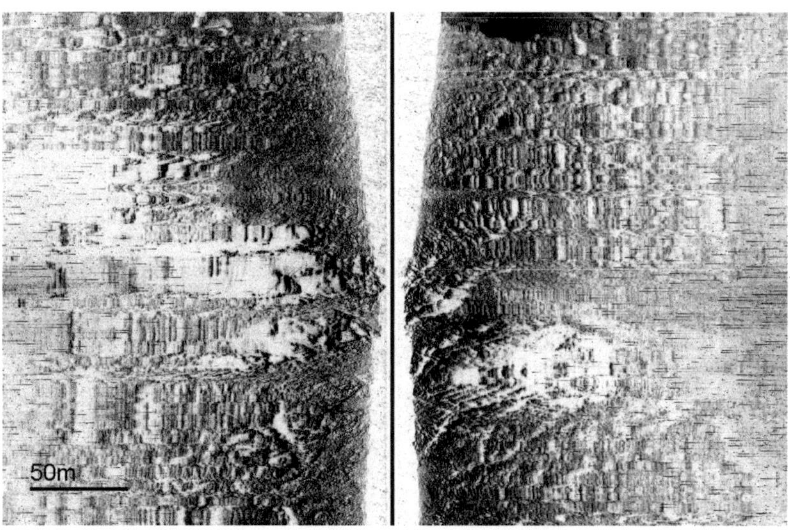


Fig. 9. *Posidonia oceanica* on rocky substrate. See at the centre of the sonograph the short distance recorded between the tow fish and the bottom and the prominent shadows that the rocky outcrops create on the left channel.

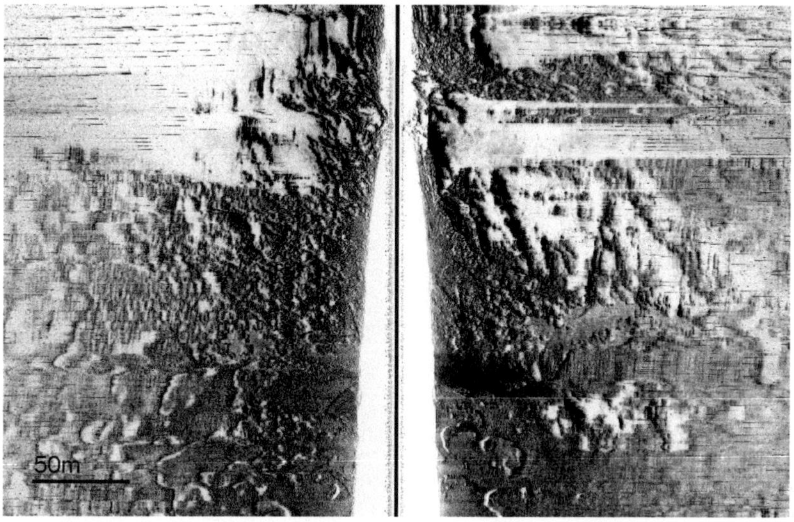


Fig. 10. Passage from *Posidonia oceanica* on rocky substrate (upper part of the sonograph, where prominent shadows occur on both the channels) to *Posidonia oceanica* on matte (middle part of the sonograph) and to degraded matte of *Posidonia oceanica* with marked intramattes and erosive features (lower part of the sonograph, especially on the left channel).

Geological attributes that can be imaged with medium-range SSS sonographs are:

- Rocky outcrops where sub-metric fractures and other structural features can also be detected (Figs. 12, 13 and 14).

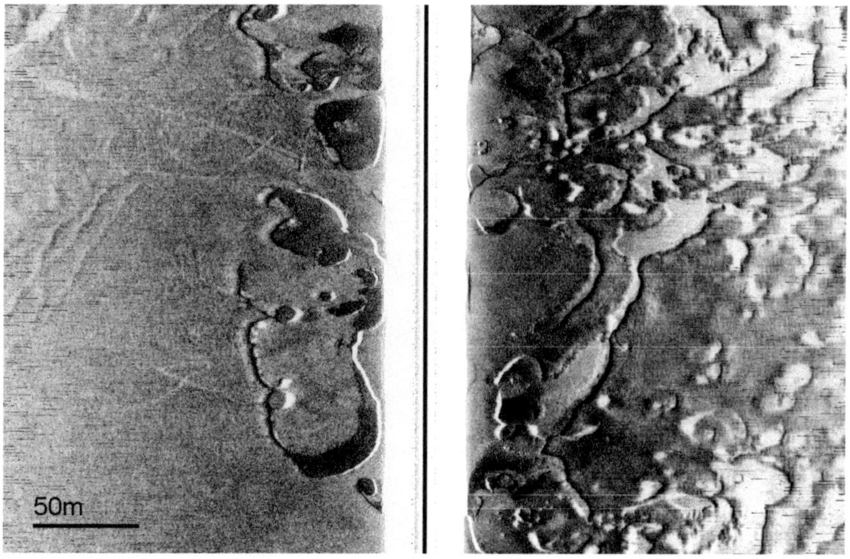


Fig. 11. Dead and degraded matte of *Posidonia oceanica*, that appears nearly compact on the left channel and highly affected by strong erosion on the right.

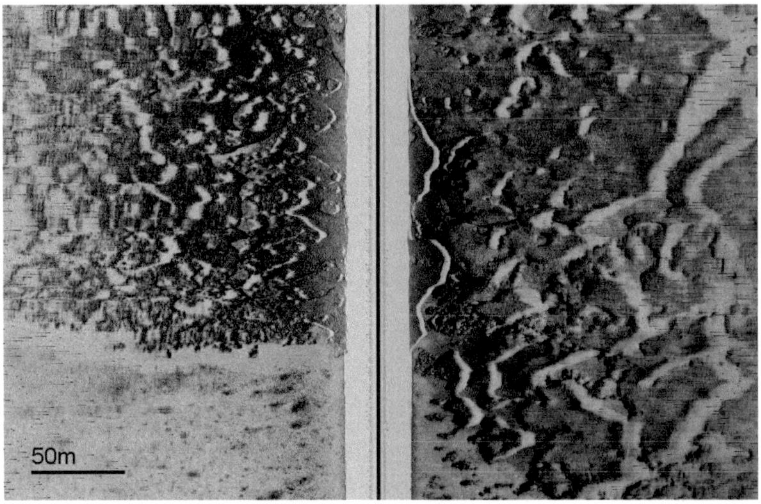


Fig. 12. Upper limit of degraded matte of *Posidonia oceanica* where erosional features are dominant.

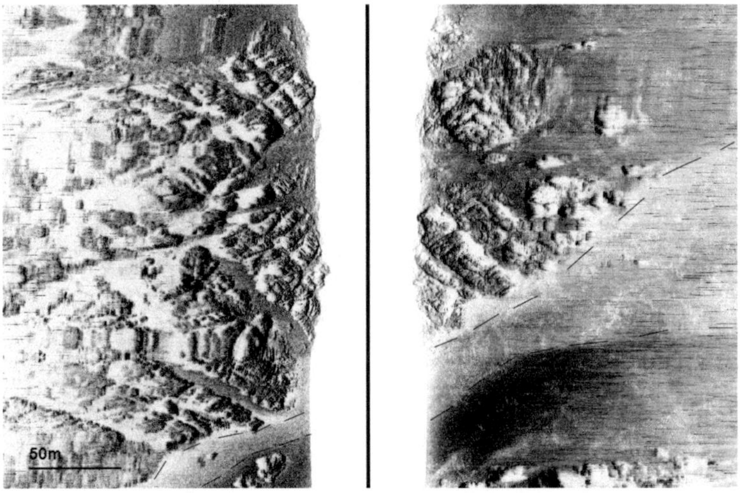


Fig. 13. Rocky outcrops where the orientation of structural lineaments, which consist of perpendicular fractures, are clearly imaged.

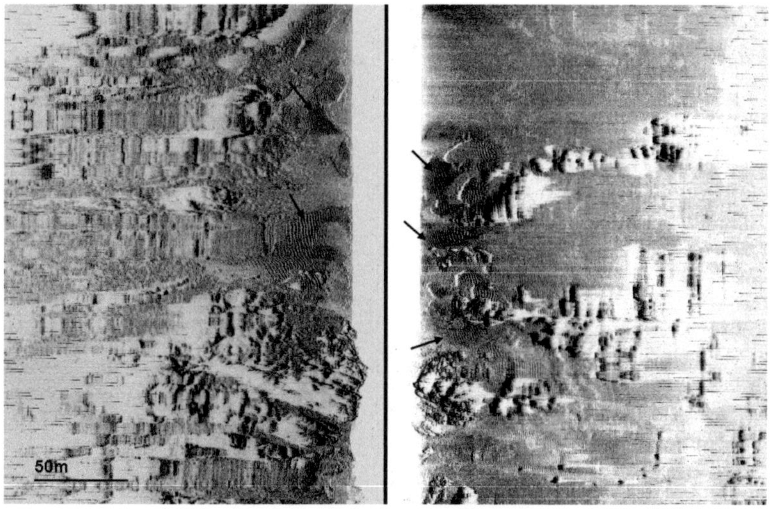


Fig. 14. Rocky outcrops and sediment. Arrows indicate well defined areas made by coarse sediment with ripples.

- Erosive features of sub-metric scales: scours, furrows etc.
- Positive reliefs of biogenic origin (e.g. biogenic formations made by coralline algae).
- Small scale mud volcanoes and pockmarks, methane-derived carbonate heterogeneous crusts and slabs, carbonate build-ups associated with seepage (Figs. 15 and 16).
- Sedimentary structures: small dunes in their different morphologies (barchan dunes, lunate dunes etc.), sediment waves, ripples, gravel/sand bars, irregular gravel/sand patches, sand/gravel ribbons, surface lineations, obstacles and scours (comet marks) etc. (Fig. 13).

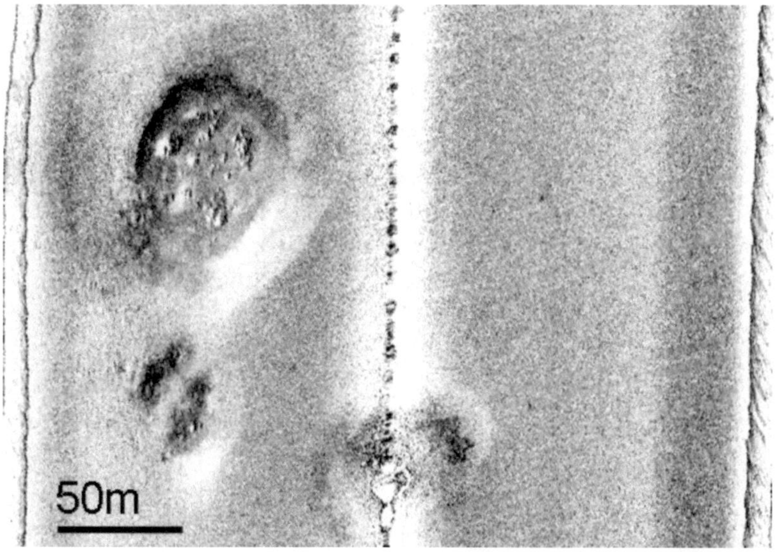


Fig. 15. Small sub-circular mud volcanoes (Savini et al., 2009). See at the top the backscatter associated to the occurrence of methane-derived carbonate crusts.

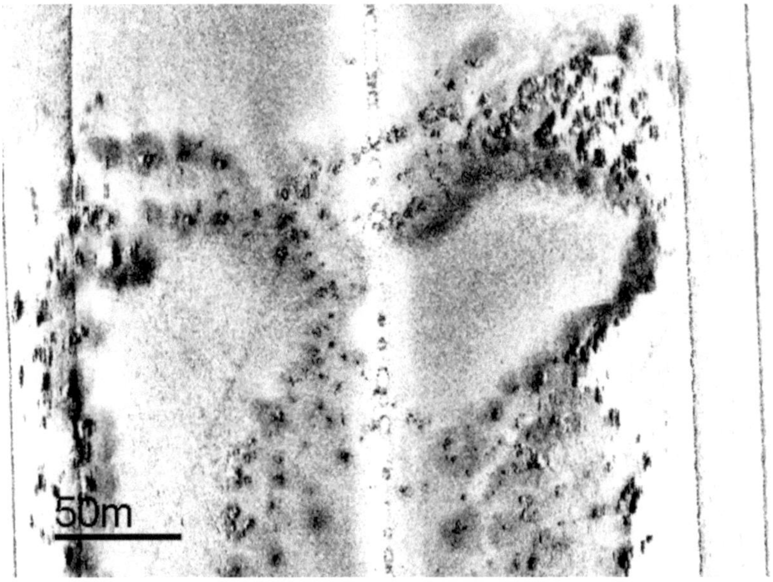


Fig. 16. Hard substrata made by methane-derived carbonate build-ups.

### 3.3 High frequency SSS imagery through small-range setting

Range between a few metres and 100 m are considered. SSS surveys conducted using a small-range setting are used in shallow water and provide high resolution sonographs. Often a high resolution setting of the SSS system is needed in a survey which aims to search for specific targets, such as small shipwrecks and other small-scale objects, or to monitor underwater pipelines. In optimal conditions, sonograph resolution can be less than 10 cm per pixel.

Operating with small-scale range, the transducer has to keep a short distance from the seafloor (a few metres), so that its safety is not compromised by articulated topography with pinnacles and/or local rocky outcrops, even if a fast recovery of the tow fish is allowed by the shallow water (because of the few metres of cable out).

While habitat mapping is mainly performed through medium-range SSS surveys, a small-range setting in a SSS survey can provide remote data useful for further habitat characterization. For instance, within the Mediterranean setting, meadows of *Posidonia oceanica* (which represent benthic habitat of conservation value, listed in the EC Habitats Directive due to its high level of biological diversity) can be imaged in detail by small-range sonographs, since most of their physiographic attributes and ecological descriptors (i.e. upper and lower depth limit of the meadow, density, bottom cover, structure of the matte, occurrence of intramattes and other erosive features, characterization of the substrate, identification of dead or highly degraded matte, human impact due to anchoring and fishery) are at metric and sub-metric scale and are well imaged through SSS sonographs acquired with small-range settings (Figs. 17, 18, 19, 20 and 21).

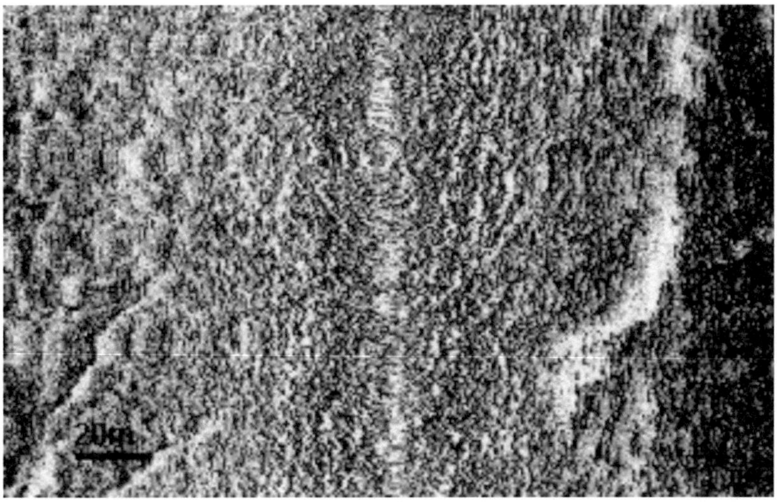

Fig. 17. *Posidonia oceanica* on matte.

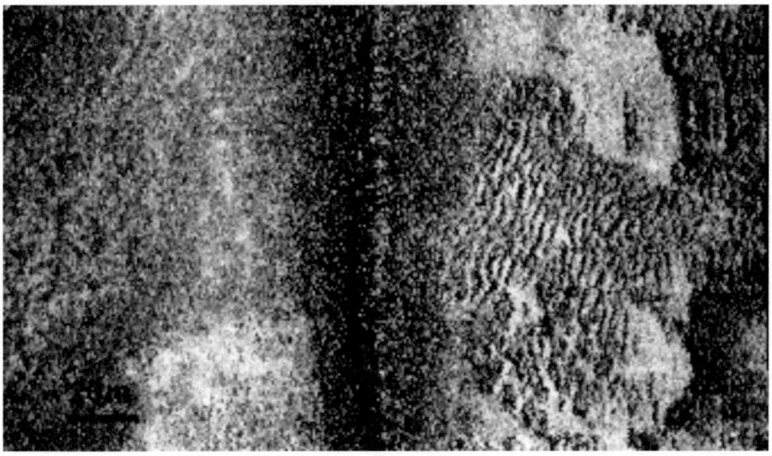

Fig. 18. Passage from *Posidonia oceanica* on matte (on the right channel) to dead/degraded matte of *Posidonia oceanica*.

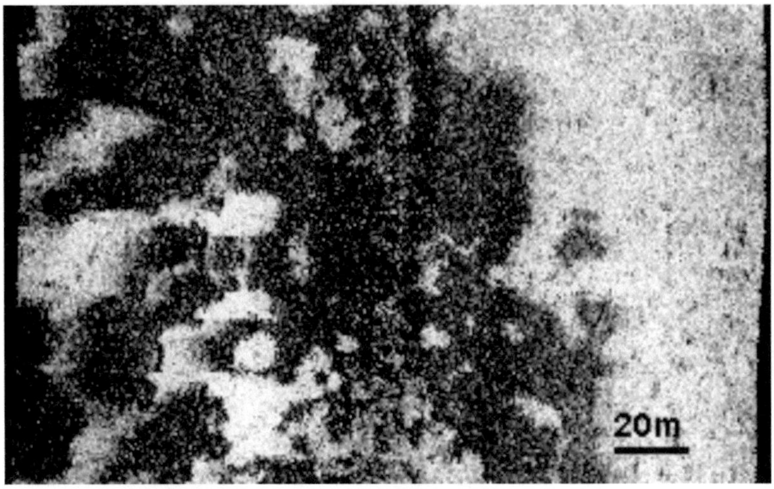


Fig. 19. Meadows of *Cymodocea nodosa* (high backscatter).

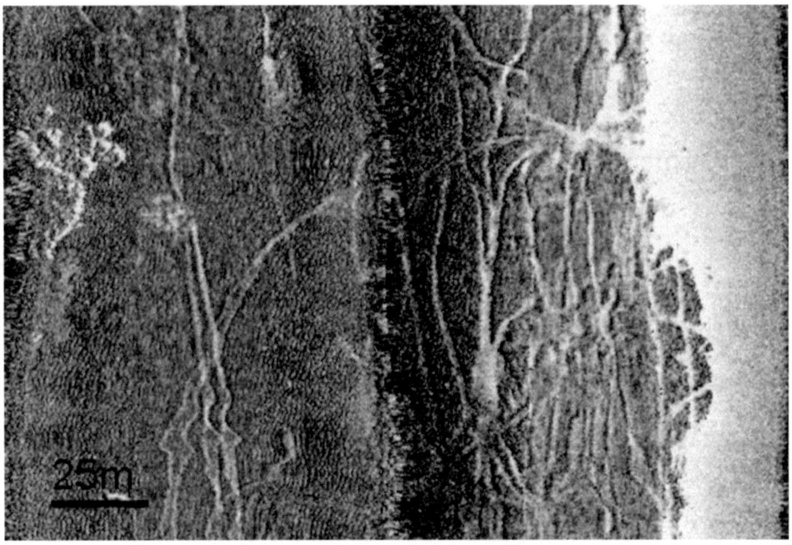


Fig. 20. Prominent track left on the matte of *Posidonia oceanica* by anchoring operations.

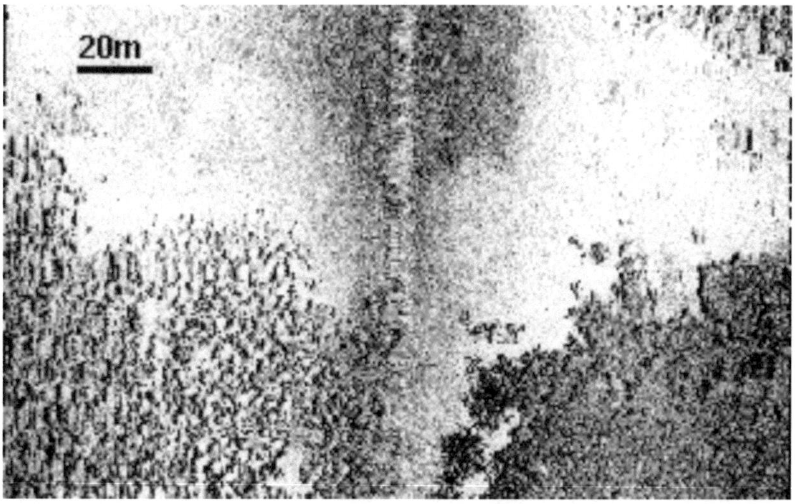


Fig. 21. *Posidonia oceanica* on sediment (on the left channel, lower part), *Cymodocea nodosa* (right channel lower part) and sediment.

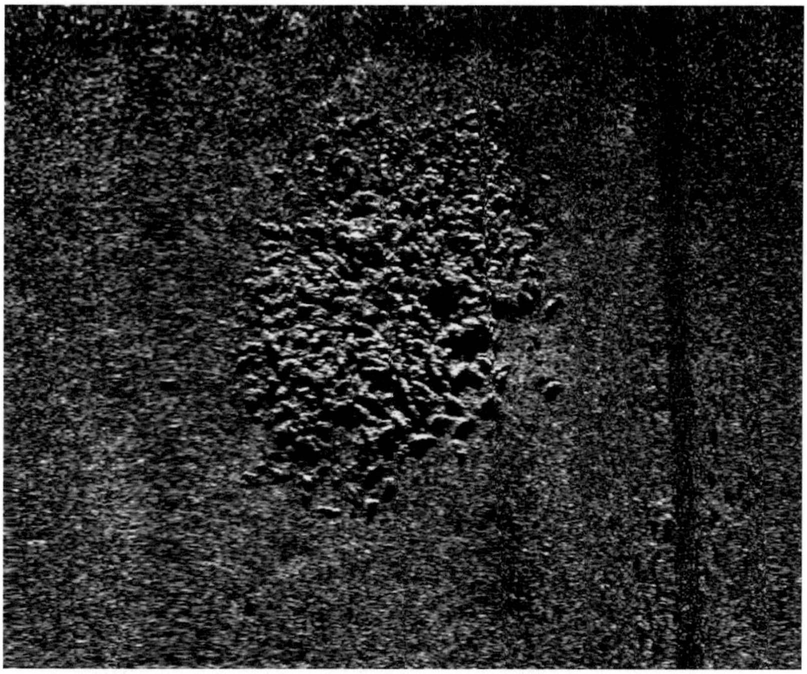

Fig. 22. Amphora on the seafloor (Savini and Petrillo, 2009). Each one is 1,3m in length.

Geological attributes that can be well imaged on small-range SSS sonographs are mostly related to small scale sedimentary structures (such as ripples and surface lineations at the centimetric scale, small comet marks etc.) and erosional bottom current related bedforms (e.g. small scours and furrows). However, as stated above, small-range SSS surveys are generally carried out to investigate/image small targets on the seafloor (Fig. 22).

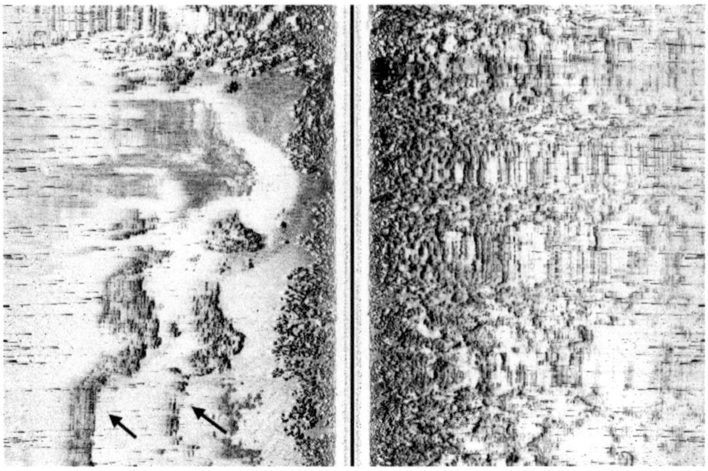

Fig. 22. Multiple (black arrows) on the left channel.

### 3.4 Peculiar SSS images

Here after are some peculiar situations that can occur during a side-scan sonar survey. Such situations occur in sonographs affected by noise that comes from the underwater environment (e.g. signals generated from other ships' propellers and/or from interference with other acoustic sources, artefacts created by objects or fishes in the water column) and particular objects/infrastructures mapped on the seafloor.

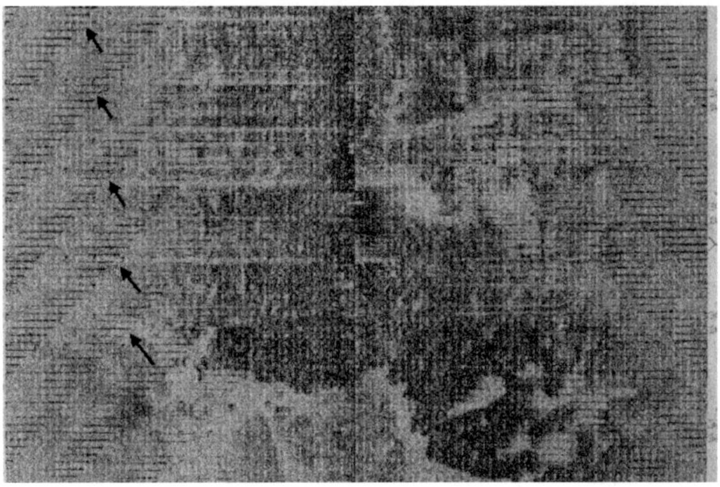

Fig. 23. Noise on the sonograph (black arrows) produced by interference with other acoustic source (in this case a single beam echosounder).

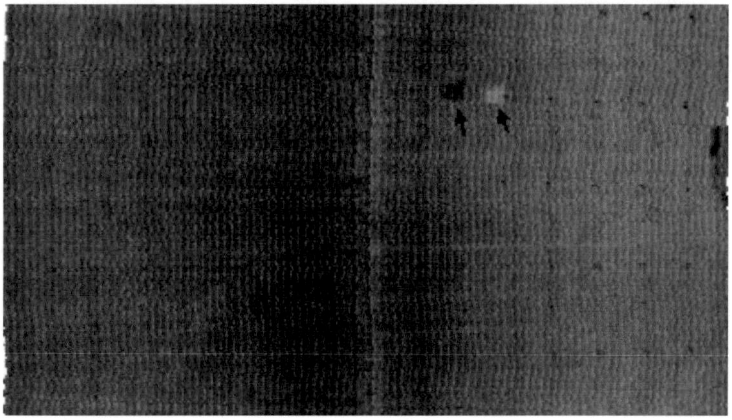

Fig. 24. A shoal of fish and its associated shadows, which appear separated on the sonograph because of the position of the shoal in the water column.

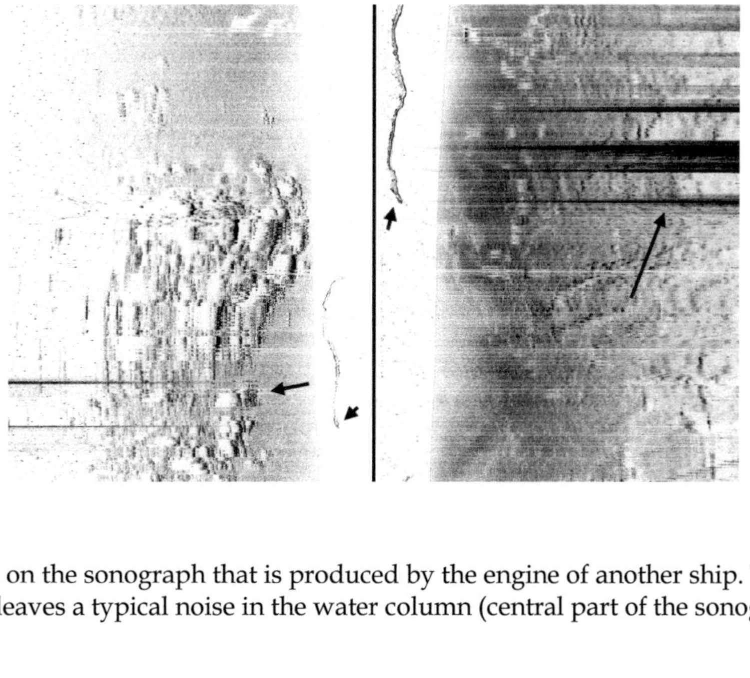

Fig. 25. Noise on the sonograph that is produced by the engine of another ship. The wake of the ship also leaves a typical noise in the water column (central part of the sonograph).

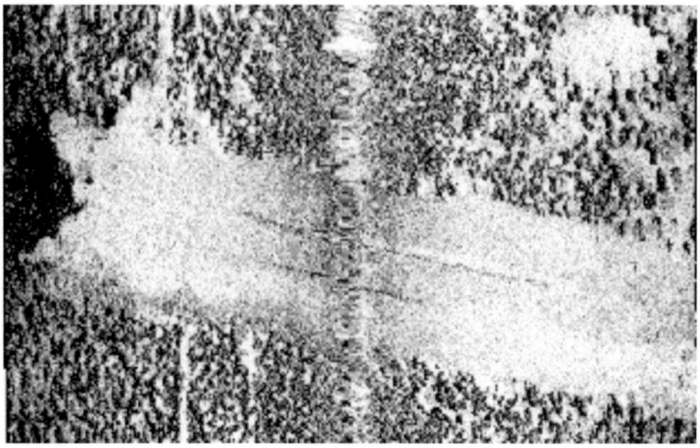

Fig. 26. Meadow of *Posidonia oceanica* interrupted by sediment where some parts of two pipelines are visible.

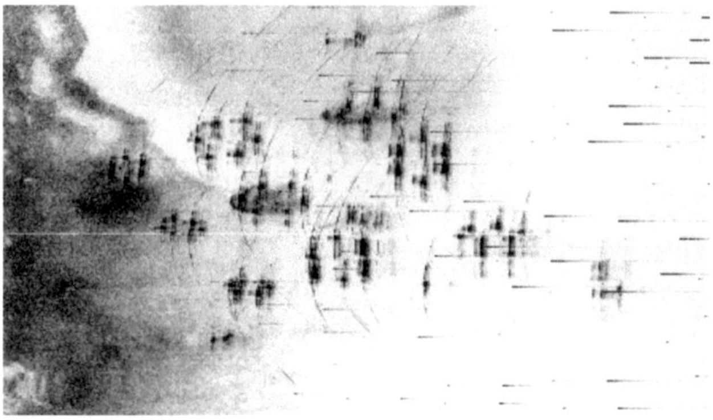

Fig. 27. Fish aquaculture.

Fig. 28. Fishing net 'at work' on the seafloor. Note the two gears which keep open the net and the noise of the engine of the fishing boat (lower part on the right).

## 4. Conclusion

This chapter aimed to provide a useful reference document representing important seafloor features and sediment characteristics on continental shelf in the Mediterranean setting. Indeed there is an exhaustive literature dealing with theoretical issues concerning SSS functioning and data processing (Fish and Carr, 1999; Blondel and Murton, 1997) and there are also some published books that can support sonograph interpretation within the oceanic environment (e.g. Blondel, 2009). Nevertheless a number of peculiar seafloor features of the Mediterranean continental shelf, often monitored in many surveys, are not so well known from literature, in terms of their peculiar acoustic backscattering. This chapter aimed to offer a reference that can help in recognising the most significant seafloor features that characterise the Mediterranean continental shelves on SSS sonographs.

## 5. Acknowledgements

This chapter benefited from funding through the 2000–2001 Italian MIUR programme 'founding for young researchers' (ministry advise n. 1707, 22.07.1998 ), by which the author was coordinator of the project 'Atlas of side-scan sonar sonographs: the acoustic response of the different geomorphological features that characterize the Mediterranean continental shelf'. The author is also grateful to captains, crew and all colleagues (among other Lidia Olivieri, Andrea Grazzini, Paolo Orrù and Tommaso Granata) met during the numerous activities carried out at sea.

## 6. References

Belderson, R.H., Kenyon, N.H., Stride, A.H. and Stubbs, A.R., (1972). *Sonographs of the Sea Floor*. Elsevier, Amsterdam, 185 p.

Blondel, P. and Murton, B.J., (1997). *Handbook of Seafloor Sonar Imagery. Chichester*. John Wiley and Sons.

Blondel, P., (2009). *The Handbook of Side-scan Sonar*. Springer Verlag

Blondel, P., Parson, L.M. Robigou, V., (1998). "TexAn: Textural Analysis of Side-scan Sonar Imagery and Generic Seafloor Characterisation." *Proceedings of Oceans '98* vol. 2: 419-423.

Boyle, F.A. and Chatiros, *N.P., (*1995*)*. A model for acoustic backscatter from muddy sediments. *Journal of the Acoustical Society of America* '98(1), 525-530.

Cochrane, G.R. and Lafferty, K.D., 2002. "Use of acoustic classification of side-scan sonar data for mapping benthic habitat in the Northern Channel Islands, California." *Continental Shelf Research* 22: 683-690.

Damuth, J.E., (1980). "Use of high frequency (3.5 - 12 kHz) echograms in the study of near bottom sedimentation processes in the deep sea." *Marine Geology* 38: 51-75.

Fish, J.P. and Carr, H.A., (1990). *Sound underwater images. A guide to the generation and interpretation of side-scan sonar data*. Orleans (Lower Cape Publishing), 188 p.

Fleming, B.W., (1976). "Side Scan Sonar: A Practical Guide," *International Hydrographic Review*, Vol.1 III, No. 1 (January 1976).

Flood, R.D., (1980). "Deep-sea sedimentary morphology: modeling and interpretation of echo-sounding profiles." *Mar. Geol.*, 38: 77--92.

Goff, J.A., Olson, H.C., Duncan, C.S., 2000. "Correlation of side-scan backscatter intensity with grain size distribution of shelf sediments, New Jersey margin." *Geo Mar. Lett.* 20, 43–49.

Holland, C. W., Hollett R. and Troiano L., (2000). "Measurement technique for bottom scattering in shallow water," *J. Acoust. Soc. Am.* 108, 997–1011.

Huvenne, V.A.I., Blondel, Ph., Henriet, J.-P., 2002. "Textural analyses of side-scan sonar imagery from two mound provinces in the Porcupine Seabight." *Mar. Geol.* 189, 323–341.

Jackson, D.R. and Briggs K.B., (1992). "High frequency bottom backscattering: roughness versus sediment volume backscattering." *J. Acous Soc. Am.* 92: 962-977.

Jones E.J.W., (1999). *Marine Geophysics.* Wiley.

Kenny, AJ, Cato, I, Desprez, M, Fader, G, Schuttenhelm, R.T.E. and Side, J., (2003). "An overview of seabed-mapping technologies in the context of marine habitat classification." *ICES Journal of Marine Science* 60:411–418.

Lo Iacono, C., Gràcia, E., Diez, S., Bozzano, G., Moreno, X., Dañobeitia, J. and Alonso, B., (2008). "Seafloor characterization and backscatter variability of the Almería Margin (Alboran Sea, SW Mediterranean) based on high-resolution acoustic data." *Marine Geology* 250 (2008) 1–18.

McQuillin, R. and Ardus, D.A., (1977). *Exploring the geology of shelf seas.* Graham & Trotman (London), 234 p.

Medwin, H. and Blue, J. E., (2005). *Sounds in the sea: from ocean acoustics to acoustical oceanography.* Cambridge University Press.

Medwin, H. and Clay, C. S., (1998). *Fundamentals of acoustical oceanography. Applications of modern acoustics.* Academic Press

Morang, A., Larson, R. and Gorman, L., (*1997*). "Monitoring the Coastal Environment, Part III, Geophysical and Research Methods." *Journal of Coastal Research*, 13, 1064-1085.

Savini, A. (2004). "Metodologie di analisi di prospezioni geofisiche in ambiente marino e loro implementazione in applicazioni S.I.T.". *PhD Thesis,* Milano-Bicocca University.

Savini A., Malinverno E., Etiope G., Tessarolo C., Corselli C. (2009). "Shallow seep-related seafloor features along the Malta Plateau (Sicily channel -Mediterranean Sea): Morphologies and geo-environmental control of their distribution". *Marine and Petroleum Geology* 26, 1831-1848.

Savini A., Petrillo M. (2009). "Seascapes of war: investigations with marine geophysics". *Europe's Deadly Century – Perspectives on 20th century conflict heritage.* Edited by Neil Forbes, Robin Page and Guillermo Pèrez. Published by English Heritage.